The First Half Century of Finite-Time Thermodynamics

The First Half Century of Finite-Time Thermodynamics

Guest Editors

Bjarne Andresen
Peter Salamon

MDPI Basel • Beijing • Wuhan • Barcelona • Belgrade • Novi Sad • Cluj • Manchester

Guest Editors
Bjarne Andresen
Niels Bohr Institute
University of Copenhagen
Copenhagen
Denmark

Peter Salamon
Department of Mathematics
and Statistics
San Diego State University
San Diego
USA

Editorial Office
MDPI AG
Grosspeteranlage 5
4052 Basel, Switzerland

This is a reprint of the Special Issue, published open access by the journal *Entropy* (ISSN 1099-4300), freely accessible at: https://www.mdpi.com/journal/entropy/special_issues/RTDUBPRS1D.

For citation purposes, cite each article independently as indicated on the article page online and as indicated below:

Lastname, A.A.; Lastname, B.B. Article Title. *Journal Name* **Year**, *Volume Number*, Page Range.

ISBN 978-3-7258-7286-2 (Hbk)
ISBN 978-3-7258-7287-9 (PDF)
https://doi.org/10.3390/books978-3-7258-7287-9

Cover image courtesy of Bjarne Andresen based on a sketch by Lena Pedersen

Contents

About the Editors

Bjarne Andresen

Bjarne Andresen was professor at the Niels Bohr Institute, University of Copenhagen 1978–2020, and is currently affiliated with the NBI as emeritus. Along with the late professor R. Stephen Berry from the University of Chicago and the other editor of this Special Issue, he started the field of finite-time thermodynamics (FTT) in 1975. He has since then participated in spreading the idea that limiting the duration of a process to a fixed time opens up a new (more realistic) arena of optimizing a system's power performance rather than its efficiency (usually at zero power). As part of that effort, he has applied the concept to a wide range of subjects. A key element of FTT is optimizing the path of the process. Prof. Bjarne Andresen has contributed to applying FTT to diverse fields, such as chemistry, physics, engineering, optimization theory, economics, and biology. Lately, he has worked with prof. Christopher Essex on extending thermodynamics to extremely long time and size scales for which they received the 2022 Entropy Best Paper Award. Early in his career, he contributed to atomic collision theory. In 1984, he helped start the Telluride Science Research Center as a secluded retreat for developing new wild ideas in science.

Prof. Bjarne Andresen has been visiting a professor at several universities, including the University of Chicago, Stanford University, San Diego State University, University of New South Wales, University of Western Ontario, and Nagpur University.

At the periphery of science, prof. Bjarne Andresen has been president of the Section for Theoretical Chemistry of the Danish Chemical Society, chairman of the committee on research and education of the Danish Confederation of Professional Associations, chairman of the Danish Forum for University Policy and Planning, and is currently treasurer of the Danish Academy of Natural Sciences.

Peter Salamon

Peter Salamon is a professor in the Mathematics Department at San Diego State University from 1980 to the present. He received his PhD in chemical physics from the University of Chicago under the guidance of the late professor R. Stephen Berry. During this period and in collaboration with the other editor of this volume, finite-time thermodynamics emerged. He has had visiting positions at the Hebrew University, the University of Heidelberg, the University of Copenhagen, among others.

His research has always revolved around thermodynamics very broadly interpreted, which enabled him to stray as far as machine learning, global optimization, and microbial ecology. His numerous collaborations earned him an Erdös number of one and the designation as a hub in the Collaborative Distance Graph linking collaborators from numerous far-flung fields to Paul Erdös.

He co-founded the Telluride Science Research Center, which gave early finite-time thermodynamics and other fledgling fields an important venue.

Preface

Finite-time thermodynamics (FTT), as an area of theoretical study and its associated applications, was initiated 50 years ago. As is often the case when new concepts crystalize, stray ideas about "what if" had appeared earlier (e.g., Chambadal in 1957, Novikov in 1958, Curzon and Ahlborn in 1975, and Weinhold in 1975), but it was only in 1976 that a more comprehensive theory started to evolve. At the same time, it became clear that the restriction of finite-time, or more generally finite resources, is pervasive for all real processes. The "finite-time" idea spread to chemistry, physics, quantum mechanics, economics, engineering (under a different name and with preciously few references to its origin), and, in more recent years, biology. It is now a solidly established field with an exceptionally wide range of applications.

In the present Special Issue of *Entropy*, we have collected 14 papers that provide a wide range of current topics for which finite-time is a crucial element, with some being very abstract, very applied, or moving in unexpected directions. To set the stage, Boye Ahlborn and Frank Curzon provide a personal view of the beginnings half a century ago. Much more FTT and FTT-inspired research is on its way. Happy reading.

Bjarne Andresen and Peter Salamon
Guest Editors

entropy

MDPI

Editorial

The First Fifty Years of Finite-Time Thermodynamics

Bjarne Andresen [1,*] and Peter Salamon [2]

[1] Niels Bohr Institute, University of Copenhagen, Jagtvej 155 A, DK-2200 Copenhagen N, Denmark
[2] Department of Mathematics and Statistics, San Diego State University, San Diego, CA 92182-7720, USA; psalamon@sdsu.edu
* Correspondence: andresen@nbi.ku.dk

1. Background

The year 1975 marked the beginning of an entirely new direction for thermodynamics with the publication of Curzon and Ahlborn's innocent-looking short paper "*Efficiency of a Carnot engine at maximum power output*" [1]. That paper not only ignited what became known as Finite-Time Thermodynamics (FTT), it is also an important story about how significant new ideas come out of casual conversations, totally unplanned, and all the trouble the originators have to go through to get their ideas published. We are very fortunate that Boye Ahlborn and Frank Curzon were willing to tell us about the birth and publication troubles of the main idea behind FTT in their paper "*The hunt for new understanding*" in this volume. The fact that overcoming the obstacles ended in success should be an encouragement to all young scientists with unconventional ideas: Persevere and you will win.

One thing is writing the "spark" that starts a new development. But that only works if the scientific environment is ready for the change. In 1975, three additional facts came together. (1) Prof. R. Stephen Berry had recently taken up his position at the University of Chicago. The terrible air quality in Chicago at the time reminded him of his earlier interest in incorporating time into thermodynamic analyses and prompted him to start pondering possible ways of reducing waste in industrial processes. (2) Prof. Peter Salamon had just started his Ph.D. studies under Prof. Berry with a desire to work out a differential geometrical formulation of thermodynamics. Finally, (3) Prof. Bjarne Andresen had just come to postdoc with Prof. Berry on atomic collision theory. This small diverse group caught the generality and importance of Curzon and Ahlborn's result restricting the duration of a thermodynamic process. Time is fully as important as a thermodynamic variable as the various free energies are, not merely as an evolution parameter used in kinetics, but as a key ingredient in calculating realistic efficiencies. For the full details, see [2]. The scientific world was ready for this new view on thermodynamic optimization, and many young researchers soon joined the effort.

To be honest, the idea of limiting the duration of a thermodynamic cycle had appeared in print earlier than the Curzon–Ahlborn paper, but for less general systems, in conference proceedings, and in local languages (see [3] for a historical account). All of these circumstances prevented the idea from being noticed. It only caught the attention of the scientific community as presented in 1975 by Curzon and Ahlborn, unaware of the previous attempts.

Our own first paper on FTT from 1977 [4] was the first to pick up the idea and coin its name; that is why we celebrate the 50th anniversary of Finite-Time Thermodynamics this year. Actually, we were in a dilemma as to whether the celebration should include a Special Issue of *Entropy*, considering that we, for unrelated reasons, collated an FTT Special

Issue just a few years ago [5]. In the end, we decided that the importance of the 1975 paper was large enough to carry another Special Issue. We hope that you agree.

Some people have argued that Finite-Time Thermodynamics was not a good name for what the topic has evolved into, e.g., that it should have been called Finite-Resource Thermodynamics instead. To a certain extent, we understand such wishes since many of the ideas behind FTT have turned out to be so general that they also apply to limitations on other quantities like heat exchange area, chemical reaction rates, memory, and capital. Furthermore, the basic theory to be extended does not have to be thermodynamics, it can also be economics, computer science, or other disciplines. Thus, "time" in FTT can be read in a very general way as anything, which is limited in a dynamic system by construction.

However, for thermodynamics, time is a very special quantity. Besides designating the direction of evolution, as it does also in many other fields, it shares with extensive variables the property that it measures one of the net changes in a process. The traditional naming of thermodynamics was always a little inappropriate since the dynamics are not specified and in fact it was traditionally what has come to be called comparative statics—enabling the comparison of the static initial and final equilibrium states. The net changes in the extensive variables traditionally gave bounds on the possible values of the process variables such as work, heat, and entropy production associated with the process. Our original goal in extending thermodynamics to finite time was to add one more net effect: the time elapsed. This additional net effect would result in more stringent limits on the process variables while keeping the comparative statics framework intact, i.e., keeping the limits dependent only on the net effects of the process. This comparative statics framework had previously been borrowed from thermodynamics for economic analyses [6] where it is oft used to this day.

The first focus of FTT was the limited time allotted to the completion of the chosen process and the influence that this has on the efficiency of that process. A second important element in FTT is the definition of a metric among the relevant thermodynamic variables from which a length of the reaction path can be calculated and in turn used to bound the dissipation in that process. Such a metric was defined in 1980 [7], curiously growing out of a re-formulation of the Maxwell relations in quantum–mechanical bra–ket notation [8], in an effort to make them more understandable. This metric, available in a wide selection of basic coordinates, only further expanded the generality of FTT, e.g., into statistical mechanics, information theory, and quantum mechanics.

2. Content

We start out this Special Issue celebrating the first 50 years of Finite-Time Thermodynamics with Boye Ahlborn and Frank Curzon's own account of how they got the idea and the trouble they had getting it published, "*Maximum Power Efficiency*".

Next, Anatoly Tsirlin, Alexander Balunov, and Ivan Sukin give a review on how the ideas of FTT are equally useful within economic theory in "*Finite-time thermodynamics: problems, approaches, results*".

Gregory Behrendt and Sebastian Deffner go beyond the usual chemical energy range and analyze "*Endoreversible Stirling cycles: plasma engines at maximal power*", showing that only the functional form of the equation of state being linear in temperature and additive in volume are important, not its further details.

Also Xiu-Hua Zhao and Yu-Han Ma go back to basics and provide a closer look at the historical Yvon engines in "*Revisiting Endo-Reversible Carnot Engine: Extending the Yvon Engine*".

 https://doi.org/10.3390/e28010049

In a more practical approach, François Lanzetta focuses on "*Compressor power and efficiency optimisation: a finite time thermodynamics approach*". In particular, he studies the influence of suction and discharge tube diameters and of gas pressures.

Stability of finite-time engines against perturbations of their assumed path is important. Julian Gonzalez-Ayala, David Pérez-Gallego, Alejandro Medina, José M. Mateos Roco, Antonio Calvo Hernandez, Santiago Velasco, and Fernando Angulo-Brown analyze the stability regions for different optimality criteria in their paper "*Linking optimization success and stability of finite-time thermodynamics heat engines*".

In "*Re-Evaluation of the Extremum Rate of Entropy Production Principles and the Fourth Law of Thermodynamics*", Anil Bhalekar and Vijay Tangde attack the issue of stability in more general terms by using Lyapunov stability theory to connect the Fourth Law of thermodynamics to the rate of entropy production and how to control it.

Related to this, Sergey Amelkin shows that any macrosystem, thermodynamic or not, described by extensive (i.e., proportional to size) variables and under the influence of intensive (i.e., independent of size) forces may be described by the same theory; "*Thermodynamic Theory of Macrosystems: Entropy Production as a Metric*".

Ricardo T. Páez-Hernández, Juan Carlos Pacheco-Paez, Juan Carlos Chimal-Eguía, Delfino Ladino-Luna, and Javier Contreras Sánchez describe the efficiency of their heat exchangers in terms of their respective "Number of Transfer Units" (NTUs) and distinguish between a number of performance situations using different objective functions in "*Transfer irreversibilities in the Lenoir cycle: FTT design criteria with e-NTU*".

Since his initial papers mentioned above [8], Frank Weinhold has used his bra–ket notation and associated reaction coordinates widely in atomic physics. In "*Thermodynamics of Intrinsic Reaction Coordinate (IRC) Chemical Reaction Pathways*" he describes many uses of such state-dependent intrinsic reaction coordinates focusing on the full reaction paths, not just the initial and final states.

Abhishek Dutta, Bitan Mukherjee, Sk Aftab Hosen, Meltem Turan, Denis Constales, and Gregory Yablonsky solve chemical kinetics rate equations using artificial intelligence in the form of so-called "physics-informed neural networks" (PINNs), i.e., the physical and chemical equations are built into their training data sets. The results for "conservatively perturbed equilibrium" systems are presented in "*A Physics-informed neural network (PINN) approach to over equilibrium dynamics in conservatively perturbed linear equilibrium systems*".

In "*Finite-Time Thermodynamics and Complex Energy Landscapes: A Perspective*", Christian Schön has written an extensive review on how FTT is being used successfully to generate efficient pathways for the synthesis of desired molecules.

Finally, Bernard Guy explores the philosophical aspects of FTT by connecting it to relativity for a geological example in "*(Finite-Time) Thermodynamics, Hyperbolicity, Lorentz Invariance: Study of an Example*".

3. Generality of FTT Ideas

Thermodynamics, originally developed by Carnot to model the efficiency of heat engines, was soon after extended by Gibbs to analyze chemical reactions. Similar extended analyses have had unbelievable applicability in other areas as well, e.g., computer science, quantum mechanics, and economics. The reason is that thermodynamics is really a theory (description) of the flow of conserved quantities, driven by a relevant force, through surroundings that resist such flow. Many natural as well as abstract setups fit this general description. Hence, it is not surprising that thermodynamic ideas may be carried over to seemingly unrelated disciplines ranging from real applications in engineering to abstract information handling in computer science, economic trading, and beyond. That, of course,

 https://doi.org/10.3390/e28010049

goes for FTT as well. This feature has also made it a lot of fun to work in this area, offering chances to interact with colleagues from many different specialties.

What ties all the many uses of FTT together is "the cost of haste". Quite universally, as soon as you try to push any dynamic system to go faster, you need to increase the gradients driving it. FTT tells you how to do it wisely, i.e., with a minimum of dissipation. Importantly, that is not necessarily the path that free-running kinetics would take. Rather, the path generally should be uniformly close to the reversible (infinitely slow) path. The driving sequence necessary to achieve the optimal path is sometimes called counter diabatic [9]. The most recent development in this direction includes the cost of measuring (checking) the actual time evolution of that path. This involves a balance between energy, speed, and accuracy [10], a very interesting combination.

4. Where Is FTT Headed?

It is hard to imagine any field where the finite-time concept will not have an important role. Humans are impatient, so anything we want to do, be it in science, technology, economics, or psychology, should be completed as quickly as possible and at as low a cost as possible, i.e., minimizing "the cost of haste". At the end of our previous Special Issue [11], we concluded with a summary entitled "Future perspectives of Finite-Time Thermodynamics" in which we were quite specific about several future avenues.

The one additional specific thermodynamic development we would hope to see in the near future is a theory for counting *compensated dissipation* in analogy to Clauisus' concept of "uncompensated heat" [12] as the work not harvested in a process but dissipated as heat. With such a measure we could examine real processes and have some clear methods for counting how much of the dissipation in the process is inescapably due to each of the different constraints acting in the process. Methods for accounting losses in this manner would be invaluable for process design and for understanding entropy production in biological processes. In this latter direction, we would hope for methods to spell out the entropy production that must be present to account for the choreography involved in the various steps of a process and in the material translocation of the reactants and products for the various steps of a complex reaction network.

Many things have been developing as we forecasted back then, but one direction has gone faster than our prior expectation: AI (artificial intelligence). Being a massively large-scale informational process, it naturally also follows thermodynamic/statistical mechanical principles and its workings can be optimized for efficiency. Not least when going to the next level, AGI (artificial general intelligence), the efficiency in time, power consumption, and accuracy of results will continue to be very important.

Other fertile directions also remain. In particular, we expect further breakthroughs based on counter diabatic driving, quantum computing, and in pushing our understanding further along biochemical pathways.

5. Lessons Particularly for Young Researchers

- Believe in what you are doing regardless of opposition from the establishment.
- Look for concepts and principles and search for equivalents outside your main field. Your results will undoubtedly apply there too, under different names. Have an open mind, but be quantitative.
- You did not work hard in research just to solve somebody else's problem, but to search for insight into unknown territory.

Conflicts of Interest: The authors declare no conflict of interest.

 https://doi.org/10.3390/e28010049

References

1. Curzon, F.L.; Ahlborn, B. Efficiency of Carnot engine at maximum power output. *Am. J. Phys.* **1975**, *43*, 22–24. [CrossRef]
2. Berry, R.S.; Salamon, P.; Andresen, B. How it all began. *Entropy* **2020**, *22*, 908. [CrossRef] [PubMed]
3. Feidt, M. The history and perspectives of efficiency at maximum power of the Carnot engine. *Entropy* **2017**, *19*, 369. [CrossRef]
4. Andresen, B.; Berry, R.S.; Nitzan, A.; Salamon, P. Thermodynamics in finite time. I. The step-Carnot cycle. *Phys. Rev. A* **1977**, *15*, 2086–2093. [CrossRef]
5. Berry, R.S.; Salamon, P.; Andresen, B. Finite-Time Thermodynamics, Entropy Special Issue. 2022. Available online: https://www.mdpi.com/journal/entropy/special_issues/FTT (accessed on 25 December 2025).
6. Samuelson, P.A. The stability of equilibrium: Comparative statics and dynamics. *Econometrica* **1941**, *9*, 97–120. [CrossRef]
7. Salamon, P.; Andresen, B.; Gait, P.D.; Berry, R.S. The significance of Weinhold's length. *J. Chem. Phys.* **1980**, *73*, 1001–1002; 5407E. [CrossRef]
8. Weinhold, F. Metric geometry of equilibrium thermodynamics. *J. Chem. Phys.* **1975**, *63*, 2479–2483. [CrossRef]
9. Deffner, S.; Jarzynski, C.; del Campo, A. Classical and quantum shortcuts to adiabaticity for scale-invariant driving. *Phys. Rev. X* **2014**, *4*, 021013. [CrossRef]
10. Klinger, J.; Rotskoff, G. Universal energy-speed-accuracy trade-offs in driven nonequilibrium systems. *Phys. Rev. E* **2025**, *111*, 014114. [CrossRef] [PubMed]
11. Andresen, B.; Salamon, P. Future perspectives of Finite-Time Thermodynamics. *Entropy* **2022**, *24*, 690. [CrossRef] [PubMed]
12. Clausius, R. Über verschiedene für die Anwendung bequeme Formen der Hauptgleichungen der mechanischen Wärmetheorie. *Ann. Phys.* **1865**, *125*, 313–356.

https://doi.org/10.3390/e28010049

entropy

MDPI

Editorial

Maximum Power Efficiency

Boye Ahlborn * and Frank Curzon

Department of Physics and Astronomy, University of British Columbia, Vancouver, BC V6T 1Z3, Canada; curzon@phas.ubc.ca
* Correspondence: ahlborn@phas.ubc.ca

New research often starts with vague, dream-like ideas, conversed on over coffee in the free flow of animated discussions about physics, the growing up of one's children, politics, and the success of the local ice hockey team.

Neither of us—Frank Curzon and myself—had any theoretical aspirations. Frank lectured on an E&M course and his students worked on Z-pinch experiments, and my students worked on pressure- or energy-driven plasma flow, like shock waves, plasma jets, and chemical detonations; I was just preparing lecture notes on thermodynamics.

One day, I casually mentioned that the Carnot efficiency is just an energy budget, one that has nothing to do with the operation of an engine that delivers mechanical power. Frank picked this casual comment up. He liked puzzles and played with physics relations for amusement, and he found solutions to questions that had not been asked before.

The authors in 1975.

Boye Ahlborn

Frank Curzon

After a few days, Frank came back with the observation that my hunch actually led to an optimum for the energy conversion when energy flows into, and out of, a heat engine at a certain rate [1]; namely, the energy transfer velocity v_{rms}, or the root mean square velocity of a gas in thermal equilibrium.

$$v_{rms} = \sqrt{3kT_D/m_b}, \quad (1)$$

The efficiency at this maximum is

$$\eta_{max} = 1 - \sqrt{T_{cold}/T_{hot}}, \quad (2)$$

and the square root contains the ratio of energy transfer velocities:

$$\sqrt{T_{cold}/T_{hot}} = \frac{v_{rms,cold}}{v_{rms,hot}} \quad (3)$$

Instead of the rms velocities, the average velocity, or the maximum of the velocity distribution could be used, since these velocities differ from the rms velocity only by a numerical constant, which drops out in the ratio of T_{cold}/T_{hot}.

Equation (2) is clearly different from the energy bookkeeping of the Carnot efficiency, and we chose the name "maximum power efficiency" in a short research note—which was rejected by two different journals. To us, however, this result looked too straightforward to be wrong, and it came as a relief when the American Journal of Physics let it go into print, although the referee noted that "while he could see nothing wrong with the derivation he had doubts about the relevance of this result".

The root mean square velocity, Equation (1), connects thermodynamic and mechanic parameters that otherwise live independent of each other in different branches of physics: temperature has neither time nor distance, and mechanical motion has no association with temperature, but both deal with energy. The gap between them is closed by fundamental physics constants and by the conservation of energy. In the case of the efficiency calculations, the Boltzmann constant, $k = 1.38 \times 10^{-23}$ [J/°K], is the missing link, which also appears in the gas law

$$p = NkT, \tag{4}$$

Connections between temperature, T, and energy also appear in radiation laws, like Stephan's law, which gives radiation intensity, W, as function of the temperature of a "black body".

$$W = \sigma T^4,\ [\text{watt/m}^2\text{K}], \tag{5}$$

Here, the connection comes in through the constant $\sigma = 5.67 \times 10^{-8}$ [K^4], and in Wien's displacement law for the maximum wavelength, λ_{max}, of a blackbody spectrum

$$\lambda_{max} T = b = 2.98 \times 10^{-3}\ [\text{m}\cdot\text{K}] \tag{6}$$

The parameters b, k, and σ are derived from the statistical mechanics of the random velocity distributions of particles in local thermal equilibrium. Actually, at low densities, a plasma can have "hot" electrons, where $T_{el} > T_{ion}$ and each group is in local thermal equilibrium.

Energy is a unique parameter that appears in all branches of physics, including fluid dynamics (plasma physics). This can be seen in the conservation of energy for a one-dimensional flow, driven by the input of some energy $Q\left[\frac{J}{kg}\right] = \left[\frac{m^2}{s^2}\right]$ which is equal to the sum of all forms of energy per unit mass (not including electrodynamic energies): the enthalpy increase $\Delta h = \Delta p + \Delta u$ (pressure p + internal energy u), the generation of kinetic energy per unit mass ½v^2, and the energy lost or gained by field in terms of W [watt/m^2] of radiation or sound:

$$Q\left[\frac{J}{kg} = \frac{m^2}{s^2}\right] = \Delta h\left[\frac{J}{kg}\right](T) + \frac{1}{2}v^2\left[\frac{m^2}{s^2}\right] + \frac{W(T)\left[\frac{watt}{m^2}\right]}{\rho_1 v_1\left[\frac{kg}{m^2}\right]} \tag{7}$$

Some examples of such flow are rockets, detonations, plasma jets, and even astrophysical radiation fronts. Another overriding principle is the conservation of mass flow J, of density ρ [kgm^3], and velocity v [m/s]: What flows in $\rho_a v_a$ must come out $\rho_b v_b$:

$$J = \rho_a v_a = \rho_b v_b = \text{const}\ [\text{kg/sm}^2] \tag{8}$$

Equation (7) is used to refer the field energy W to the mass flow J, thus generating the unit J/kg. Between Q and any two of these energy forms, a conversion efficiency could be defined.

A "hunch" is like a spark that flashes up when steel hits the grinding wheel, and lights a fire in nearby fuel. Finite time thermodynamics was certainly not in sight at the time of our morning coffee efficiency-speculations. Obviously, the impact of a communication does not depend on the size of the research program, but rather on taking an unconventional look at existing questions.

Conflicts of Interest: The authors declare no conflict of interest.

Reference

1. Curzon, F.L.; Ahlborn, B. Efficiency of Carnot engine at maximum power output. *Am. J. Phys.* **1975**, *43*, 22–24. [CrossRef]

entropy

MDPI

Article

Finite-Time Thermodynamics: Problems, Approaches, and Results

Anatoly M. Tsirlin [1,*], Alexander I. Balunov [1,2,*] and Ivan A. Sukin [1]

[1] System Analysis Research Center, Ailamazyan Program Systems Institute of RAS, 152021 Pereslavl-Zalessky, Russia; ivsukin@gmail.com
[2] Department of Cybernetics, Yaroslavl State Technical University, 150023 Yaroslavl, Russia
* Correspondence: tsirlin@sarc.botik.ru (A.M.T.); balunovai@ystu.ru (A.I.B.)

Abstract: In this manuscript, the typical problems of "finite-time thermodynamics", their general methodology, and the general features of their solutions are considered. We also consider the role of minimal dissipation processes, the properties of the irreversibility index, and the consequences of its existence. A generalization of the Carathéodory theorem for averaged optimization problems corresponding to cyclic processes and the properties of optimal solutions following from it are given. The existence of the irreversibility index for economic macrosystems and their analogies to and differences from thermodynamic systems are proven.

Keywords: irreversible thermodynamics; optimization; problem statements; thermodynamic balances; Carathéodory theorem; averaging; entropy production

1. Introduction. History of the Emergence of "Finite-Time Thermodynamics" and the Validity of Its Name

The development of thermodynamics, starting with the work of S. Carnot, is closely connected with extremal problems on the ultimate possibilities of thermodynamic systems. Carnot would hardly have obtained a solution to the problem of the ultimate efficiency of a heat engine if he had posed it in a mathematically rigorous manner, such as the following: *find a law $T(t)$ of change in the temperature of the working fluid of a heat engine receiving heat from a source with temperature T_+ and giving the heat to a source with temperature T_- such that the ratio ρ of the work produced to the heat taken from the hot source is maximal.* After all, one of the variables sought for in this problem is the duration of the cycle τ. The set of admissible values of this variable is limited only by the condition of non-negativity and therefore is not closed. Consequently, according to the Weierstrass theorem, which Carnot could not have known, the problem may not have a solution. Indeed, the maximum of ρ does not exist. The upper limit of the efficiency supremum is achieved in the limit as τ tends to infinity.

This feature was discovered to be characteristic of other extreme problems in thermodynamics (on the minimum work for separation of mixtures, on the efficiency of cycles for systems with sources of finite capacity, etc.). The solutions to these problems are reversible processes in which the exchange flows are arbitrarily close to zero; therefore, an arbitrarily long time is required to exchange a finite amount of substance or energy.

Note that a reversible Carnot heat engine can also have finite power but only in cases where the heat exchange coefficients between the heat sources and the working fluid of the heat engine (the dimensions of the engine) are arbitrarily large. In this case, the Carnot efficiency is the ratio of the "reversible power" to the heat flow taken from the hot source.

Apparently, the first problem in optimization thermodynamics was the "maximum power problem", about a form of the heat engine cycle receiving heat from a source of infinite capacity with temperature T_+ and giving off heat to a source with temperature T_-, for which the power of the heat engine would be maximum. This was considered and solved soon after Carnot's work; however, this result was not noticed. Much later, the problem of the maximum power cycle was solved in [1,2]. At that time, nuclear power began to develop. The cost of building nuclear power plants is very high, and the cost of fuel is much less than for thermal power plants. Under these conditions, obtaining maximum power is much more important than achieving maximum efficiency.

In the problem of maximum power, instead of the quantities of heat and work, heat and work flows appear within the kinetics of heat transfer when the dimensions of the heat engine are arbitrarily large. The authors of the above-mentioned and numerous other studies often solved this problem independently of each other but according to the same scheme:

1. Heat exchange flows in contact with each of the sources were assumed to be proportional to the difference in temperature between the source and the working fluid (Newtonian kinetics).
2. The desired cycle was assumed a priori to consist of two isotherms and two adiabats, similar to the Carnot cycle, and the temperatures of the working fluid in contact with the sources were selected based on the condition of maximum power, taking into account the law of conservation of energy and the fact that the entropy flow coming from a hot source must be equal to the entropy flow given to a cold source. The latter meant that the processes inside the working fluid were assumed to be reversible.

It was revealed that the maximum power of a heat engine in contact with reservoirs having temperatures T_+ and T_- is equal to

$$N_{max} = \frac{\alpha_1 \alpha_2}{\alpha_1 + \alpha_2} \left(\sqrt{T_+} - \sqrt{T_-} \right)^2. \tag{1}$$

where α_1 and α_2 are heat exchange coefficients in contact with sources. The efficiency of the maximum power cycle does not depend on the heat exchange coefficients and is equal to

$$\eta = 1 - \sqrt{\frac{T_-}{T_+}}. \tag{2}$$

These works did not answer the following questions:

1. For what heat transfer kinetics does the maximum power cycle consist of two isotherms and two adiabats?
2. Is the efficiency of the maximum power cycle always independent of the heat transfer coefficients?
3. What is the shape of the heat engine cycle that has the maximum efficiency at a fixed power?

Answers to these questions were given in the works [3,4], in which the methods of averaged optimization developed in [5] were used. It was found that the following statements are true:

1. For any heat exchange kinetics satisfying the natural condition that the direction of heat flow coincides with the sign of the temperature difference of the contacting bodies, the maximum power cycle consists of two isotherms and two adiabats.

2. The efficiency corresponding to this cycle for kinetics that differ from Newtonian depends on the kinetic coefficients.
3. The cycle of a heat engine with maximum efficiency at a fixed power for arbitrary kinetics can consist of three isotherms and three adiabats. There, a condition was obtained under which the number of isotherms is equal to two.

The development of optimization thermodynamics from the «maximum power problem» to its current state began after the publication of the work of Curzon and Ahlborn [2]. This happened largely because the talented researchers P. Salamon, B. Andresen and K. H. Hoffmann ended up at the Department of Chemistry at the University of Chicago, headed by R.S. Berry. They realized that in real thermodynamic processes, thermal, mass transfer, chemical, etc., the factor of limited duration plays a crucial role. It is important to evaluate the capabilities of systems with an additional restriction on the duration of processes, i.e., in the class of irreversible processes. Thus, a new field of thermodynamics arose, which they called «Finite-time thermodynamics» (see [6–9] and many other publications).

The name «Finite-Time Thermodynamics» (FTT) has taken hold and is hard to abandon. However, this name cannot be considered successful. The overwhelming majority of problems on optimal irreversible processes do not contain time at all. The processes occur in stationary or cyclically operating systems. In the latter case, the indicators are averaged over the cycle time and time is not included in the problem after averaging.

Three classes of problems are typical:

1. Optimal periodic processes of limited duration
2. Optimal cyclic processes
3. Optimal processes in open stationary systems.

At the same time, for any class of problems, including problems on periodic processes, their conditions lead to non-zero flows due to their limited duration.

The correct name for this area of thermodynamics would be «Thermodynamics with non-zero flows». It emphasizes the features of all problems arising in it.

In order to be able to use the apparatus of equilibrium thermodynamics when solving problems that arise, the system is divided into several equilibrium subsystems, and irreversibility is determined only by the processes that occur when equilibrium subsystems come into contact with each other.

Here, thermodynamic systems should be understood as all systems consisting of a large number of interacting elements, each of which cannot be controlled. Control of such systems is possible only at the macro level and this control imposes the same conditions on all microelements. Such systems are called *macrosystems*. Examples of macrosystems may be economic systems, where microelements are sellers and buyers, individual «households.»; social systems, where microelements are taxpayers, transport passengers, and others. An important class of macrosystems are *segregated* systems [10], whose elements (households, units in fluidized bed apparatuses, biomass in biochemical reactors, etc.) evolve under the influence of the environment. The parameters of the environment are determined by external influences and the average effect of the units.

All macrosystems are characterized by the division of the variables characterizing them into intensive and extensive. That implies the existence of processes that proceed spontaneously in one direction, without the expenditure of external resources, and in the other requiring the involvement of external resources for their implementation, i.e., the processes in such systems are irreversible. The irreversibility index for any processes in closed macrosystems increases. However, the form and the fact of the existence of such an index require proof for each type of macrosystem. For thermodynamic systems, such an indicator is entropy, for economic systems, the «utility function», the proof of the existence of which is given below. The irreversibility indicator is always the product of the flow and the driving force causing this flow. The sign of the flow coincides with the sign of the driving force, so their product is non-negative.

In macrosystems, there is a special subsystem, an *intermediary*. The intermediary (a reseller in economics, a working fluid in thermodynamics) reduces the growth rate of the irreversibility indicator and, due to this, extracts the target resource from the system (capital in economics, work in thermodynamics). For the intermediary to operate, a difference in the values of intensive variables in the subsystems is necessary. In economics, a difference in prices, in thermodynamics, a difference in temperatures or chemical potentials.

In the following sections, we will present the formulations of typical FTT problems, then the general methodology, and the results of their solution. Let us consider the mathematical features of solving problems of resource exchange and optimization of cyclic processes.

The present work is a review of some results obtained by the authors in solving FTT problems and in developing mathematical methods necessary for such a solution. Our goal is to show by examples that FTT has developed as a complete section of thermodynamics with its own methodology and class of problems, that it has allowed formulating and solving problems that could not even be posed within the framework of classical thermodynamics, that the FTT methodology is also applicable to other types of macrosystems, such as microeconomics, that FTT required the development of averaged optimization methods, and that there are unsolved problems in FTT.

Let us list these results:

1. Processes of minimum dissipation.
2. Construction of attainability sets of thermodynamic systems.
3. Extraction of maximum work from a non-uniform system in a limited time.
4. Limit capabilities of heat exchange systems.
5. Microeconomic analogies.
6. Optimality conditions for averaged problems and Lyapunov-type equations.

2. Statements and Solutions of Problems of FTT

2.1. The Problem of Maximum Possible Productivity

Each non-isolated thermodynamic system exchanges energy and material flows with its environment. In steady-state conditions, these flows can be stationary or periodic. In the latter case, the flow intensity is understood as its average intensity over a period. The mechanism of system operation (kinetics of heat and mass transfer, chemical reactions, etc.) establishes a connection between the incoming and outgoing flows. From the outgoing flows, a target flow can be identified or formed. Its intensity is the system's productivity. The incoming flows form a cost flow. For a heat engine, the target flow is power. The cost flow is heat taken from a hot source.

The most studied types of macrosystems are thermodynamic. In the first section, we will present the formulations, methodology, and results of solving typical problems for such systems. In the second section we will focus on systems of an economic nature, and in the third we will prove the conditions that determine the type of optimal solution to problems with averaging, characteristic of macrosystems.

The problem of the maximum possible productivity of a thermodynamic system of arbitrary nature under certain constraints and the value of the cost flow that ensures this maximum productivity is a direct generalization of the maximum power problem.

This raises the following questions:

1. For which systems is productivity limited from above, and for which systems, by increasing the cost flow, can productivity be made arbitrarily large, and therefore the problem of maximum productivity has no solution?
2. There is a system of two or more thermodynamic reservoirs and a working fluid that contacts each of them in a steady state or alternately and produces a target flow. How should the contacts of the working fluid be organized to obtain the maximum value of the target flow? What should be considered the efficiency of such a system?
3. What will change if in problem 2, instead of reservoirs, there are sources of finite capacity? In particular, what is the maximum work that can be extracted in a closed thermodynamic system with a fixed duration? This problem coincides with the problem of calculating the exergy of a system in the particular case where the duration of the process is unlimited.
4. How to organize thermodynamic processes so that for a given average flow intensity increase in entropy of the system is minimal (minimum dissipation processes)?
5. In particular, what criterion should be used to evaluate the heat exchange process? How to organize the heat exchange process of two vector flows so that for a given heat load and total heat exchange coefficient the entropy production is minimal?
6. How to construct the region of realizable modes of a thermodynamic system with limited kinetic coefficients in a space along the axes of which the flow intensities are plotted.

The above formulations, of course, do not exhaust the problems in FTT, but they allow us to judge the nature and applied focus of the problems that arise. Note that in most of them the duration of the process does not figure.

2.2. General Methodology for Solving Problems of FTT, Dissipation

Thermodynamic systems are characterized by two types of variables: intensive and extensive. The first type do not change when combining subsystems if before combining they were equal in each subsystem (temperature, pressure, concentration ...), while the second type are are added when combining such subsystems (volume, number of moles, internal energy...).

In this case, subsystems can be either *passive*, in which case intensive variables are determined by extensive variables and the equation of state, or *active* (the working fluid of a heat engine, or an absorption desorption cycle), in which case intensive variables (they are controls) are selected to achieve one or another goal.

Let us first characterize the general scheme by which these and similar problems can be solved.

- The first step in studying the ultimate possibilities of thermodynamic systems is to compile balance relations for matter, energy, and entropy. The last of these relations will include a term characterizing the irreversibility of processes: entropy production rate σ. It is equal to the rate of growth of the entropy of the system. This term is equal to zero if all processes in the system are reversible, and greater than zero for irreversible processes. The non-negativity of dissipation, due to the equations of thermodynamic balances, determines a certain set of realizability in the parameter space of input and output flows.
- If additional conditions of finite duration of processes, given average flow intensity, and limited kinetic coefficients are imposed on the system, then the dissipation value that is minimally possible under these restrictions is found. In any real system $\sigma \geq \sigma_{min}$, which narrows the set of realizability. Now this set takes into account the kinetics of the processes as well as the dimensions of the installation through the heat and mass transfer coefficients.

- The third step is to obtain from the balance equations the relationship between one or another indicator of the system's efficiency and the dissipation σ. As a rule, natural efficiency indicators monotonically deteriorate with the growth of σ and reach their limit values in a reversible process, which leads to estimates similar to the Carnot efficiency for processes of very different nature.
- Since in a complex system dissipation additively depends on dissipation in each of the elementary processes, an important stage of the study is to identify the conditions of minimal dissipation. Optimal organization of processes in a complex system comes down to coordinating individual processes of minimal dissipation with each other.

2.3. Thermodynamic Balances

Thermodynamic balances are a system of equations of material, energy, and entropy balances [11,12]. For simplicity, we will consider them for an open system. Some flows enter the system from outside, others are generated in the system. One of such flows generated in the system is entropy production, which is non-negative. This circumstance turns the entropy balance equation into an inequality. Together with the other equations, it identifies a region of feasibility in the space of flows, the boundary of which corresponds to reversible processes. If, however, in one way or another it was possible to solve the problem of the minimum possible entropy production σ_{min} under the restrictions imposed on the process, then the equations of thermodynamic balances with the condition $\sigma \geq \sigma_{min}$ identify a region whose boundary is determined by processes of minimal dissipation. This region lies inside the region bounded by reversible processes.

Open system. Thermodynamic balances establish a connection between the flows of each substance, energy, and entropy that the system exchanges with the environment, as well as the occurrence of these quantities in the system and the rate of change of their quantity. We will sum up all flows further, considering the incoming flows as positive, and the outgoing flows as negative. We will divide the flows into convective and diffusive, marking the latter with the index d. Unlike the convective flow, the diffusion flow depends on the difference between the intensive variables of the system under study at the point where it enters or leaves and the intensive variables of the environment. In addition, we will use the following notations: j—flow index, e_j, v_j—internal energy and volume of one mole of the corresponding flow, and P_j—its pressure, $h_j = e_j + P_j v_j$—molar enthalpy, h_{dj}—enthalpy in the flow entering by diffusion, q_j—j-th heat flow, N_a—power produced by the system.

Let us present the general form of balance equations:

$$\frac{dE}{dt} = \sum_j g_j h_j + \sum_j q_{dj} + \sum_j q_j - N_a, \tag{3}$$

$$\frac{dN_i}{dt} = \sum_j g_j x_{ij} + \sum_j g_{dj} x_{dj} + \sum_\nu \alpha_{i\nu} W_\nu, \tag{4}$$

$$\frac{dS}{dt} = \sum_j g_j s_j \sum_j \frac{q_{dj} - \sum_i g_{dj}\mu_{dij}}{T_{dj}} + \sum_{i\nu} \frac{\mu_{i\nu} n_{i\nu}}{T_\nu} + \sum_j \frac{q_j}{T_j} + \sigma. \tag{5}$$

Here $n_{i\nu} = -\alpha_{i\nu} W_\nu$ is the intensity of formation of the i-th substance in the ν-th reaction, T_ν is the temperature of the ν-th reaction. If there are no diffusion flows, then

$$\frac{dE}{dt} = \sum_j g_j h_j + \sum_j q_j - N_a, \tag{6}$$

$$\frac{dN_i}{dt} = \sum_j g_j x_{ij} + \sum_\nu \alpha_{i\nu} W_\nu, \tag{7}$$

$$\frac{dS}{dt} = \sum_j g_j s_j + \sum_j \frac{q_j}{T_j} + \sum_{i\nu} \frac{\mu_{i\nu} n_{i\nu}}{T_\nu} + \sigma, \tag{8}$$

where the number of heat flows includes heat flows released or absorbed during chemical reactions, which depend on the reaction rate.

If we consider a stationary mode of the process, i.e., when $dE/dt = dN_i/dt = dS/dt = 0$, then the written equations from differential equations are transformed into final relations. When considering a cyclic process, balances can be written not for each moment in time, but on average for the cycle of the installation. Since the state of the system is the same at the beginning and end of the cycle, the total change in energy, amount of substance and entropy of the working fluid for the cycle are zero. In this case, balances are also reduced to a system of relations linking the average values of the terms on the right-hand sides of the equations for the cycle.

For **closed systems** consisting of several equilibrium subsystems, thermodynamic balances have the form

$$\dot{E}_0 = \sum_i \dot{E}_i, \quad \dot{N}_0 = \sum_i \dot{N}_i, \quad \dot{S}_0 = \sum_i \dot{S}_i,$$

where i is the subsystem number, and index 0 refers to the system as a whole. In turn, $\dot{E}_i$, $\dot{N}_i$ and $\dot{S}_i$ are determined by the relations (3)–(5).

3. Processes of Minimal Dissipation

The requirement of minimum entropy production in processes of a given average intensity leads to the problem of such an organization of the process, in which the entropy production associated with it will be minimal [13]. Let us present a scheme for solving this problem for the scalar case.

Let two bodies contact, characterized by their extensive variables Y (volume, internal energy, entropy, number of moles of substance, …) and intensive u (temperature, composition, pressure, …). The difference in intensive variables leads to the appearance of an exchange flux J. The entropy increase over time L is equal to the integral of the product of the flux J and the driving force X, which, like J, depends on the intensive variables. Moreover, J and X always have the same sign. Dissipation is equal to the average rate of entropy increase. In this case, the intensive variables themselves depend on the extensive ones due to the equation of state. We will assume that the intensive variable u_2 of the second of the two contacting systems can be changed in an optimal way, and u_1 changes due to the influence of the exchange flow on it. The average value of the flow is given. The problem will take the form:

$$\overline{\sigma} = \frac{1}{L}\int_0^L J(u_1, u_2) X(u_1, u_2) dl \to \min_{u_2 \in V} \tag{9}$$

under conditions

$$\frac{du_1}{dl} = \varphi(u_1, u_2), \quad u_1(0) = u_{10}, \tag{10}$$

$$\frac{1}{L}\int_0^L J(u_1, u_2) dl = \overline{J}. \tag{11}$$

In these equations l is the duration of contact.

The value of L can be either fixed or subject to optimal selection.

The optimality conditions for this problem, as well as the optimality conditions for vector flows characterized by Onsager kinetics, are given in [14,15]. For the most important case, when

$$\varphi(u_1, u_2) = c(u_1)J(u_1, u_2) \tag{12}$$

we obtain

$$J^2(u_1, u_2) = \lambda_2\left(\frac{\partial J(u_1, u_2)}{\partial u_2} : \frac{\partial X(u_1, u_2)}{\partial u_2}\right). \tag{13}$$

4. Form of the Realizability Region

Let the organization of processes in the system correspond to the conditions of minimal dissipation and the corresponding production of entropy σ_{min}, being substituted into the entropy balance equation, distinguishes the realizability region in the space of flows. The following statement is true [15]:

If the system productivity monotonically depends on the flow of mechanical, electrical power, or separation power generated by the system, and the flow of costs monotonically depends on the heat flow, then the region of realized modes in the productivity-costs plane is limited from above (Figure 1a). If productivity is determined by the heat flow, and costs are determined by mechanical or electrical power, then productivity monotonically increases with increasing costs (Figure 1b).

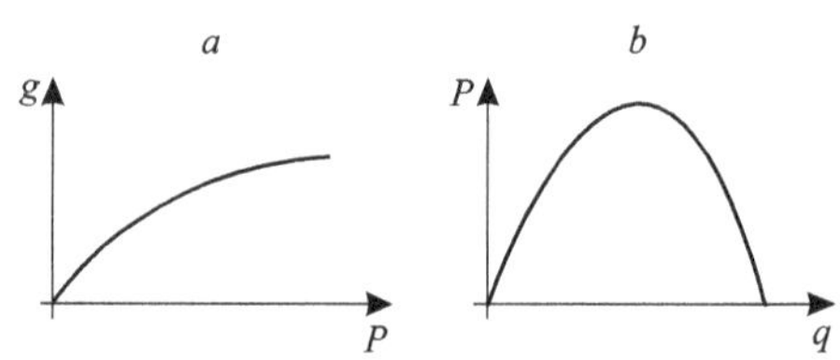

Figure 1. The form of the realizability set of a thermodynamic system for cases where the cost flow is mechanical or electrical energy (**a**) and when the cost flow is thermal energy (**b**). g is the target flow, P is the flow of expended (**a**) or produced (**b**) mechanical energy, q is the flow of heat.

The first type includes heat engines, rectification processes, absorption refrigerators, etc. The second type includes compression refrigerators and heat pumps, membrane separation systems, electric heaters, etc.

5. Maximum Work Problem

Let there be a non-uniform thermodynamic system consisting of several subsystems (reservoirs, subsystems of finite capacity), the intensive variables of which at the initial moment of time differ from each other. In addition, the system has a working body that can establish contact with each of the subsystems. The contact function $U(t)$ takes the value of one in the presence and zero in the absence of contact. The coefficients of heat and mass transfer for such contact are limited. It is required to organize the contacts of the working body with the subsystems in such a way as to extract the maximum work from the system in a given time τ. The extracted work is equal to the change in the internal energy of the system, and since the initial state is given, the goal of the solution is the maximum decrease in the internal energy in a limited time. In this case, the state of the working body at the beginning and end of the process is usually assumed to be the same.

The result of solving the problem are two statements [16]:

Statement 1: In a thermodynamic system consisting of reservoirs and a working fluid with a given initial state, for any laws of heat and mass transfer, the maximum work extracted during time τ corresponds to a process for which

- *the vector of intensive variables u and contact functions U on the interval $(0, \tau)$ is piecewise constant, and the number of values it takes does not exceed $r + m + 2$, where r is the number of conditions imposed on the final state of the subsystems and m is the dimension of the concentration vector;*
- *at the beginning and end of the process, the intensive variables of the working fluid change abruptly to some optimal values corresponding to optimal pressures;*
- *the entropy of the system grows on the interval $(0, \tau)$ as a piecewise linear function.*

Statement 1 follows from the properties of solutions of averaged problems presented in the previous section. Depending on the given boundary conditions, the maximum work can be greater or less than zero. In the latter case, it corresponds to the minimum of the expended work. We emphasize that such a structure of the optimal process is characteristic of any kinetics of heat and mass transfer.

Corollary: When no restrictions are imposed on the composition and entropy of the working fluid and the increments of the extensive variables of the reservoirs at $t = \tau$, the entropy of the system in the optimal process for any laws of heat and mass transfer increases at a constant rate, and the working fluid contacts the same reservoirs throughout the process.

In the presence of subsystems of finite capacity, the problem of maximum work turns out to be an optimal control problem with integer variables $U(t)$. In this case, it is true that

Statement 2: On each interval of constancy of the contact function between the working fluid and the subsystem of finite capacity, the law of change of the vector $u(t)$ of intensive variables of the working fluid in the optimal process must satisfy the conditions of minimal dissipation.

6. Ideal Organization of Heat Exchange Systems

In multi-flow heat exchange systems with given input temperatures and heat capacities of hot flows, total heat exchange coefficient and heat load for Newton's law of heat exchange, the minimum of entropy production corresponds to such an organization in which at each point of contact the ratio m of absolute temperatures of the heated and heating flows is the same, as is the temperature $\overline{T}$ of cold flows at the system outlet [17].

The following relationships are valid:

$$\left.\begin{array}{l}
\overline{T} = \frac{\sum\limits_i T_{i0} W_i - \overline{q}}{\sum\limits_i W_i}, \\
q^*(T_{i0}) = W_i(T_{i0} - \overline{T}), \\
\alpha^*(T_{i0}) = \frac{\overline{\alpha} W_i(\ln T_{i0} - \ln \overline{T})}{\sum\limits_i W_i(\ln T_{i0} - \ln \overline{T})}, \\
m = 1 - \frac{1}{\overline{\alpha}} \sum\limits_i W_i(\ln T_{i0} - \ln \overline{T}), \\
\overline{\sigma}^* = \overline{\alpha} \frac{(1-m)^2}{m}, \\
\alpha^*(T_{i0}) = q^*(T_{i0}) = W_i = 0, \quad T_{i0} \le \overline{T}.
\end{array}\right\}$$

Here T_{i0}, W_{i0} are the temperature and heat capacity of the i-th hot stream, q is the heat flux, σ is the entropy production rate, and $\overline{\alpha}$ is the total heat transfer coefficient for the system. The last equality indicates that hot streams whose temperature does not exceed $\overline{T}$ should be excluded from the system.

7. Existence of the Irreversibility Index and Thermodynamic Analogy for Economic Macrosystems

In the economy, economic agents (EA) function at the micro level, each of which is uncontrollable. Together, they form a macrosystem that is largely analogous to the thermodynamic one [18].

7.1. Existence of the Welfare Function and Its Properties

The analogy between thermodynamic and microeconomic systems is obvious and has attracted the attention of many researchers [19–22]. The reason for this analogy is that both systems consist of a large number of individually uncontrolled elements, economic agents (EA), i.e., they are macrosystems. In order to make such an analogy strict, it is necessary to introduce a variable for microeconomic systems that is an indicator of the irreversibility of the processes occurring in them. The proof of the existence of such an indicator is given in [16]. Here it is given with minor abbreviations.

In exchange processes, the EA buys and sells resources, simultaneously changing the stocks of resources and capital. Let us introduce the function U, the differential of which

$$dU = dM + \sum_i p_i dN_i. \tag{14}$$

We will call this function *capitalization* or *total capital* of the EA, since its change takes into account the change in both the basic resource M and the *tied capital* $F = \sum_i p_i N_i$. The vector p depends on the stocks of resources. The values p_i have the dimension of the cost of the i-th resource in units of the basic resource.

Let some firm carry out an equilibrium exchange with the EA, exchanging some types of resources for others through purchases and sales. The exchange occurs reversibly and in such a way that the initial and final states of the EA in the space with coordinates N_i coincide. If in such a process the firm could extract a certain amount of the basic resource, this would mean that the possibilities of its extraction are not limited, since it can be obtained only from one EA, without causing any changes in its state or in the state of the environment. From the impossibility of such an economic "perpetual motion machine" it follows that $r = \text{const}$

$$\oint \sum_i p_i(N, M)\, dN_i = 0.$$

From this, in turn, it follows that there exists a function $Q(N, r)$, whose partial derivatives with respect to N_i are equal to p_i, and the differential has the form

$$dQ = \sum_i p_i\, dN_i + \frac{\partial Q}{\partial r}\, dr.$$

Here r is the valuation of the EA capital.

The expression (14) can be rewritten as

$$dU = dM + dQ - \frac{\partial Q}{\partial r}\, dr = d(M + Q) - \frac{\partial Q}{\partial r}\, dr.$$

Denoting $M + Q = Y$, and $-\frac{\partial Q}{\partial r} = \gamma$, we obtain that

$$dU = dY + \gamma\, dr. \tag{15}$$

Expression (15) depends on three variables: Y, γ and r. These variables are related to each other. If this were not so, then on the plane with coordinates Y, r it would be possible to move from a given initial state to any other with constant capitalization by choosing γ. In reality, the possibilities of such a transition are limited.

Indeed, let point 1 correspond to the initial state on the plane Y, r, and point 2 to the final state, such that with reversible exchange with the environment with a constant value of r, the capitalization of the EA decreases $(dU < 0)$. If it were possible to get to the same point at $dU = 0$, then the EA could organize a cyclic exchange process in which its state would change from point 1 to point 2 along the trajectory with $dU = 0$, and would return from point 2 to point 1 with a constant value of r and an increase in capitalization $(dU > 0)$. So it could increase its capitalization by any amount in a reversible exchange, which is impossible. Thus, γ is a function of Y and r. The differential of the total capital is thus a Pfaffian form of two variables, which always has an integrating factor.

Recall that a Pfaffian form is a first-order differential form, i.e., the sum of products of functions of several variables by the differentials of these variables,

$$dK = \sum_{i=1}^{n} F_i(x)\, dx_i.$$

If $n = 2$ and the functions F_i are differentiable, then there will always be an integrating factor $f(x)$ such that $dS = f(x)dK$ is a total differential, i.e., S depends on x, and $\oint dS = 0$. The factor $f(x)$ is usually not unique.

Before finding the integrating factor $p_0(Y, r)$, let us clarify the dependence of γ on these variables. To do this, consider a system consisting of two EAs with identical estimates of the basic resource $(r_1 = r_2 = r_0)$. Capitalization U and variable Y and the basic resource reserves for the system are additive, so that $U = U_1 + U_2$, $Y = Y_1 + Y_2$. So that $Q = Q_1 + Q_2$. Accordingly, $\gamma = \frac{\partial Q}{\partial r} = \gamma_1 + \gamma_2$. Thus, the dependence of γ on Y satisfies the superposition principle: $\gamma(Y_1 + Y_2, r) = \gamma(Y_1, r) + \gamma(Y_2, r)$ and has the form

$$\gamma(Y, r) = G(r)Y + l(r), \tag{16}$$

Thus, for the capitalization differential we obtain the expression

$$dU = dY + (G(r)Y + l(r))\, dr.$$

The integrating factor can be any function p_0 such that

$$dS = p_0 dU = p_0 dY + p_0(G(r)Y + l(r))dr$$

is a total differential. Since the mixed derivative of the function S does not depend on the order of differentiation, the function p_0 must satisfy the condition

$$\frac{\partial S}{\partial r \partial Y} = \frac{\partial S}{\partial Y \partial r} \rightarrow \frac{\partial p_0}{\partial r} = \frac{\partial [p_0(G(r)Y + l(r))]}{\partial Y}. \tag{17}$$

We will look for p_0 as a function of r, then the condition (17) leads to the equation

$$\frac{\partial p_0}{\partial r} = p_0 G(r). \tag{18}$$

One of the solutions to this equation is the function $p_0(r) = \exp[G(r)]$. It uniquely depends on r and can also serve as an estimate of the base resource.

Thus, there is some function of state (extensive variables) $S(N, M)$ such that its differential has the form

$$\oint dS = 0. \tag{19}$$

In a process with non-zero flows, when prices do not coincide with estimates, the function S (the irreversibility index) always increases.

The estimates of resources can be expressed in terms of the function S as

$$p_0 = \frac{\partial S}{\partial M}, \quad p_i = \frac{\partial S}{\partial N_i} \Big/ \frac{\partial S}{\partial M}, \quad i = 1, 2, \ldots \tag{20}$$

In this case, the assessment of the basic resource p_0 is positive for all economic agents, and p_i can be negative if the i-th resource requires, for example, utilization or storage costs. Cash can be considered one of the resources with a reserve of N_j, the assessment of which $p_j = 1/p_0$. The function $S(N, M)$ is called the welfare function or, more briefly, *welfare*.

When exchanging resources between economic agents, the conditions of voluntariness must be observed, which consist in the fact that none of the welfare functions S_ν decreases (the exception is associated exchange, charity). The conditions of voluntariness make it impossible to directly exchange one type of resource if its assessments by the EAs in contact with each other are of the same sign. Such an exchange becomes possible only if there is an intermediary.

If the welfare function is measured in national currency, then the value p_0 characterizes the value for the EA of the international currency and has the dimension [unit of national currency/unit of the basic resource]. The assessment p_0 of the basic resource in the processes of exchanging money by an economic agent on the foreign exchange market plays the same role as the assessment p in the exchange of resources.

The description of economic systems becomes formally close to the relations of thermodynamics if we introduce an *'economic temperature'*,

$$T = \frac{1}{p_0}.$$

This notation was introduced in [23], the interpretation of this quantity, called *liquidity*, and its properties were given considerable attention in [24].

When the scale of the EA changes without changing its properties, the stocks of resources and capital are proportional to the scale. It is natural to assume that the welfare function changes in the same way, i.e., it is an extensive variable like N and M. In this case, the S function is homogeneous of the first order, and its derivatives with respect to N and M (estimates) are homogeneous functions of the zeroth order. According to Euler's theorem for homogeneous functions, it can be written in the form

$$S(N, M) = p_0(M, N)\left(\sum_i p_i(M, N) N_i + M\right). \tag{21}$$

The dependence $p(N, M)$ can be found by formulas (20) and experimentally by the behavior of the EA in exchange processes.

If the existence of a welfare function S is postulated, then the estimates are determined through an extremal problem in which they enter as parameters:

$$S(N, M) \to \max \Big/ \left(\sum_i p_i N_i + M\right) = U, \tag{22}$$

where the value of U is fixed.

The Lagrange function of the problem (22) is

$$L = S(N, M) - p_0 \left(\sum_i p_i N_i + M \right),$$

where p_0 is an indefinite Lagrange multiplier. The stationarity conditions for L lead to the relations

$$p_i(N, M) = \frac{\partial S}{\partial N_i} \Big/ \frac{\partial S}{\partial M}, \tag{23}$$

which coincides with the expression (20). The function S is often assumed to be twice continuously differentiable, monotonically increasing in each of the arguments, strictly convex upward and equal to zero at the origin. Its partial derivatives tend to infinity when the corresponding variable tends to zero.

As a consequence of these assumptions, the solution to the problem (22) is unique and positive, and p_i decreases with an increase in N_i. With this description, the EA is similar to a thermodynamic subsystem of finite capacity. For an economic reservoir, the resource estimates p_i and p_0 are constant, and the welfare function is linear.

Note that in the general case, the welfare function of a system is not equal to the sum of the welfare functions of its constituent subsystems. Moreover, for each subsystem, the welfare function can have its own dimension. In contrast, the capitalization of each of the subsystems, the volume of the basic resource and the associated capital have the same dimension and their sum makes sense for the system as a whole under the conditions of internal equilibrium of each of the EA included in it.

7.2. Differential Relations Between Estimates, an Analogue of the Gibbs-Duhem Equation

Let's write the differential of the function S,

$$dS = p_0 \left(dM + \sum_{i=1}^{n} p_i dN_i \right) = p_0 dU. \tag{24}$$

Solving (24) with respect to dM, we obtain

$$dM = \frac{dS}{p_0} - \sum_{i=1}^{n} p_i dN_i. \tag{25}$$

From equality (21) it follows that

$$M = \frac{S}{p_0} - \sum_{i=1}^{n} p_i N_i, \tag{26}$$

$$dM = \frac{dS}{p_0} + Sd\left(\frac{1}{p_0}\right) - \sum_{i=1}^{n} (p_i dN_i + N_i dp_i). \tag{27}$$

Comparing equalities (27) and (25), we obtain a relationship linking the estimates of resources and capital

$$Sd\left(\frac{1}{p_0}\right) - \sum_{i=1}^{n} N_i dp_i = 0. \tag{28}$$

Similarly, comparing the differential S found from equality (21) with the expression (24), we obtain

$$Mdp_0 + \sum_{i=1}^{n} N_i d(p_0 p_i) = 0. \tag{29}$$

Conditions (28) and (29) follow from the existence of the function S and its homogeneity. They are economic analogues of the Gibbs-Duhem equations. For example, it follows from them that if the state of the system changes so that the resource estimates are constant, then the capital estimate p_0 is also unchanged.

Due to the symmetry of the matrix of second derivatives for a twice differentiable function, the sensitivity of estimates to changes in resource stocks and capital are related by the equalities

$$\frac{\partial(p_0 p_i)}{\partial N_j} = \frac{\partial(p_0 p_j)}{\partial N_i} = \frac{\partial^2 S}{\partial N_i \partial N_j}, \tag{30}$$

$$\frac{\partial p_0}{\partial N_j} = \frac{\partial(p_0 p_j)}{\partial M} = \frac{\partial^2 S}{\partial M \partial N_j}. \tag{31}$$

From conditions (30) and (31) it follows that

$$\frac{\partial p_i}{\partial N_j} + p_i \frac{\partial p_j}{\partial M} = \frac{\partial p_j}{\partial N_i} + p_j \frac{\partial p_i}{\partial M}, \quad i,j = 1,\ldots,n. \tag{32}$$

Equalities (30) and (31) are the economic analogue of Maxwell's relations.

7.3. Dissipation of Capital

Let us return to the cyclical process of interaction of a firm with one EA and require that the average intensity of exchange processes be fixed. Then the firm when purchasing a resource will be forced to raise prices compared to the estimates p_i, and sell at prices that are lower than these estimates. The capitalization of the EA will increase, since

$$\Delta U = \oint \sum (p_i(N,M) - c_i)\, dN_i \geq 0, \tag{33}$$

and the firm will suffer losses in comparison with the reversible process in the amount of ΔU.

We will call the intensity of the firm's losses due to irreversibility,

$$\sigma(t) = \sum_i n_i(p_i, c_i)(p_i - c_i) \geq 0, \tag{34}$$

the *dissipation of capital* associated with the irreversibility of the resource exchange process. In this case, it has the meaning of trading costs.

The condition (33) of non-decreasing capitalization (and therefore welfare) in economic exchange is an analogue of the Clausius integral. Similarly, the law according to which at the contact of two EA, the resource passes from the EA for which its valuation is lower to the EA for which its valuation is higher, and at the same time the total value of the tied capital does not decrease ($\Delta(F_1 + F_2) \geq 0$), is an analogue of the second law of thermodynamics. This allows us to construct an irreversible microeconomics, in many ways similar to finite-time thermodynamics.

7.4. The Second Law of Microeconomics

The laws of conservation of matter and energy in microeconomics are the laws of conservation of resources. Here we will focus on the economic analogy of the second law of thermodynamics.

There are several formulations of the second law of thermodynamics, each of which can be considered a consequence of the others. Let us discuss the analogs of some of these formulations in microeconomics.

Among the numerous formulations of the second law, we will highlight two: the formulation of Clausius with Planck's clarification: «*Heat by itself cannot pass from a cold body to a hotter body without leaving other changes*», and also the formulation of Leontovich: «*It is impossible to construct a device, as a result of the action of which positive work would be produced only due to the cooling of one body without any other changes*».

In microeconomics, the following statements correspond to the above formulations:

1. *A scalar resource flow cannot pass from an EA with a higher valuation to an EA with a lower valuation without leaving other changes.*
2. *It is impossible to extract capital by exchanging resources with one EA without any other changes.*

M. Planck formulated the following statement as a consequence of the second law of thermodynamics: «*In an irreversible process in a closed thermodynamic system, entropy can only increase and the exergy of the system can only decrease. The equilibrium state of such a system corresponds to the maximum entropy and the minimum performance under conditions corresponding to the imposed restrictions*». Similarly, for economic systems resource exchange processes in isolated microeconomic systems proceed in such a direction that the total tied capital of economic agents increases and reaches a maximum, and the potential ability of extracting the basic resource (profitability) decreases and reaches a minimum, under the restrictions imposed on the system, which include the conditions of voluntariness.

7.5. Analogies Between Thermodynamic and Microeconomic Systems and the Variables Characterizing Them

The notations adopted here and listed in Table 1 are: T_{-} and T—temperatures of the reservoir and the system in contact with it, p_{-}—valuation of the resource on the market, c—price of the resource set by the monopolist firm, N—stock of the resource, U—internal energy of the system and total capital, q and n—flows of heat and resource, M and F—basic resource and tied capital.

Table 1. Equivalent nomenclatures for thermodynamic and microeconomic systems.

Thermodynamic System		Microeconomic System	
Name	**Designation**	**Name**	**Designation**
Reservoir (reversible heat exchange)	T_{-}	Economic reservoir	p_{-}
Reservoir (irreversible heat exchange)	$q = \alpha(T - T_{-})$	Monopoly market	$n = \alpha(c - p_{-})$
Amount of substance	N	Reserve of resource	N
System with finite capacity, chem. potential	$\mu(N)$	EA, resource assessment	$p(N)$
Heat engine, temperature	$T(t)$	Intermediary firm, price	$c(t)$
Free energy, work	A	Basic resource	M
System performance	E	System profitability	E
System entropy	S	Tied capital	F
Entropy production	σ	Capital dissipation	σ
Internal energy	U	Total capital	$U = M + F$

7.6. Microeconomic Balances

7.6.1. Open System

Let us write the balance equations for a heterogeneous economic system exchanging resource and capital flows with its environment. In this case, we assign index i to the i-th type of resource, and index j to the j-th subsystem. External resource and capital flows entering the system will be considered positive, and those leaving it will be considered negative. These flows are divided into two categories: those forced into the system and changing their intensity over time under the influence of external factors, and those dependent on prices set by external sellers and buyers, and on resource valuations in the subsystem under consideration. The first of these, by analogy with thermodynamics, will be called *convective* and marked with index k, and the second *diffusive* (index d). In addition, the subsystem can transform some types of resources into others.

The balance equation for the i-th resource is

$$\dot{N}_i = \sum_j \left(n_{ij}^k(t) + n_{ij}^d(p_j, c_j) + W_j(p_j)\alpha_{ij} \right), \quad i = 1, 2, \dots \tag{35}$$

Here the summation is carried out over all subsystems; $W_j(p_j)$—the intensity of the resource transformation process in the j-th subsystem, and the coefficients $\alpha_{ij} > 0$ if the i-th resource arises in the j-th subsystem and $\alpha_{ij} < 0$ if this resource is spent, they determine how much of the i-th resource arises (is spent) per unit of time; c_j—the price vector during the exchange of the j-th subsystem with the environment.

The balance equation for the basic resource is

$$\dot{M} = \sum_j \left(m_j^k(t) - \sum_i c_{ij} n_{ij}^d(p_j, c_j) \right) \tag{36}$$

and the balance sheet equation for tied capital is

$$\dot{F} = \sum_{i,j} p_{ij}(N_j, M_j)(n_{ij}^k(t) + n_{ij}^d(p_j, c_j)) + \sigma, \tag{37}$$

where capital dissipation σ is equal to

$$\sigma = \frac{1}{2} \sum_j \sum_\nu n_{j\nu}(p_j, p_\nu)(p_j - p_\nu) + \sum_j W_j(p_j) A_j. \tag{38}$$

Here p_j and p_ν are vectors of resource estimates for the j-th and ν-th subsystems with components p_{ij} and $p_{i\nu}$, $A_j = \sum_i \alpha_{ij} p_{ij}$, and $n_{j\nu} = -n_{\nu j}$ is a vector function of resource flow.

The first term in (38) is a change in associated capital due to resource exchange, and the second is due to the transformation of resources. p_j and p_ν are vectors of resource estimates for contacting subsystems with components p_{ji} and $p_{\nu i}$. Finally, $n_{j\nu} = -n_{\nu j}$—vector function of the exchange flow.

The value $\sigma(p) \geq 0$, so that in a heterogeneous open system in the absence of convective flows, the bound capital in the outgoing flows is not less than in the flows, entering the system. The equal sign corresponds to a homogeneous system.

As in thermodynamics, the condition $\sigma(p_1, p_2) \geq 0$ together with the balance relations (35)–(37) distinguishes the boundary of the feasibility region of an economic system in the class of reversible processes. The conditions imposed on the intensity of a particular flow allow us to determine the value $\sigma_{\min} > 0$, achievable under these conditions. In this case, the area of feasibility narrows, since instead of the inequality $\sigma \geq 0$, the inequality $\sigma \geq \sigma_{\min}$.

If $\Delta p_{jv} = p_j - p_v$ is small, and the kinetic function n_{iv} is differentiable with respect to the set of arguments, then σ is a positive-definite quadratic form of the variables Δp^j_{iv}. In the steady state, the right-hand sides of Equations (35)–(37) are equal to zero. In cyclic mode, when

$$N(0) = N(\tau), \quad M(0) = M(\tau), \quad F(0) = F(\tau)$$

the integrals of the right-hand sides of these equations are equal to zero.

The condition of voluntary exchange imposes restrictions on the set of possible states of an open system. To find these restrictions, we write down the balance relations for the resource and capital for j-th subsystems;

$$\dot{N}_{ij} = n^k_{ij}(t) + n^d_{ij}(p_j, c_j) + \sum_v n_{ijv}(p_j, c_{j\,nu}) + W_j(p_j)\alpha_{ij}, i = 1, 2, \ldots \tag{39}$$

$$\dot{M}_j = m^k_j(t) - \sum_i [c_{ij} n^d_{ij}(p_j, c_j) + \sum_v c_{ij\,nu} n_{ijv}(p_j, c_{jv})]. \tag{40}$$

Here c_{jv} is a vector of intermediate prices with components c_{jvi}, which is determined from the continuity condition of flows,

$$\tilde{n}_{vji}(p_j, c_{jv}) = -\tilde{n}_{jvi}(c_{jv}, p_v) = n_{vji}(p_j, p_v), \quad i = 1, \ldots, m. \tag{41}$$

For any subsystem due to the conditions of voluntariness,

$$\dot{U}_j = \dot{S}_j / p_{0j} = \sum_i (p_{ij} \dot{N}_{ij} + \dot{M}_j) \geq 0,$$

which, taking into account (39) and (40), leads to restrictions on estimates and flows in the system:

$$\begin{aligned} &\sum_i \Bigg[p_{ij}[n^k_{ij}(t) + W_j(p_j)\alpha_{ij}] + \sum_v [n_{ijv}(p_j, c_{jv})(p_{ij} - c_{ijv})] + \\ &+ n^d_{ij}(p_j, c_j)(p_{ij} - c_{ij}) \Bigg] + m^k_j(t) \geq 0, \quad j = 1, 2, \ldots. \end{aligned} \tag{42}$$

7.6.2. Isolated System

For an isolated system, there are no external flows and the balance relations (35)–(37) take the form

$$\dot{N}_i = \sum_j W_j(p_j)\alpha_{ij}, \quad N_i(0) = \sum_j N_{ij}(0), \tag{43}$$

$$\dot{M} = \sum_j \dot{M}_j = 0, \tag{44}$$

$$\dot{F} = \sum_j \dot{F}_j = \sigma \geq 0. \tag{45}$$

In equilibrium, tied up capital is maximal, and the flows n_{ij} and the rates of resource transformation $W_j(p_j)$ are equal to zero. In this case, the equilibrium distribution of the basic resource $\overline{M}$ depends on the kinetics of resource exchange between subsystems.

In thermodynamics, the equilibrium conditions can be obtained from the problem about the maximum of the total entropy of the system for a given total value of one or another extensive variable of the subsystems. For example, if the total volume of the subsystems is given, and the remaining extensive variables are unchanged, then the maximum of the total entropy corresponds to such a distribution of volumes, in which the derivatives of the entropy of each system with respect to its volume are the same, and this leads to equality of pressures. Similarly, the distribution of the total amount of thermal energy in equilibrium leads to equality of temperatures, etc.

In microeconomics, the conditions for the equilibrium distribution of resources can also be derived from the requirement of the maximum total tied capital of the system, given the total stock of resources. This leads to a distribution in which the derivatives of the tied capital of each subsystem with respect to the stock of its allocated resource (resource valuation) are the same.

For each j-th subsystems of the isolated system

$$\dot{N}_{ji} = \sum_{\nu} n_{\nu ji}(p_j, p_\nu) + W_j(p_j)\alpha_{ij}, \tag{46}$$

$$\dot{M}_j = -\sum_{\nu,i} \tilde{n}_{\nu ji}(p_j, c_{j\nu})c_{j\nu i} \quad \nu, j = 1, \ldots, n. \tag{47}$$

Here $c_{j\nu}$ are intermediate prices determined by conditions (41). Thus, the price vector, and hence the right-hand side of the Equation (47) depends on the type of kinetics $\tilde{n}_{\nu j}$ of demand and supply of resources.

After determining from (41) the intermediate price $c_{j\nu}(p_j, p_\nu)$ and substituting it into $\tilde{n}_{j\nu}$ and $\tilde{n}_{\nu j}$, these functions turn out to be the same and equal to the kinetic function with components $n_{j\nu}(p_j, p_\nu)$, appearing in expressions (38) and (46).

As an example, let us consider the kinetics of the exchange of two EAs linear with respect to the difference between the price and the valuation and calculate the dissipation of capital σ in this case:

$$n_1(p_1, c) = a_1(p_1 - c), \tag{48}$$

$$n_2(p_2, c) = a_2(p_2 - c). \tag{49}$$

Let us find $c(p_1, p_2)$ using the condition $-n_1 = n_2 = n$:

$$a_1(p_1 - c) + a_2(p_2 - c) = 0,$$

where

$$c = \frac{a_1 p_1 + a_2 p_2}{a_1 + a_2}, \tag{50}$$

$$n(p_1, p_2) = -n_1(p_1, c(p_1, p_2)) = \bar{a}(p_2 - p_1), \tag{51}$$

$$\bar{a} = \frac{a_1 a_2}{a_1 + a_2}. \tag{52}$$

Then the resource exchange dissipation is

$$\sigma(p_1, p_2) = (p_2 - p_1)\bar{a}(p_2 - p_1) = \bar{a}(p_2 - p_1)^2 = \frac{n^2(p_1, p_2)}{\bar{a}}. \tag{53}$$

The equilibrium distribution of the basic resource $\overline{M}$ is determined by the exchange kinetics, and the equilibrium stocks of resources $\overline{N}$ also depend on $\overline{M}$, since they satisfy the equality

$$p_j(\overline{M_j}, \overline{N_j}) = p_\nu(\overline{M_j}, \overline{N_j}) = \lambda, \quad \forall j, \nu. \tag{54}$$

In a number of cases, the set Q of values of the vector $\overline{M}$ that can be reached from a given initial state of the system for various functions $\tilde{n}(p, c)$ of demand and supply is of interest.

For each j-th subsystem, the minimum capital increase $\Delta M_j = \overline{M_j} - M_{j0}$ is achieved when the exchange with other subsystems is conducted at prices $c_{j\nu}$ arbitrarily close to p_j, i.e., reversibly, so that the value $\overline{M_j}^{\min}$ can be found by the condition $S_j(\overline{N_j}, \overline{M_j}) = S_j(N_{j0}, M_{j0})$.

The maximum value of $\overline{M_j}^{\max}$ corresponds to an exchange at which $c_{j\nu}$ are arbitrarily close to p_ν. Thus, in the space with coordinates M_j, it is possible to construct a parallelepiped with boundaries $\overline{M_j}^{\min} \leq \overline{M_j} \leq \overline{M_j}^{\max}$. The section of this parallelepiped by the plane

$$\sum_j \overline{M_j} = \sum_j M_{j0} \tag{55}$$

selects the set Q of all possible equilibrium distributions of the basic resource between subsystems for different kinetic functions.

8. Generalization of Carathéodory's Theorem and the Structure of Optimal Processes in Macrosystems

Exchange flows in macrosystems depend on intensive variables of subsystems, and the rate of change of extensive variables is determined by these flows. This leads to problems in which extensive variables are not included in the right-hand sides of differential equations that determine their change. For example, the change in the internal energy of two contacting subsystems is characterized by the differential equation

$$\frac{dU_1}{dt} = q(T_1, T_2). \tag{56}$$

The internal energy itself is not included in the right-hand side of this equation. If the values of the internal energy are subject to constraints $U_1(0) = u_{10}, U_1(\tau) = u_{1\tau}$, then these constraints correspond to the condition imposed for a given value of τ on the average value of the current

$$\frac{1}{\tau}\int_0^\tau q(T_1, T_2) = \frac{1}{\tau}\Delta U_1.$$

The Equation (56) are called Lyapunov-type equations by L.I. Rozonoer [3]. When solving problems, these equations can be discarded, replacing them with conditions imposed on the average rate of change of state variables, which leads to problems of averaged optimization. The same is typical for cyclically occurring processes in systems with intermediaries. In such systems, their efficiency indicators are averaged over a time equal to the duration of the cycle, which also leads to averaged problems. Below we present a proof of the theorem that defines the structure and type of optimal solutions to such problems.

8.1. On the Relationship Between Time Averaging and Set Averaging

The mean value of a continuous scalar function $f(x(t)), t \in [0, \tau], x \in V \subset R^n$ can be calculated over time as

$$\overline{f_t(x)} = \frac{1}{\iota}\int_0^\tau f(x(t))dt \tag{57}$$

or over a set as

$$\overline{f_p(x)} = \int_V f(x)p(x)dx. \tag{58}$$

The function $p(x)$ is called the distribution density. In the case where $x(t)$ is a random function, $p(x)$ is the distribution density of a random variable. It is non-negative and its integral on V is equal to one. In our case, $x(t)$ is a deterministic function. Let us dwell in more detail on the properties of the function $p(x)$, such that the results of averaging by Formulas (57) and (58) are the same.

We will assume that the variable x is scalar, the set V here and below is bounded and closed, and introduce the function $\theta(x_0), x_0 \in V$, equal to the total duration of those time intervals t for which $x(t) \leq x_0$. Obviously, this function does not exceed τ. By $P(x_0)$ we denote the ratio $\frac{\theta(x_0)}{\tau}$, i.e., the fraction of the interval $[0, \tau]$ for which $x(t) \leq x_0$. This function increases monotonically with x_0, changing from zero to one. It is similar to the distribution function of a random variable.

The distribution density is

$$p(x_0) = \frac{dP(x_0)}{dx_0} = \frac{1}{\tau}\frac{d\theta(x_0)}{dx_0} = \frac{1}{\tau}\frac{1}{\sum_{\nu}\left|\frac{dx_\nu}{dt}\right|_{x_\nu = x_0}}. \tag{59}$$

Here the interval θ increases with x_0 for any sign of the derivative for those values x_ν of the function $x(t)$ for which it is equal to x_0.

If for some value x_0 the function $x(t)$ is constant during a fraction γ of the interval $[0, \tau]$, then the function $P(x_0)$ experiences a jump of magnitude γ at this point, and the distribution density at it is equal to $\gamma\delta(x - x_0)$.

Example 1—Linear functions. Let $x(t) = \frac{ht}{\tau}$. Then by Formula (59) we obtain $p(x) = \frac{1}{h} = const$. The same distribution density corresponds to all triangles with base $[0, \tau]$ and height h.

Example 2—Piecewise constant functions. These functions take discrete values x_i, each during a fraction γ_i of the interval $[0, \tau]$. Any such function, according to Formula (59), corresponds to the distribution density

$$p(x_0) = \sum_i \gamma_i \delta(x - x_i), \quad \gamma_i > 0, \quad \sum_i \gamma_i = 1. \tag{60}$$

The order in which a piecewise constant function assumes one or another of the possible values does not matter.

From these examples it is clear that to each function $x(t)$ there corresponds a distribution density of its values $p(x)$, defined on V, and to each distribution density there corresponds an arbitrary number of functions $x(t)$, for which $\overline{f_p(x)} = \overline{f_t(x)}$. An exception is the distribution density of the form $p(x) = \delta(x - x_1)$. In this case, the corresponding function $x(t) = x_1 = const$ on the entire interval $[0, \tau]$ and it is unique.

Let us consider the case when the function f depends on several, for simplicity on two, variables $x_1(t)$ and $x_2(t)$. In this case, the distribution function $P(x^0)$ of the values of the vector x is a fraction of the interval $[0, \tau]$ for which two inequalities are satisfied: $x_1(t) \leq x_1^0$ and $x_2(t) \leq x_2^0$. This function monotonically increases with the growth of each of the arguments, when the first component of the vector x^0 is maximum ($p_1(x_1) = 1$), it is equal to $P(x_2^0)$, and its derivative is equal to the distribution density $p(x_1^{max}, x_2) = p_2(x_2)$; similarly in the case when $x_2 = x_2^{max}, \quad p(x_2^{max}, x_1) = p_1(x_1)$. The functions $x_1(t)$ and $x_2(t)$ are independent of each other so that $p(x_1, x_2) = p_1(x_1)p_2(x_2)$.

The desired solution to the averaged optimization problem is the distribution density $p^*(x)$ of the vector x on the set V of its admissible values. To implement this solution in time, one must find one of the possible functions $x(t)$ that have the distribution $p^*(x)$. The solution to this last problem is significantly simplified by the features of the optimal solutions $p^*(x)$, proven in the next section.

8.2. On the Form of the Optimal Solution of Averaged Optimization Problems

We will denote the averaging operation by a bar drawn over the averaged function or vector. Thus,

$$\overline{x} = \int_V xp(x)dx, \quad \overline{f(x)} = \int_V f(x)p(x)dx.$$

The simplest problem of averaged optimization is the problem of maximizing the average value of a scalar function $f(x)$ for a given average value of its argument:

$$\overline{f(x)} \to \max \Big/ \overline{x} = x_0, x \in V \subset R^n \tag{61}$$

or in more detailed notation

$$\int_V f(x)p(x)dx \to \max \Big/ \int_V xp(x)dx = x_0, p(x) \geq 0, \int_V p(x)dx = 1. \tag{62}$$

The unknown in this problem is $p(x)$ (the distribution density of the vector of unknown variables). This function is non-negative and its integral on the set V is equal to one.

Carathéodory's Theorem on Convex Hulls of Functions

Carathéodory's theorem [25] on convex hulls of sets states that any element of the convex hull CoD of a compact set D in a Euclidean space of dimension n can be represented as an element of a simplex that has at most $n+1$ vertices (base points), each of which belongs to D.

In particular, a subgraph of a function $f(x)$ can be a set D. The convex hull of a function is called the convex hull of a subgraph. A function that depends on n variables is the boundary of a set in the space R^{n+1} of dimension n. The base points are certainly on this boundary, which means that their number does not exceed $n+1$. Below we will call Carathéodory's theorem the theorem on convex hulls of functions.

The ordinate of the convex hull of the function $f_0(x)$ at the point x_0, which belongs to the convex hull of the set of definition of the function, is the value of the problem:

$$\overline{f_0(x)} \to \max_{p(x)} \Big/ \begin{array}{l} \overline{x_i} = x_{i0}, \quad i = \overline{1,n}, \\ x \in V \subset R^n \end{array} \tag{63}$$

where V is compact.

According to Carathéodory's theorem, the optimal solution to this problem is

$$p^*(x) = \sum_{j=0}^{n} \gamma_j \delta(x - x^j), \quad \gamma_j \geq 0, \quad \sum_{i=0}^{n} \gamma_j = 1.$$

That is, the optimal distribution is concentrated at no more than the $(n+1)$-th base point.

This fact allows us to rewrite the problem as a nonlinear programming (NP) problem:

$$\sum_{j=0}^{n} \gamma_j f_0(x^j) \to \max \Big/ \begin{array}{l} \sum\limits_{j=0}^{n} \gamma_j x^j - x_0, \\ x^j \in V \subset R^m, \quad \sum\limits_{j=0}^{n} \gamma_j = 1, \quad \gamma_j \geq 0. \end{array} \tag{64}$$

The variables are the base vectors x^j and the vector of weight coefficients γ. We use the Kuhn-Tucker theorem [25] to solve it: If y^* is a solution to a nonlinear programming problem

$$f(y) \to \max \Big/ \varphi_i(y) \leq 0, \quad y_j \geq 0, i = 1, 2, \ldots, m, j = 1, \ldots, n, \tag{65}$$

then there exists a nonzero vector of factors

$$\lambda = \lambda_0, \ldots, \lambda_m \ \ (\lambda_0 \text{ is equal to 0 or 1}, \ \ \lambda_i \leq 0 \text{ for } i > 0),$$

such that for the Lagrange function

$$R = \lambda_0 f(y) + \sum_{i=1}^{m} \lambda_i \varphi_i(y)$$

the following conditions hold:

$$\left(\frac{\partial R}{\partial y_j}\right)_{y=y^*} = 0 \ \text{ if } y_j^* > 0; \qquad \left(\frac{\partial R}{\partial y_j}\right)_{y=y^*} \leq 0 \ \text{ if } y_j^* = 0, \tag{66}$$

and

$$\lambda_i = 0 \ \text{ if } \varphi_i(y^*) < 0; \qquad \lambda_i \leq 0 \ \text{ if } \varphi_i(y^*) = 0. \tag{67}$$

For problem (64), the Lagrange function has the form

$$R = \sum_{j=0}^{n} \gamma_j \Big[f_0(x^j) + \sum_{i=1}^{n} \lambda_i x_i^j - \Lambda \Big], \tag{68}$$

where Λ is the Lagrange multiplier corresponding to the condition that the sum of the weight coefficients is equal to one.

The Kuhn–Tucker local unimprovability conditions with respect to weight coefficients lead to the requirements

$$R^0(x_j, \lambda) = f_0(x^j) + \sum_{i=1}^{n} \lambda_i x_i^j < \Lambda \ \text{ if } \gamma_j = 0, \tag{69}$$

$$R^0(x_j, \lambda) = f_0(x^j) + \sum_{i=1}^{n} \lambda_i x_i^j = \Lambda \ \text{ if } \gamma_j > 0, \ j = 0, \ldots, n+1.$$

Here R^0 is the Lagrange function of problem (64) in the absence of averaging. Such a problem will be called *original*. Thus, we have

Theorem 1. *For each of the basic values x included in the optimal solution to the problem of the convex hull of the function f_0 with a non-zero weight, the Lagrange function of the original problem is maximal, and the number of such points does not exceed $n+1$.*

8.3. Generalization of Carathéodory's Theorem

In the NP problem with averaging of functions defining relationships between variables, it is required to achieve the maximum of the mean value of the function $f_0(x)$ on the set V of admissible values of x, provided that the mean value of the vector function $f(x) = (f_1(x), \ldots f_i(x), \ldots f_m(x))$ is zero. Formally

$$\overline{f_0(x)} \to \max \Big/ \overline{f_i(x)} = 0, \quad i = 1, \ldots, m, \quad x \in V \in R^n, \tag{70}$$

yielding.

Theorem 2. *1. The optimal distribution density in problem (70) has the form*

$$p^*(x) = \sum_{j=0}^{m} \gamma_j \delta(x - x^j), \quad \gamma_j \geq 0, \sum_{j=0}^{m} \gamma_j = 1. \tag{71}$$

2. There is a nonzero vector

$$\lambda = \lambda_0, \dots \lambda_i, .., \lambda_m, \quad \lambda_0 = (0; 1),$$

such that at each base point x^j the Lagrange function of the original problem

$$R = \sum_{i=0}^{m} \lambda_i f_i(x) \tag{72}$$

is maximal with respect to $x \in V$.

Proof. To prove this statement, we introduce the concept of *reachability function* of the problem (70):

$$f_0^*(C) = \max f_0(x) \Big/ f_k(x) = C_k, \quad k = 1, \dots, m, \quad x \in V. \tag{73}$$

This function is defined algorithmically on the set

$$V_c = \{C \in R^m : f(x) = C, x \in V \subset R^n\}.$$

It may not be smooth and upper semi-continuous.

The following is true: For those values of x for which $f(x) = C$, $p^*(x)$ is certainly equal to zero if $f_0(x) \neq f_0^*(C)$. Thus, the solution of the averaged problem can include with non-zero weight only those values of $x = x^*(C)$ for which the value of $f_0(x)$ coincides with the ordinate of the attainability function. Otherwise, it would be possible to change the distribution density so that the average value of $f_0(x)$ increases.

Since for each C the value f_0 coincides with the ordinate of the reachability function, problem (70) can be rewritten in the form

$$\overline{f_0^*(C)} \to \max \Big/ \overline{C_k} = 0, \quad k = 1, \dots, m, \quad C \in V_c \subset R^m. \tag{74}$$

This is the problem of the ordinate of the convex hull of the reachability function at zero. According to the Carathéodory theorem on convex hulls, its optimal solution is

$$p^*(C) = \sum_{j=0}^{m} \gamma_j \delta(C - C^j), \quad \gamma_j \geq 0, \sum_{j=0}^{m} \gamma_j = 1. \tag{75}$$

Since each basic value C^j corresponds to the value $x^{j*}(C^j)$, the optimal distribution density in problem (70) has the form (71). Thus the first assertion of Theorem 2 is proved.

The proof of the second assertion completely repeats the analogous proof for the problem of the ordinate of the convex hull of a function, with the difference that the Lagrange function of the unaveraged problem has the form (72). We emphasize that the number of base points does not depend on the dimension of the vector x, but is determined by the dimension m of the vector function f. In the special case when $f_j(x) = x_j$, we obtain the theorem on the convex hull of a function. □

Note that here and below, the conditions in the form of the maximum principle do not require the functions defining the averaged problem to be smooth in x; the set V may not be simply connected.

Example 1. Consider a system consisting of an electric motor, a pump it rotates, and a tank. The motor consumes power n on which the pump performance g depends. The dependence $g(n)$ is shown in Figure 2. It is required to find the mode for which, for a given average expended power $\overline{n}$, the average productivity $\overline{g}$ is maximum.

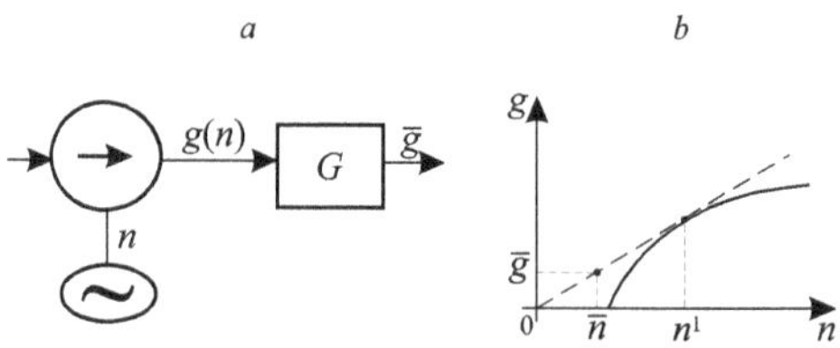

Figure 2. A system consisting of a pump and a smoothing tank (**a**); the dependence of the flow rate on the expended power (**b**).

This is a problem about the ordinate of the convex hull of the function $g(n)$ at the point $\overline{n}$. The number of base points is equal to two, one of them is the origin, and the second n^1, is determined by the condition of the maximum of the Lagrange function $R = g(n) + \lambda n$ at it, the same as at $n = 0$. Eliminating λ from the conditions for the maximum of the Lagrange function and the requirement that this maximum be zero, we arrive at the equation for n^1:

$$\frac{g(n)}{n} = \frac{dg(n)}{dn}.$$

There are as many optimal implementations of this solution in time as you like. For each of them the pump power takes the value zero for one fraction of the period, and n^1 for the remaining fraction, and the fraction of the interval τ for which $n = n^1$ is equal to $1 - \frac{\overline{n}}{n^1}$. The maximum value of the interval τ is determined by the value of the capacity G, it is equal to

$$\tau_{max} = \frac{2G}{g(n^1)}.$$

The value of the problem is

$$\overline{g}^* = g(n^1)\left(1 - \frac{\overline{n}}{n^1}\right).$$

It does not depend on G. When the capacity tends to zero, the sliding mode becomes the optimal solution.

8.4. Averaged Problem with Deterministic Variables

In averaged problems there can be two types of variables, random and deterministic. There is no averaging over the variables of the second type. Let us consider the NP problem, in which there is no averaging over some variables.

A problem with averaging over some variables will take the form

$$\overline{f_0(x,y)} \to \max \Big/ \overline{f_j(x,y)} = 0, \;\; x \in V \subset R^n, \;\; y \in V_y \subset R^K, \;\; j = 1, \ldots, m. \tag{76}$$

The functions $f_0, \ldots, f_m$ are continuous and continuously differentiable with respect to the set of arguments, the bar corresponds to averaging over $x \in V$, and the sets V and V_y are closed and bounded.

For any y, this problem is an averaged nonlinear programming problem (70), and therefore, by Theorem 2, the optimal density function of x is concentrated at most at the $(m+1)$-th base point, so that $p^*(x) = \sum_0^m \gamma_j \delta(x - x^j)$, and there exists a nonzero vector λ such that at each of these points the Lagrange function of the original problem

$$R = \sum_{j=0}^{m} \lambda_j f_j(x, y), \;\; x \in V \subset R^n, \;\; y \in V_y \subset R^K \tag{77}$$

is maximal with respect to x.

The Lagrange function of problem (76), in which the density function x is $p^*(x)$, has the form

$$\overline{R^*} = \sum_{j=0}^{m} \lambda_j \sum_{i=0}^{m} \gamma_i f_j(x^i, y), \;\; x^i \in V \subset R^n, \;\; y \in V_y \subset R^K. \tag{78}$$

For any density function $p(x)$ of randomized variables, problem (76) is a nonlinear programming problem and, according to the Kuhn-Tucker theorem, there exists a nonzero vector λ with components $\lambda_0 = (0; 1)$, λ_j, $j = 1, \ldots, m$, such that for the function (77) the optimal solution satisfies the conditions of local non-improvability with respect to y, in particular, the stationarity conditions

$$\frac{\partial \overline{R^*}}{\partial y_l} \delta y_l \leq 0, \;\; l = 1, \ldots K. \tag{79}$$

Here δy_l is the admissible variation of y_l.

9. Conclusions

In FTT the specificity of problems on the optimal exchange of resources and problems with cyclically changing variables of the working fluid or mediator leads to averaged problems of nonlinear programming. Their optimal solutions are of a switching nature between the basic values of the variables. The maximum number of basic values is one more than the number of averaged conditions, and in each of them the Lagrange function of the non-averaged problem reaches a maximum.

The presented results of solving the problems of "Thermodynamics at finite time" illustrate the nature of its applications for systems of various natures, but do not pretend to be an exhaustive presentation. Not all the problems of FTT have been solved. In particular, the problem of the conditions of minimal dissipation for vector flows whose kinetics differs from Onsager's awaits its solution. It is not clear what to consider the efficiency of a machine receiving energy from several sources. It is not clear what the irreversibility indicator looks like for macrosystems of a social nature, etc.

Author Contributions: Methodology, A.M.T.; software, A.I.B. and I.A.S.; validation, A.I.B. and I.A.S.; formal analysis, A.M.T. and A.I.B.; investigation, A.M.T. and I.A.S.; resources, A.M.T.; data curation, A.I.B. and I.A.S.; writing—original draft preparation, A.M.T., A.I.B. and I.A.S.; writing—review and editing, A.M.T. and A.I.B.; visualization, I.A.S. and AIB; supervision, AMT; project administration, AIB. All authors have read and agreed to the published version of the manuscript.

Funding: This research received no external funding.

Data Availability Statement: No new data were created or analyzed in this study. Data sharing is not applicable to this article.

Acknowledgments: The authors would like to express their deep gratitude to the editors, Bjarne Andreessen and Peter Salamon, with whom long-term collaboration made it possible to obtain many of the results presented in the article and many not presented in the article. We, like them, get great pleasure from doing FTT.

Conflicts of Interest: The authors declare no conflict of interest.

References

1. Novikov, I.I. The efficiency of atomic power stations. *J. Nuclear Energy* **1958**, *7*, 125–128.
2. Curzon, F.L.; Ahlborn, B. Efficiency of a Carnot engine at maximum power output. *Am. J. Phys.* **1975**, *43*, 22–24. [CrossRef]
3. Rozonoer, L.I.; Tsirlin, A.M. Optimal control of thermodynamic processes. *Autom. Remote Control* **1983**, *44*, 55–62. 209–220. 314–326.
4. Berry, R.S.; Kazakov, V.A.; Sieniutycz, S.; Szwast, Z.; Tsirlin, A.M. *Thermodynamic Optimization of Finite Time Processes*; Wiley: Chichester, UK, 1999.
5. Tsirlin, A.M. *Methods of Averaged Optimization and Its Applications*; Fizmatlit: Moscow, Russia, 1997. (In Russian)
6. Andresen, B.; Salamon, P.; Berry, R.S. Thermodynamics in finite time: Extremals for imperfect heat engines. *J. Chem. Phys.* **1977**, *66*, 1571–1577. [CrossRef]
7. Andresen, B. *Finite-Time Thermodynamics*; University of Copenhagen: Copenhagen, Denmark, 1983.
8. Andresen, B.; Salamon, P.; Berry, R.S. Thermodynamics in finite time. *Phys. Today* **1984**, *37*, 62–70. [CrossRef]
9. Salamon, P.; Hoffmann, K.H.; Schubert, S.; Berry, R.S.; Andresen, B. What conditions make minimum entropy production equivalent to maximum power production? *J. Non-Equilib. Thermodyn.* **2001**, *26*, 73–83. [CrossRef]
10. Tsirlin, A.M. Mathematical models and control for systems with segregation. *J. Math. Sci.* **2013**, *195*, 427–438. [CrossRef]
11. Sukin, I.A.; Balunov, A.I.; Tsirlin, A.M. *Methods of Irreversible Thermodynamics in Problems of Conceptual Design of Heat Transfer and Separation Processes*; RAS: Moscow, Russia, 2021. (In Russian)
12. Balunov, A.; Sukin, I.; Tsirlin, A. Entropic balance conditions and optimization of distillation column system. *Entropy* **2021**, *23*, 1468. [CrossRef] [PubMed]
13. Tsirlin, A.M. *Minimal Dissipation Processes in Irreversible Thermodynamics*; Lan: Saint-Petersberg, Russia, 2022. (In Russian)
14. Tsirlin, A.M.; Mironova, V.A.; Amelkin, S.A.; Kazakov, V.A. Finite-time thermodynamics: Conditions of minimal dissipation for thermodynamic process with given rate. *Phys. Rev. E* **1998**, *58*, 215. [CrossRef]
15. Tsirlin, A.M.; Kazakov, V.A. Maximal work problem in finite-time thermodynamics. *Phys. Rev. E* **2000**, *62*, 307. [CrossRef] [PubMed]
16. Tsirlin, A.M. *Mathematical Models and Optimal Processes in Macrosystems*; Nauka: Moscow, Russia, 2006. (In Russian)
17. Tsirlin, A.M.; Gagarina, L.G.; Balunov, A.I. The Synthesis of Heat-Exchange Systems Integrated with a Technology Process. *Theor. Found. Chem. Eng.* **2021**, *55*, 439–450. [CrossRef]
18. Tsirlin, A.M.; Balunov, A.I. On the Analogy of Processes in Thermodynamic and Microeconomic Systems. *Processes* **2023**, *11*, 2974. [CrossRef]
19. Samuelson, P.A. Extension of the Le Chatelier Principle. *Econometrica* **1960**, *28*, 368–379. [CrossRef]
20. Samuelson, P.A. Maximum Principle in Analytical Economics. *Am. Econ. Rev.* **1972**, *B2*, 249–262.
21. Sieniutycz, S.; Salamon, P. (Eds.) *Finite-Time Thermodynamics and Thermoeconomics*; Taylor & Francis: New York, NY, USA, 1990.
22. Tsirlin, A.M.; Kazakov, V.A. Extremal principles and limiting possibilities of open thermodynamic and economic systems. In *Variational and Extremum Principles in Macroscopic Systems*; Sieniutycz, S., Farkas, H., Eds.; Kluwer Academic Publishers: Dordrecht, The Netherlands, 2004.
23. Rozonoèr, L.I. A Generalized Thermodynamic Approach to Resource Exchange and Allocation. II, III. *Autom. Remote Control* **1973**, *34*, 915–927. 1272–1289.
24. Martinas, K. About Irreversibility in Economics. *Open Syst. Inf. Dyn.* **2000**, *7*, 349–364. [CrossRef]
25. Hadley, J. *Nonlinear and Dynamic Programming*; Mir: Moscow, Russia, 1967. (In Russian)

entropy

MDPI

Article

Endoreversible Stirling Cycles: Plasma Engines at Maximal Power

Gregory Behrendt [1,2,*] and Sebastian Deffner [1,2,3]

[1] Department of Physics, University of Maryland, Baltimore County, Baltimore, MD 21250, USA; deffner@umbc.edu
[2] Quantum Science Institute, University of Maryland, Baltimore County, Baltimore, MD 21250, USA
[3] National Quantum Laboratory, College Park, MD 20740, USA
* Correspondence: gregorb2@umbc.edu

Abstract: Endoreversible engine cycles are a cornerstone of finite-time thermodynamics. We show that endoreversible Stirling engines operating with a one-component plasma as a working medium run at maximal power output with the Curzon–Ahlborn efficiency. As a main result, we elucidate that this is actually a consequence of the fact that the caloric equation of state depends only linearly on temperature and only additively on volume. In particular, neither the exact form of the mechanical equation of state nor the full fundamental relation are required. Thus, our findings immediately generalize to a larger class of working plasmas, far beyond simple ideal gases. In addition, we show that for plasmas described by the photonic equation of state, the efficiency is significantly lower. This is in stark contrast to endoreversible Otto cycles, for which photonic engines have an efficiency larger than the Curzon–Ahlborn efficiency.

Keywords: plasma engine; endoreversible Stirling cycle

1. Introduction

Among all the various states of matter, plasmas have a unique place. In particular, quantum plasmas exist only in the most extreme environments, such as the interior of stars, in the early universe, or more mundanely in highly intense laser fields [1]. In its simplest form, a plasma is a superheated gas, in which electrons have been stripped from atoms, creating a mixture of positively charged ions and negatively charged electrons [1]. It provides the circumstances for nuclear fusion, which is a nuclear reaction where two light atomic nuclei combine to form a heavier one, releasing a tremendous amount of energy [2]. The potential technological applications of nuclear fusion are tremendous, which is why its realization is considered one of the "14 Grand Challenges for Engineering in the 21st Century" [3].

From a thermodynamic point of view, nuclear reactions in plasmas can also be understood as heat engine cycles [4,5]. In fact, the electrostatic interactions within a plasma behave very akin to the usual mechanical pressure in typical gases [6]. Interestingly, the somewhat natural engine cycle for plasma engines is the Stirling cycle [7]. More broadly, technological applications of the Stirling cycle appear rather promising; see, for instance, the recent review [8]. The natural question thus arises what the efficiency of plasma Stirling engines is at maximal power output.

In the present work, we answer this question within the framework of endoreversible thermodynamics [9]. In endoreversible thermodynamics one assumes that all processes are slow enough that the system *locally equilibrates*, yet the processes are too fast for the system to reach a state of equilibrium with the environment. More specifically, imagine an engine, whose working medium is at equilibrium at temperature T. However, T is not equal to the temperature of the heat bath, T_{bath}, and thus there is a temperature gradient at the boundaries of the engine. Now further imagine that the engine undergoes a slow, cyclic state transformation, where slow means that the working medium remains *locally* in equilibrium at all times. Then, from the point of view of the environment the device undergoes an irreversible cycle. Such state transformations are called *endoreversible* [9], which means that locally the transformation is reversible, but globally irreversible.

In a seminal work, Curzon and Ahlborn showed [10] that the efficiency of a Carnot engine undergoing an endoreversible cycle at maximal power is given by

$$\eta_{\text{CA}} = 1 - \sqrt{\frac{T_c}{T_h}}, \tag{1}$$

where T_c and T_h are the temperatures of the cold and hot reservoirs, respectively. Since its discovery the Curzon–Ahlborn efficiency (1) has received a great deal of attention. Also, see even earlier studies by Moutier [11] and Novikov [12], and related works by Rozonoer and Tsirlin [13–15]. More recently, it has been found, e.g., that endoreversible Otto [16] and Brayton [17] engines operating with ideal gases also have the same efficiency. However, it has also been shown that whether or not a finite-time Carnot cycle really assumes η_{CA} is determined by the "symmetry" of the dissipation [18] and on the specific form of the fundamental relation [19–31].

In the present work, we focus on plasma engines that run in endoreversible Stirling cycles. It is interesting to note that there are also several accounts in the literature of the fact that endoreversible Stirling cycles at maximum power operating with classical, ideal gases are described by η_{CA} [32–34]. Given that it is often a good assumption (in first order approximation) that plasmas can also be described as ideal gases [35], one is tempted to conclude that, clearly, plasma engines at maximal power also have the Curzon–Ahlborn efficiency. However, given that previous treatments of the endoreversible Stirling make, sometimes implicitly, often explicitly the assumption that the working medium is a regular, classical, ideal gas, it is not immediately obvious that plasmas do not require a separate treatment.

Therefore, we start with a detailed discussion on the necessary conditions under which the Curzon–Ahlborn efficiency arises. Particular emphasis is placed on a comprehensive and pedagogical derivation, which then also leads to immediate generalizations. In fact, we will see that any gases that are described by caloric equations of states that are linear in temperature and at most additive in volume lead to the Curzon–Ahlborn efficiency. Neither the mechanical equation of state nor the full fundamental relation is required.

This means, in particular, that endoreversible Stirling engines whose working plasmas require a second-order virial expansion also have the same efficiency. In the simplest case, their corresponding gas law is given by the van der Waals equation of state.

As an example of plasma engines that fall not within this class, we then analyze an engine that operates with an electron–positron–photon plasma. For high enough temperatures [36] such plasmas can be described by the photonic equation of state, which permits an almost completely analytical treatment. We find that photonic Stirling engines have a significantly smaller efficiency at maximal power. This is in stark contrast to Otto engines, in which case photonic working mediums lead to higher efficiency [25].

2. One-Component Plasmas—Modified Ideal Gas

We start with a Stirling engine that operates with a one-component plasma. For such situations one commonly assumes that the plasma can be treated as an "effective ideal gas" [35]. A classical, ideal gas is comprised of uniform, non-interacting, identical particles, whose caloric equation of state is proportional to temperature and independent of the volume:

$$E = \varsigma T\,. \tag{2}$$

The constant ς depends on natural constants and the number of degrees of freedom. For instance, for a classical ideal gas in three spatial dimensions we have [37], $\varsigma = 3/2N\,k_B$, where N is the number of particles and k_B is Boltzmann's constant.

A one-component plasma is comprised of non-interacting, identical particles. At the fundamental level, a one-component plasma is nothing but a collection of uniform ions, that have been immersed in an equally and oppositely charged background, neutralizing the total charge of the plasma, and thus preventing any potential interparticle-Coulomb interactions [35]. Thus, it follows that the corresponding caloric equation of state for the one-component plasma is still directly proportional to the temperature, and all "quantum" modifications can be accounted for by an effective number of degrees of freedom along with an arrangement of phenomenological constants that are unique to that plasma. We can write

$$E = E_0 + f\,T \tag{3}$$

where E_0 is a constant off-setting the background energy, and f quantifies the effective degrees of freedom [35]. As we will see shortly, Equation (3) is a necessary and sufficient condition to obtain the Curzon–Ahlborn efficiency.

2.1. *Endoreversible Cycle and Efficiency*

In complete analogy to the endoreversible Carnot [10], Otto [16,24–31], and Brayton cycles [17], we now construct the endoreversible Stirling cycle.

The Stirling cycle is a 4-stroke process comprising isothermal expansion, isochoric cooling, isothermal compression, and isochoric heating. The corresponding PV- and TS-diagrams for a one-component plasma (3) are depicted in Figure 1.

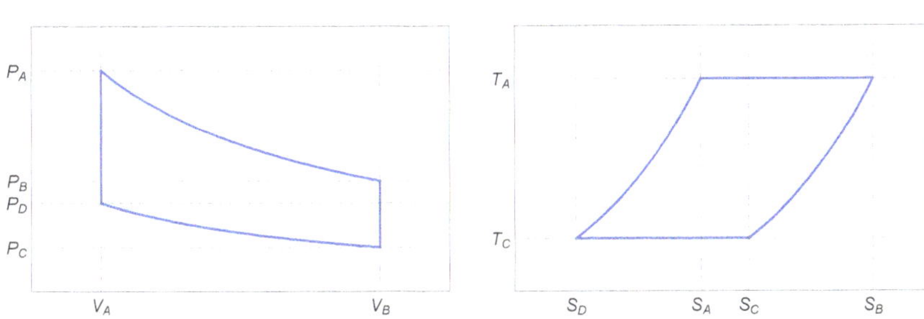

Figure 1. Schematic PV- and TS-diagrams of the Stirling cycle for the one-component plasma (3) as a working medium.

2.1.1. $A \to B$: Isothermal Expansion

During the hot isotherm, the plasma is in contact with a heat reservoir at temperature T_h. However, as usual in endoreversible thermodynamics [9], we assume that the plasma has not fully equilibrated with the hear reservoir, and rather has a temperature "slightly" below T_h, namely we have $T_{h,p} < T_h$. Fourier's law in Newton's form then dictates that the heat flux is linear in the temperature gradient [10], and we can write

$$Q_{\rm AB} = \alpha\, t_{AB} \left(T_h - T_{h,p}\right), \tag{4}$$

where α is the thermal conductivity and t_{AB} is the duration of the stroke.

Now note that for one-component plasmas the internal energy is constant for isothermal processes; cf. Equation (3). Thus, we immediately have

$$\Delta E = E_B - E_A \quad \text{and} \quad W_{AB} = -Q_{AB}, \tag{5}$$

namely the work produced during the isothermal expansion is equal and opposite in sign to the heat absorbed from the heat reservoir.

2.1.2. $B \to C$: Isochoric Cooling

During the isochoric stroke, the plasma is disconnected from the heat reservoirs. Thus, this stroke is identical to the ideal cycle. In any case, we have

$$W_{BC} = 0 \quad \text{and} \quad Q_{BC} = E_B - E_C = f\,(T_B - T_C) \tag{6}$$

where we used the caloric equation of state (3).

2.1.3. $C \to D$: Isothermal Compression

In complete analogy to the hot isotherm, during the cold, isothermal compression the plasma is in contact with a cold reservoir at temperature T_c. However, we again assume that the plasma has not fully equilibrated with the reservoir, and that its temperature, $T_{c,p}$, is slightly above T_c. Again, employing Fourier's law, we can write

$$Q_{CD} = \beta\, t_{CD}\,(T_c - T_{c,p}) \tag{7}$$

where β is the thermal conductivity for the cold stroke, and t_{CD} is the duration. As before, we can also write

$$\Delta E = E_D - E_C \quad \text{and} \quad W_{CD} = -Q_{CD}, \tag{8}$$

which follows from the caloric equation of state (3).

2.1.4. $D \to A$: Isochoric Heating

The cycle is completed with another isochoric stroke. Again, the plasma is disconnected from the heat reservoirs. We have

$$W_{DA} = 0 \quad \text{and} \quad Q_{DA} = f\,(T_D - T_A) \tag{9}$$

2.1.5. Endoreversible Efficiency

We are now interested in the efficiency at maximal power. To this end, consider that the plasma absorbs heat from the hot reservoir during the hot isotherm $A \to B$. Thus, we can write

$$\eta \equiv -\frac{W_{\rm cyc}}{Q_{AB}} = 1 + \frac{Q_{CD}}{Q_{AB}}, \tag{10}$$

where we used that the work produced during the entire cycle is $W_{\text{cyc}} = -(Q_{AB} + Q_{CD})$. Using Equations (4) and (7), we also have

$$\eta = 1 + \frac{\beta\, t_{CD}}{\alpha\, t_{AB}} \frac{T_c - T_{c,p}}{T_h - T_{h,p}}, \tag{11}$$

which appears to suggest that the efficiency depends on the stroke times (see also Appendix A).

However, the expression for the efficiency (11) can be further simplified by employing the entropy balance over one cycle. In fact, we have

$$0 = \Delta S_{\text{cyc}} = \Delta S_{AB} + \Delta S_{BC} + \Delta S_{CD} + \Delta S_{DA}\,. \tag{12}$$

Along the isothermal strokes, $A \to B$ and $C \to D$, the entropy is simply given by the heat divided by temperature,

$$\Delta S_{AB} = \frac{Q_{AB}}{T_{h,p}} = \frac{\alpha\, t_{AB}\left(T_h - T_{h,p}\right)}{T_{h,p}} \quad \text{and} \quad \Delta S_{CD} = \frac{Q_{CD}}{T_{c,p}} = \frac{\beta\, t_{CD}\,(T_c - T_{c,p})}{T_{c,p}}, \tag{13}$$

where we again used Equations (4) and (7).

For the isochoric strokes, we again exploit *only* the caloric equation of state (3). In its differential form, along the isochor, $đW = 0$, we have $dE = đQ = T\,dS$. Thus, the total change in entropy can be written as

$$\Delta S = \int_{S_i}^{S_f} dS = \int_{T_i}^{T_f} dT\,\frac{f}{T} = f \ln\left(\frac{T_f}{T_i}\right). \tag{14}$$

Consequently, we obtain

$$\Delta S_{BC} = -\Delta S_{DA} = f\, \ln\left(\frac{T_{c,p}}{T_{h,p}}\right), \tag{15}$$

and thus,

$$\frac{\alpha\, t_{AB}\left(T_h - T_{h,p}\right)}{T_{h,p}} = \frac{\beta\, t_{CD}\,(T_{c,p} - T_c)}{T_{c,p}}. \tag{16}$$

In other words, the efficiency (11) of an endoreversible Stirling cycle operating with a one-component plasma simply becomes

$$\eta = 1 - \frac{T_{c,p}}{T_{h,p}}, \tag{17}$$

which is identical to the ideal efficiency [37] replacing the temperatures of the heat reservoirs with the corresponding temperatures of the plasma.

2.2. Efficiency at Maximal Power

As stated above, we are now interested in the efficiency at maximal power output. Inspecting Equation (17) we need to to determine the temperatures, $T_{h,p}$ and $T_{c,p}$, that maximize the power. As usual, we write

$$P \equiv -\frac{W_{\text{cyc}}}{\tau_{\text{cyc}}} = \frac{Q_{AB} + Q_{CD}}{\gamma\,(t_{AB} + t_{CD})}, \tag{18}$$

where γ is a real constant. It will prove convenient to introduce the variables

$$x \equiv T_h - T_{h,p} \quad \text{and} \quad y \equiv T_{c,p} - T_c \tag{19}$$

which is identical to the notation introduced by the original treatment by Curzon and Ahlborn [10].

After a few lines of simple algebra, we obtain

$$P(x,y) = \frac{\alpha\beta\, xy\,[(T_h - x) + (T_c + y)]}{\gamma[\alpha\, x\,(T_h - x) + \beta\, y\,(T_c + y)]}\,, \tag{20}$$

where we once again employed Equations (4) and (7). The maximum of $P(x,y)$ is determine using standard calculus, namely solving $\partial_x P(x,y) = 0$ and $\partial_y P(x,y) = 0$ for x and y.

Denoting the solutions by x^* and y^*, we find

$$x^* = \frac{\sqrt{\beta}}{\sqrt{\alpha} + \sqrt{\beta}}\left(T_h - \sqrt{T_h T_c}\right) \quad \text{and} \quad y^* = \frac{\sqrt{\alpha}}{\sqrt{\alpha} + \sqrt{\beta}}\left(-T_c + \sqrt{T_h T_c}\right). \tag{21}$$

Substituting these solutions into the expression for the efficiency (17) and simplifying the expression we finally obtain

$$\eta = 1 - \sqrt{\frac{T_c}{T_h}}\,. \tag{22}$$

That is, the efficiency at maximal power of an endoreversible Stirling engine operating with a one-component plasma as working medium is given by the Curzon–Ahlborn efficiency. Our result (22) corroborates earlier findings for classical ideal gases [32–34]. Thus, engines operating in endoreversible Stirling cycles with one-component plasmas and classical ideal gases have the same efficiency.

2.3. Generalization to Second Order Virial Expansion

Before we move on to more intricate working mediums, we emphasize that the present analysis is entirely based on the fact that the caloric equation of state is of the form of an ideal gas (3). In particular, we did not need to require the mechanical equation of state to be given by the ideal gas law, nor did we need the full fundamental relation.

Interestingly, equations of state of the form (3) are *not* restricted to *classical* ideal gases. For instance, classical harmonic oscillators [16] and rubber bands [37] are also described by equations of state that are linear in temperature (and independent of volume).

For the present purposes, it is even more interesting to observe that the van der Waals gas is described by [37]

$$E = a_0\, T - \frac{b_0}{V^2}\,, \tag{23}$$

where a_0 and b_0 are phenomenological constants. It is a simple exercise to show that the van der Waals equation of state is obtained from the virial expansion in second order [38]. Since the dependence on the volume V in Equation (23) is only additive, the changes in entropy during isochoric strokes remain identical to above (14). Consequently, we obtain the same expressions for efficiency (17), power (20), and efficiency at maximal power (22).

It is worth emphasizing that for many real plasmas virial coefficients have been determined; see, for instance, [39–43]. Our present results remain valid for any endoreversible Stirling cycle for plasma, for which a description of up to second order in the virial expansion is a good description.

3. Relativistic Electron–Positron–Photon Plasma

In the preceding section we analyzed the performance of endoreversible plasma engines that are described by equations of state which are (i) linear in temperature and (ii) at most additive in volume. The situation becomes more involved if these conditions are not met.

To this end, we conclude the analysis with another type of plasma, namely the relativistic electron–positron–photon plasma. Rather recently, it was shown in Ref. [36] that in the high-temperature limit, the corresponding caloric equation of state can be written as the familiar photonic gas law [37]:

$$E = \epsilon V T^4 , \tag{24}$$

where ϵ is a constant collecting natural constants, such as the speed of light, the electron mass, and $\hbar$; see Ref. [36]. In principle, we could also work with the more involved expression at finite temperature, but this would necessitate a fully numerical treatment. We choose to continue with Equation (24) as it allows an almost entirely analytical treatment.

Note that Equation (24) does not have the simple, linear form of Equation (3) that we exploited above. Therefore, the natural question arises whether engines with photonic plasmas operate at higher or lower efficiency than those with one-component plasmas. Note that for the same question, but for Otto cycles it was found that photonic gases significantly outperform classical ideal gases [25].

3.1. *Endoreversible Stirling Cycle*

To answer this question, we repeat the construction of the endoreversible Stirling cycle paying special attention to the required modifications. In Figure 2, we plot the corresponding PV- and TS-diagrams schematically.

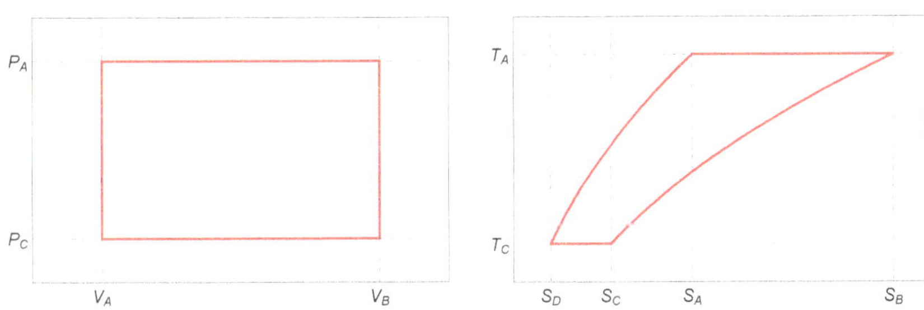

Figure 2. Schematic PV- and TS-diagrams of the Stirling cycle for the photonic gas (24) as a working medium.

3.1.1. $A \to B$: Isothermal Expansion

As before, we assume that during the hot isotherm the plasma is slightly colder than the hot heat reservoir, and that the heat flux is given by Fourier's law. Hence, we can write again $Q_{AB} = \alpha t_{AB} x$. However, in this case we can no longer assume that the internal energy is constant, as Equation (24) depends multiplicatively on the volume. Hence, we have to write for the work

$$W_{AB} = \Delta E_{AB} - Q_{AB} = \epsilon \, T_{h,p}^4 \, (V_B - V_A) - \alpha t_{AB} x \, , \tag{25}$$

which explicitly depends on the change in volume as well as the stroke time.

It is a standard exercise [37] to show that the entropy of the photonic gas reads

$$S = \frac{4}{3}\epsilon V T^3 . \tag{26}$$

And, as discussed above, for isothermal processes, we simply have $Q_{AB} = T_{h,p}\,\Delta S_{AB}$. Therefore, we can also write

$$\alpha t_{AB} x = \frac{4}{3}\,T_{h,p}^4\,(V_B - V_A)\,, \tag{27}$$

which will become useful shortly.

3.1.2. $B \to C$: Isochoric Cooling

For the isochoric strokes, we again have that $W_{BC} = 0$ and

$$Q_{BC} = \Delta E_{BC} = \epsilon V_B (T_{c,p}^4 - T_{h,p}^4)\,. \tag{28}$$

However, as seen above, the work and heat of the isochoric strokes will not be required for further analysis.

3.1.3. $C \to D$: Isothermal Compression

In complete analogy to the hot isotherm, we now can write $Q_{CD} = -\beta t_{CD} y$, and

$$W_{CD} = \Delta E_{CD} - Q_{CD} = -\epsilon\, T_{c,p}^4\,(V_B - V_A) - Q_{CD}\,. \tag{29}$$

Moreover, again employing the expression for the entropy (26) and $Q_{CD} = T_{c,p}\,\Delta S_{CD}$ we also have

$$\beta t_{CD} y = \frac{4}{3}\,T_{c,p}^4\,(V_B - V_A)\,. \tag{30}$$

3.1.4. $D \to A$: Isochoric Heating

For completeness, we also collect the work, $W_{DA} = 0$, and heat,

$$Q_{DA} = \Delta E_{DA} = \epsilon V_A (T_{h,p}^4 - T_{c,p}^4)\,, \tag{31}$$

during the isochoric heating stroke.

3.1.5. Endoreversible Efficiency

Before we start analyzing the power, we again first derive an expression for the efficiency:

$$\eta \equiv -\frac{W_{\text{cyc}}}{Q_{AB}} = -\frac{W_{AB} + W_{CD}}{Q_{AB}}\,. \tag{32}$$

Substituting Equations (25), (27), (29), and (30) into the definition (32) we obtain,

$$\eta = \frac{1}{4}\left[1 - \left(\frac{T_{c,p}}{T_{h,p}}\right)^4\right]. \tag{33}$$

Comparing the latter result (33) with the endoreversible efficiency for one-component plasmas (17) we note that endoreversible Stirling efficiency for photonic gases is always smaller than Equation (17); cf. Figure 3.

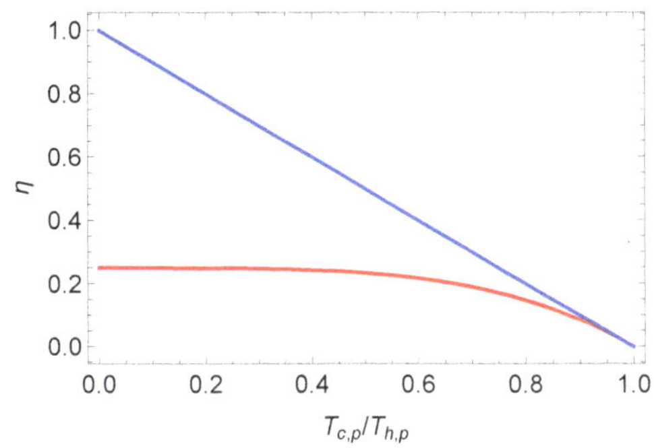

Figure 3. Endoreversible Stirling efficiency for the one-component plasma (22) (blue line) and the photonic gas (33) (red line).

3.2. *Efficiency at Maximal Power*

The obvious question now is what this means for the efficiency at maximal power. To this end, we again consider the power output

$$P = \frac{Q_{AB} + Q_{CD}}{\gamma\,(t_{AB} + t_{CD})}\,, \tag{34}$$

which in the present case can be written as

$$P(T_{c,p}, T_{h,p}) = \frac{\alpha\beta}{4\gamma}\frac{(T_h - T_{h,p})(T_c - T_{c,p})(T_{h,p}^4 - T_{c,p}^4)}{\beta T_{h,p}^4(T_c - T_{c,p}) - \alpha T_{c,p}^4(T_h - T_{h,p})}\,. \tag{35}$$

We immediately observe that Equation (35) is significantly more involved than the power for one-component plasmas (20), and thus we have to resort to a numerical analysis.

In Figure 4, we depict the solution for one set of parameters. We also observe that at maximal power, the efficiency for photonic gases is significantly below the Curzon–Ahlborn efficiency (22). We emphasize again that this is in contrast to endoreversible Otto cycle, in which the photonic case has a higher efficiency [25].

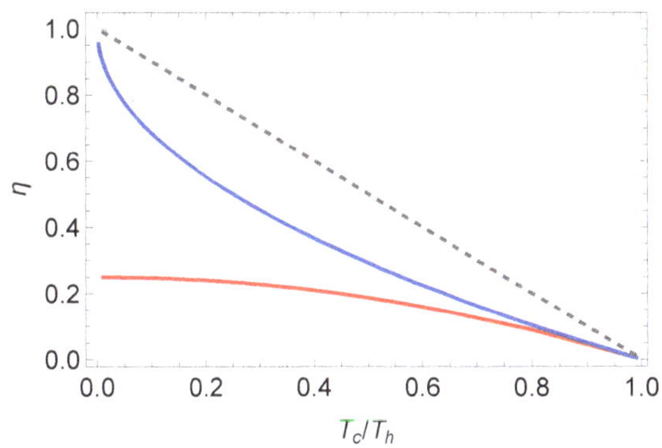

Figure 4. Efficiency at maximal power for the photonic equation of state (24) (red line) together with the Curzon–Ahlborn efficiency (22) (blue line), and the Carnot efficiency, $\eta_C = 1 - T_c/T_h$ (gray, dashed line). Parameters are $\alpha = 1$, $\beta = 1$, and $\gamma = 1$.

4. Concluding Remarks

In the present work, we analyzed the efficiency at maximal power of endoreversible Stirling cycles. As working mediums, we considered one-component plasmas with caloric equations of state, extensions in second order virial expansion, and photonic gases. We found that for all thermodynamics systems, whose caloric equation of state only depends linearly on temperature and at most additively on volume, the efficiency at maximal power is given by the seminal Curzon–Ahlborn efficiency. Interestingly, the efficiency for photonic working mediums is significantly smaller, which is in stark contrast to endoreversible Otto cycles.

Particular emphasis was put on a comprehensive, self-contained, and pedagogical presentation of the treatment. While our work is purely theoretical and it is unlikely that our results will lead directly to the experimental realization of a practically useful plasma Stirling engine, the conceptual mathematical steps of our analysis could be applied to more realistic scenarios. The only necessary "ingredient" is the caloric equation of state for any considered plasma, which then step-for-step leads to the corresponding expression for the efficiency at maximal power.

Author Contributions: G.B. and S.D. contributed equally to all aspects of the study. All authors have read and agreed to the published version of the manuscript.

Funding: S.D. acknowledges support from the John Templeton Foundation under Grant No. 62422.

Institutional Review Board Statement: Not applicable.

Data Availability Statement: The original contributions presented in this study are included in the article. Further inquiries can be directed to the corresponding authors.

Conflicts of Interest: The authors declare no conflicts of interest.

Appendix A. Equal Strokes Times—Trivial Case

In this appendix, we briefly discuss a trivial case for the one-component plasma (3). As previously conducted for Otto [16] and Brayton [17] cycles, one is tempted to make a "simplifying" assumption. Namely, to set the stroke times equal to each other, $\tau \equiv t_{AB} = t_{CD}$. In this case, we can set Equation (18) as

$$P = \frac{1}{2\gamma}\left(\alpha\, x - \beta\, y\right), \tag{A1}$$

which is linear in x and y. In this case, we can determine the maximal power simply from the maximal and minimal values of x and y, respectively. Recognizing that at no instant the plasma can be colder than the cold reservoir, we can write

$$x_{\max} = T_h - T_c \quad \text{and} \quad y_{\min} = 0\,. \tag{A2}$$

Substituting the latter into Equation (11) we obtain

$$\eta = 1 - \frac{\beta\, y_{\min}}{\alpha\, x_{\max}} = 1\,. \tag{A3}$$

References

1. Conde, L. *An Introduction to Plasma Physics and its Space Applications*; IOP Publishing: Bristol, UK, 2020; Volume 2, pp. 2053–2563. [CrossRef]
2. Morse, E. *Nuclear Fusion*; Springer: Berlin/Heidelberg, Germany, 2018.

3. Department of Energy, U.S.A. Fusion energy Sciences. Available online: https://www.energy.gov/science/fes/fusion-energy-sciences (accessed on 18 June 2025).
4. Segantin, S.; Bersano, A.; Falcone, N.; Testoni, R. Exploration of power conversion thermodynamic cycles for ARC fusion reactor. *Fusion Eng. Des.* **2020**, *155*, 111645. [CrossRef]
5. Terahara, Y.; Tanabe, K. Process design of a thermochemical cycle for hydrogen production compatible with nuclear fusion heat sources. *Fusion Eng. Des.* **2023**, *194*, 113868. [CrossRef]
6. Avinash, K. Plasma heat pump and heat engine. *Phys. Plasmas* **2010**, *17*, 082105. [CrossRef]
7. Avinash, K.; Chaudhary, S. Stirling like engine using plasma electric fields. *Proc. Indian Natl. Sci. Acad.* **2014**, *80*, 1099–1104. [CrossRef]
8. Minale, T.A.; Lanzetta, F.; Bégot, S.; Getie, M.Z. Review on the technological advancement of Stirling cycle heat pumps. *Energy Rep.* **2024**, *12*, 3504–3518. [CrossRef]
9. Hoffmann, K.H.; Burzler, J.M.; Schubert, S. Endoreversible thermodynamics. *J. Non-Equilib. Thermodyn.* **1997**, *22*, 311. [CrossRef]
10. Curzon, F.L.; Ahlborn, B. Efficiency of a Carnot engine at maximum power output. *Am. J. Phys.* **1975**, *43*, 22. [CrossRef]
11. Moutier, J. *Éléments de Thermodynamique*; Gauthier-Villars: Paris, France, 1872.
12. Novikov, I. The efficiency of atomic power stations (a review). *J. Nucl. Energy (1954)* **1958**, *7*, 125–128. [CrossRef]
13. Rozonoer, L.I.; Tsirlin, A.M. Optimal control of thermodynamic processes. I. *Avtomatika i Telemekhanika* **1983**, *1*, 70–79.
14. Rozonoer, L.I.; Tsirlin, A.M. Optimal control of thermodynamic processes. II. *Avtomatika i Telemekhanika* **1983**, *2*, 88–101.
15. Rozonoer, L.I.; Tsirlin, A.M. Optimal control of thermodynamic processes. III. *Avtomatika i Telemekhanika* **1983**, *3*, 50–64.
16. Deffner, S. Efficiency of Harmonic Quantum Otto Engines at Maximal Power. *Entropy* **2018**, *20*, 875. [CrossRef]
17. Ferketic, E.E.; Deffner, S. Boosting thermodynamic performance by bending space-time. *EPL (Europhys. Lett.)* **2023**, *141*, 19001. [CrossRef]
18. Esposito, M.; Kawai, R.; Lindenberg, K.; Van den Broeck, C. Efficiency at Maximum Power of Low-Dissipation Carnot Engines. *Phys. Rev. Lett.* **2010**, *105*, 150603. [CrossRef]
19. Leff, H.S. Thermal efficiency at maximum work output: New results for old heat engines. *Am. J. Phys.* **1987**, *55*, 602. [CrossRef]
20. Rezek, Y.; Kosloff, R. Irreversible performance of a quantum harmonic heat engine. *New J. Phys.* **2006**, *8*, 83. [CrossRef]
21. Abah, O.; Roßnagel, J.; Jacob, G.; Deffner, S.; Schmidt-Kaler, F.; Singer, K.; Lutz, E. Single-Ion Heat Engine at Maximum Power. *Phys. Rev. Lett.* **2012**, *109*, 203006. [CrossRef]
22. Deffner, S.; Campbell, S. *Quantum Thermodynamics*; Morgan & Claypool Publishers: San Rafael, CA, USA, 2019. [CrossRef]
23. Kloc, M.; Cejnar, P.; Schaller, G. Collective performance of a finite-time quantum Otto cycle. *Phys. Rev. E* **2019**, *100*, 042126. [CrossRef] [PubMed]
24. Myers, N.M.; Deffner, S. Bosons outperform fermions: The thermodynamic advantage of symmetry. *Phys. Rev. E* **2020**, *101*, 012110. [CrossRef]
25. Smith, Z.; Pal, P.S.; Deffner, S. Endoreversible Otto Engines at Maximal Power. *J.-Non-Equilib. Thermodyn.* **2020**, *45*, 305–310. [CrossRef]
26. Myers, N.M.; McCready, J.; Deffner, S. Quantum Heat Engines with Singular Interactions. *Symmetry* **2021**, *13*, 978. [CrossRef]
27. Myers, N.M.; Abah, O.; Deffner, S. Quantum Otto engines at relativistic energies. *New J. Phys.* **2021**, *23*, 105001. [CrossRef]
28. Myers, N.M.; Deffner, S. Thermodynamics of Statistical Anyons. *PRX Quantum* **2021**, *2*, 040312. [CrossRef]
29. Myers, N.M.; Peña, F.J.; Negrete, O.; Vargas, P.; De Chiara, G.; Deffner, S. Boosting engine performance with Bose–Einstein condensation. *New J. Phys.* **2022**, *24*, 025001. [CrossRef]
30. Myers, N.M.; Peña, F.J.; Cortés, N.; Vargas, P. Multilayer Graphene as an Endoreversible Otto Engine. *Nanomaterials* **2023**, *13*, 1548. [CrossRef]
31. Peña, F.J.; Myers, N.M.; Órdenes, D.; Albarrán-Arriagada, F.; Vargas, P. Enhanced Efficiency at Maximum Power in a Fock–Darwin Model Quantum Dot Engine. *Entropy* **2023**, *25*, 518. [CrossRef]
32. Erbay, L.B.; Yavuz, H. Analysis of the Stirling heat engine at maximum power conditions. *Energy* **1997**, *22*, 645. [CrossRef]
33. Blank, D.A.; Davis, G.W.; Wu, C. Power optimization of an endoreversible stirling cycle with regeneration. *Energy* **1994**, *19*, 125–133. [CrossRef]
34. Kaushik, S.; Kumar, S. Finite time thermodynamic analysis of endoreversible Stirling heat engine with regenerative losses. *Energy* **2000**, *25*, 989–1003. [CrossRef]
35. Slattery, W.L.; Doolen, G.D.; DeWitt, H.E. Improved equation of state for the classical one-component plasma. *Phys. Rev. A* **1980**, *21*, 2087–2095. [CrossRef]
36. Faussurier, G. Equation of state of the relativistic electron–positron–photon plasma at arbitrary temperature and degeneracy. *Phys. Plasmas* **2024**, *31*, 102702. [CrossRef]
37. Callen, H. *Thermodynamics and an Introduction to Thermostastistics*; Wiley: New York, NY, USA, 1985.

38. Peliti, L. *Statistical Mechanics in a Nutshell*, 2nd ed.; Princeton University Press: Princeton, NJ, USA, 2024.
39. Gervois, A.; Pomeau, Y.; Résibois, P. Non equilibrium density expansion in the dilute one component plasma. *Phys. Lett. A* **1976**, *55*, 343–344. [CrossRef]
40. Alastuey, A.; Perez, A. Virial Expansion of the Equation of State of a Quantum Plasma. *EPL (Europhys. Lett.)* **1992**, *20*, 19. [CrossRef]
41. Omarbakiyeva, Y.A.; Fortmann, C.; Ramazanov, T.S.; Röpke, G. Cluster virial expansion for the equation of state of partially ionized hydrogen plasma. *Phys. Rev. E* **2010**, *82*, 026407. [CrossRef] [PubMed]
42. Mattiello, S.; Cassing, W. Shear viscosity of the Quark–Gluon Plasma from a virial expansion. *Eur. Phys. J. C* **2010**, *70*, 243–249. [CrossRef]
43. Röpke, G.; Schörner, M.; Redmer, R.; Bethkenhagen, M. Virial expansion of the electrical conductivity of hydrogen plasmas. *Phys. Rev. E* **2021**, *104*, 045204. [CrossRef]

entropy

MDPI

Article

Revisiting Endoreversible Carnot Engine: Extending the Yvon Engine

Xiu-Hua Zhao [1] and Yu-Han Ma [1,2,3,*]

[1] School of Physics and Astronomy, Beijing Normal University, Beijing 100875, China
[2] Key Laboratory of Multiscale Spin Physics (Ministry of Education), Beijing Normal University, Beijing 100875, China
[3] Graduate School of China Academy of Engineering Physics, Beijing 100193, China
* Correspondence: yhma@bnu.edu.cn

Abstract: Curzon and Ahlborn's 1975 paper, a pioneering work that inspired the birth of the field of finite-time thermodynamics, unveiled the efficiency at maximum power (EMP) of the endoreversible Carnot heat engine, now commonly referred to as the Curzon–Ahlborn (CA) engine. Historically, despite the significance of the CA engine, similar findings had emerged at an earlier time, such as the Yvon engine proposed by J. Yvon in 1955 that shares the exact same EMP, that is, the CA efficiency η_{CA}. However, the special setup of the Yvon engine has circumscribed its broader influence. This paper extends the Yvon engine model to achieve a level of generality comparable to that of the CA engine. With the power expression of the extended Yvon engine, we directly explain the universality that η_{CA} is independent of the heat transfer coefficients between the working substance and the heat reservoirs. A rigorous comparison reveals that the extended Yvon engine and CA engine represent the steady-state and cyclic forms of the endoreversible Carnot heat engine, respectively, and are equivalent.

Keywords: finite-time thermodynamics; endoreversible heat engine; Curzon–Ahlborn efficiency; Yvon engine

1. Introduction

The seminal paper by Curzon and Ahlborn [1], published in 1975, is a landmark work in finite-time thermodynamics [2]. By analyzing an endoreversible Carnot engine with finite temperature differences between the working substance and the reservoirs, the authors optimized the power output of the engine and obtained the efficiency at maximum power (EMP), which is a more practical efficiency bound than the Carnot efficiency [3]. The EMP formula they derived,

$$\eta_{\mathrm{CA}} = 1 - \sqrt{T_c/T_h} \tag{1}$$

with T_c (T_h) being the temperature of the cold (hot) reservoir, is commonly referred to as Curzon–Ahlborn (CA) efficiency. In the following decades, studies on finite-time heat engines have advanced rapidly [2,4–8]. Notable topics include the nature of CA efficiency [9–11] and the microscopic realization of CA engines [12,13], various forms of endoreversible thermodynamic cycles [14–17], different theoretical frameworks for finite-time cycles [10,18–22] and their thermodynamic bounds [15,23–27], and the connections and shared principles among diverse heat engine models [28–30].

In fact, Curzon and Ahlborn's work was not the first to explore the performance of finite-time heat engines or derive the CA efficiency [31–34]. As early as 1955, Yvon presented the same formula as Equation (1) while investigating the optimization of actual power plants modeled as endoreversible engines [35]. Two years later, Chambadal [36] and Novikov [37] presented discussions that are quite similar to that of Yvon and reached the same conclusion. Yvon's approach, though yielding the same EMP expression, differed from Curzon and Ahlborn's in two key aspects as follows: (i) a finite temperature difference was assumed only between the working substance and the hot reservoir, with no temperature difference on the cold side; and (ii) instead of explicitly considering cycle time, finite heat flux and work rate were directly used to characterize the finite-time nature of practical engines. These distinctions render the relationship between the Yvon and CA engines less straightforward, obfuscating the grounds for their having the same EMP.

The simplicity of Yvon's optimization procedure (shown in Section 2) inspires us to extend their model to encompass more general finite-time engines. The extended Yvon engine presented in Section 3 provides a unified and straightforward framework for studying endoreversible Carnot engines, bridging the gap between Yvon's original analysis and the broader scope of Curzon and Ahlborn's work. Specifically, we relax the assumption of equal temperatures on the cold side in the Yvon engine, and optimize the work rate (power) with respect to the endoreversible temperatures of the working substance under the endoreversible condition. The derived EMP aligns with the CA efficiency, yet the optimization process is significantly more concise than that of the CA engine. Moreover, our approach naturally reproduces the trade-off relation between power and efficiency beyond the maximum-power regime [15], a topic of considerable interest in contemporary thermodynamic research [8,38]. In Section 4, we provide a rigorous comparison between the extended Yvon engine and the CA engine models to demonstrate their equivalence in describing the endoreversible Carnot heat engine. The conclusion, outlook, and some remarks of this study are presented in Section 5.

2. Yvon Engine

We start with reviewing Yvon's pioneering treatment of the finite-time Carnot heat engine [33,35]. Yvon's report [35] at the 1955 United Nations Conference on the Peaceful Uses of Atomic Energy offered a simple solution for maximizing the output power of the steam engine in nuclear reactors. Specifically, as shown in Figure 1a, the fluid at temperature T_h from the reactor core serves as the hot reservoir transferring heat to the hot end of the working substance in the engine, where the temperature of this end is maintained at $T_m < T_h$. The Yvon engine generates power through the turbine shafts and isothermally releases heat to the condenser where the working substance and the condenser are at the same lowest temperature $T_c < T_m$. This assumes that the heat flux from the fluid to the working substance obeys Newton's law of cooling, namely,

$$\dot{Q} = \Gamma(T_h - T_m), \tag{2}$$

where Γ is a constant that depends on the thermal conductivity and area of the wall separating the working substance from the hot fluid, and the overdot notation stands for the rates. This heat transfer law, proportional to the temperature difference between the reservoir and the working substance, is valid for small temperature differences [39–41], which is the scope of our study. The heat engine efficiency is defined as $\eta \equiv P/\dot{Q}$, where P is the output power of the engine. With the assumption that the working substance undergoes a reversible transformation between T_m and T_c, namely, the heat engine is equivalent to operating with a Carnot cycle between a hot reservoir of temperature T_m and a cold reservoir of temperature T_c, the efficiency of the engine reads as

$$\eta^{(Y)} = 1 - T_c/T_m, \tag{3}$$

which is now known as the endoreversible assumption [42]. Hence, the power of the Yvon engine follows as

$$P^{(Y)} = \eta^{(Y)}\dot{Q} = \Gamma\left(T_h + T_c - T_m - \frac{T_h T_c}{T_m}\right) \leq \Gamma\left(\sqrt{T_h} - \sqrt{T_c}\right)^2, \tag{4}$$

where the AM-GM inequality, $(a+b)/2 \geq \sqrt{ab}$ for nonnegative a and b, has been applied. The power output is optimized to determine a practical operating regime for real engines, where high power is required to perform real-world tasks. Associated with the third equality in Equation (4), the maximum power is achieved when

$$T_m = T_m^* \equiv \sqrt{T_h T_c}, \tag{5}$$

and the corresponding efficiency at maximum power $\eta^{(Y)}|_{T_m=T_m^*} = \eta_{CA}$ is easily checked.

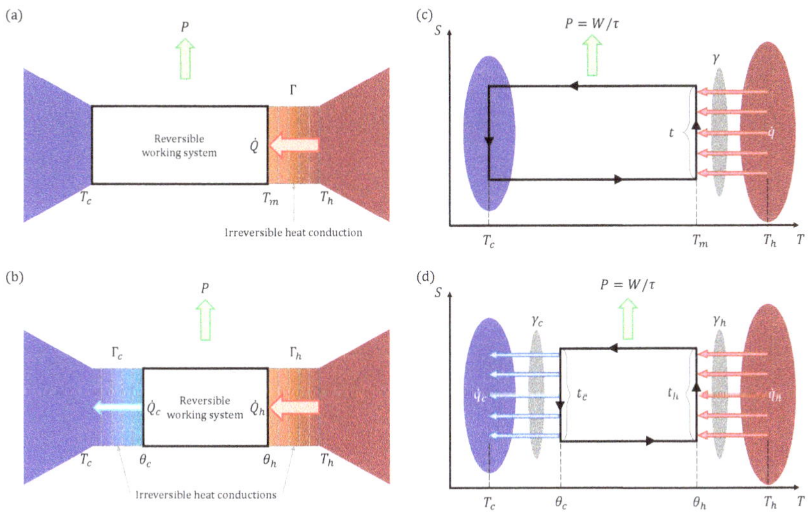

Figure 1. Steady-state (**a**,**b**) and cyclic (**c**,**d**) endoreversible heat engines. (**a**) In the original Yvon engine, finite heat flux (denoted as $\dot{Q}$) occurs only at the high-temperature end, where there is a temperature difference between the working substance and the hot reservoir. (**b**) The extended Yvon engine introduces temperature differences, and thus, heat fluxes, between the working substance and both the hot and cold reservoirs. (**c**,**d**) show the entropy (S)–temperature (T) diagrams of the endoreversible Carnot engine cycles with finite heat fluxes (denoted as $\dot{q}$) along the high-temperature and both isothermal branches, respectively.

3. Extended Yvon Engine

In the Yvon engine, heat flux is only considered when the working substance absorbs heat from the hot reservoir. In this section, we extend the model to include heat flux on the cold side, resulting from the temperature difference between the working substance and the cold reservoir, as illustrated in Figure 1b. This extended Yvon engine is consistent with the CA engine [Figure 1d], though it uses a different representation method based on energy fluxes rather than examining each branch of a thermodynamic cycle.

The heat flux from the hot reservoir at temperature T_h to the working substance at temperature $\theta_h < T_h$ follows

$$\dot{Q}_h = \Gamma_h (T_h - \theta_h), \tag{6}$$

where the constant Γ_h characterizes the heat conductivity during heat absorption. Similarly, the heat flux from the working substance, whose temperature decreases to $\theta_c < \theta_h$, to the cold reservoir at temperature $T_c < \theta_c$ is

$$\dot{Q}_c = \Gamma_c (\theta_c - T_c), \tag{7}$$

with Γ_c being a constant during heat release. As a result of the energy conservation law, the engine's output power is

$$P = \dot{Q}_h - \dot{Q}_c = \eta \dot{Q}_h. \tag{8}$$

In this case, the endoreversible assumption reads as

$$\frac{\dot{Q}_h}{\theta_h} = \frac{\dot{Q}_c}{\theta_c}, \tag{9}$$

which indicates that there is no entropy generation in the working substance for this steady-state engine, and the entropy flow from the hot reservoir to the working substance is balanced by the entropy flow from the working substance to the cold reservoir [43]. Correspondingly, $\eta = 1 - \theta_c/\theta_h$ is the endoreversible efficiency. According to Equations (6), (7), and (9), θ_h can be expressed with the temperature ratio θ_h/θ_c as

$$\theta_h = \frac{\Gamma_h T_h}{\Gamma_h + \Gamma_c} + \frac{\Gamma_c T_c}{\Gamma_h + \Gamma_c} \frac{\theta_h}{\theta_c}, \tag{10}$$

substituting which into Equation (8), we obtain

$$P = \frac{\Gamma_h \Gamma_c T_h}{\Gamma_h + \Gamma_c} \left(1 + \frac{T_c}{T_h} - \frac{\theta_c}{\theta_h} - \frac{T_c}{T_h} \frac{\theta_h}{\theta_c} \right). \tag{11}$$

Similar to Equation (4), by utilizing the AM-GM inequality, it is easy to find that, for given $\Gamma_{h,c}$ and $T_{h,c}$,

$$P \leq \frac{\Gamma_h \Gamma_c T_h}{\Gamma_h + \Gamma_c} \left(1 - \sqrt{\frac{T_c}{T_h}} \right)^2 \equiv P_{\max}, \tag{12}$$

where the maximum power $P_{\max}$ is achieved with the optimal endoreversible temperatures θ_h^* and θ_c^*, which satisfy

$$\frac{\theta_c^*}{\theta_h^*} = \sqrt{\frac{T_c}{T_h}}. \tag{13}$$

Consequently, the EMP of the engine $\eta_{\mathrm{MP}} = 1 - \theta_c^*/\theta_h^* = \eta_{\mathrm{CA}}$. In the limiting case of $\Gamma_h/\Gamma_c \to 0$, it follows from Equations (10) and (13) that

$$\theta_h^* = \sqrt{T_h T_c} = T_m^*, \;\; \theta_c^* = T_c, \tag{14}$$

and the original Yvon engine is recovered. With nonvanishing Γ_h/Γ_c, the maximum power of the extended Yvon engine [Equation (12)] is lower than that of the original Yvon engine [Equation (4)], despite operating at the same efficiency, due to the additional dissipation resulting from the finite-rate heat transfer at the low-temperature end.

Furthermore, in Equation (11), replacing the temperature ratios with efficiencies, where $T_c/T_h = 1 - \eta_C$ and $\theta_c/\theta_h = 1 - \eta$, yields

$$P = \frac{T_h}{\Gamma_h^{-1} + \Gamma_c^{-1}} \frac{\eta(\eta_C - \eta)}{(1 - \eta)}, \tag{15}$$

which is illustrated in Figure 2 for $\eta_C = 0.1, 0.7, 0.9$ with normalized axes. This relation determines the efficiency at an arbitrary power or the power at an arbitrary efficiency of the endoreversible Carnot engine. The explicit dependence of power on efficiency, which provides a comprehensive optimization regime for thermal machines, is now referred to as the trade-off or constraint relation between power and efficiency [23–25,44,45], and this constitutes one of the key focuses within the realm of finite-time thermodynamics [30]. For more details and the related progress on this issue, please refer to Ref. [8] along with the references encompassed therein.

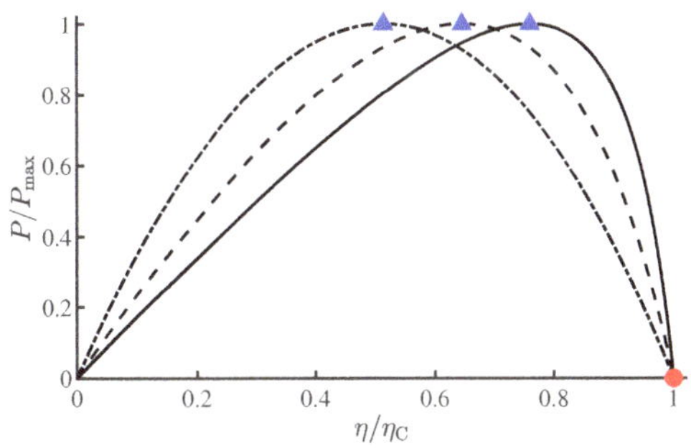

Figure 2. Trade-off relations between power and efficiency for the Curzon-Ahlborn engine with $\eta_C = 0.1$ (dash-dotted curve), $\eta_C = 0.7$ (dashed curve), and $\eta_C = 0.9$ (solid curve). The triangles and circle mark the maximum power and maximum efficiency, respectively.

4. Comparison of Curzon–Ahlborn and Extended Yvon Engines

Both the Yvon engine and its extended version are steady-state heat engines, while the CA engine adopts a cyclic representation, taking into account the duration of each thermodynamic process in the cycle. Although the previous section has demonstrated the consistency of the EMP obtained by the extended Yvon and CA engines, their strict correspondence needs further clarification. In this section, we will establish the strict correspondence between the parameters of the two models. As the cyclic counterpart of the extended Yvon engine [Figure 1b], the CA engine is depicted in Figure 1d in the entropy–temperature ($S - T$) diagram. Meanwhile, Figure 1c represents the limiting case of the CA heat engine, where the temperature difference at the low-temperature end approaches zero, corresponding to the original Yvon engine illustrated in Figure 1a.

In Curzon and Ahlborn's original derivation [1], engine power is expressed not through energy fluxes but as the total work $W = Q_h - Q_c$ divided by the duration $\tau = \xi(t_h + t_c)$ of an endoreversible Carnot cycle, namely,

$$P^{(CA)} = \frac{Q_h - Q_c}{\xi(t_h + t_c)}. \tag{16}$$

Here, t_h and t_c represent the durations of the heat absorption and release processes [see Figure 1d], respectively, while the time taken to complete the adiabatic transitions,

$(\xi-1)(t_h+t_c)$, is assumed to be proportional to the duration of the isothermal processes with $\xi>1$. The exchanged heat, $Q_{h(c)}=\int \dot{q}_{h(c)}dt$, in the isothermal processes is

$$Q_h=\gamma_h(T_h-\theta_h)t_h,\ Q_c=\gamma_c(\theta_c-T_c)t_c, \tag{17}$$

where $T_{h,c}$ and $\theta_{h,c}$ have the same meanings as in the extended Yvon model. However, the heat transfer coefficients $\gamma_{h,c}$ are different from $\Gamma_{h,c}$, as will be discussed later. As the result of the endoreversible assumption and the cyclic condition, the entropy variation of the working substance in a cycle satisfies

$$\Delta S=\frac{Q_h}{\theta_h}-\frac{Q_c}{\theta_c}=0. \tag{18}$$

Consequently, this cyclic engine's efficiency $\eta\equiv(Q_h-Q_c)/Q_h=1-\theta_c/\theta_h$. Comparing the first equality in Equation (8) with Equation (16) and Equation (9) with Equation (18), we obtain the relation between $\gamma_{h,c}$ and $\Gamma_{h,c}$ as

$$\Gamma_{h(c)}=\frac{t_{h(c)}}{\xi(t_h+t_c)}\gamma_{h(c)}. \tag{19}$$

This equation shows that the heat transfer coefficients in the steady-state engine differ from those in the cyclic engine by a time-proportional factor, which is determined by the ratio of the corresponding process duration to the total cycle duration.

According to Equation (19), the overall coefficient appearing in the upper bound of Equation (12) satisfies

$$\frac{\Gamma_h\Gamma_c}{\Gamma_h+\Gamma_c}=\frac{\gamma_h\gamma_c}{\xi(1+t_c/t_h)(\gamma_h t_h/t_c+\gamma_c)}\leq\frac{\gamma_h\gamma_c}{\xi(\sqrt{\gamma_h}+\sqrt{\gamma_c})^2}, \tag{20}$$

where $\gamma_h t_h/t_c+\gamma_c t_c/t_h\geq 2\sqrt{\gamma_h\gamma_c}$ has been used, and the equality is saturated with the optimal time ratio

$$\frac{t_h^*}{t_c^*}=\sqrt{\frac{\gamma_c}{\gamma_h}}. \tag{21}$$

This indicates that when the process durations are taken into account, cyclic engines have an additional parameter to be optimized, i.e., process time allocation t_c/t_h, compared to steady-state engines. Substituting Equation (20) into Equation (12), the maximum power of the CA engine [1] is exactly recovered as

$$P_{\max}^{(\mathrm{CA})}=\frac{\gamma_h\gamma_c T_h}{\xi(\sqrt{\gamma_h}+\sqrt{\gamma_c})^2}\left(1-\sqrt{\frac{T_c}{T_h}}\right)^2. \tag{22}$$

On the other hand, by combining $\gamma_{h,c}$ and $t_{h,c}$ into $\Gamma_{h,c}$ through Equation (19), the power of the extended Yvon engine given in Equation (11) can be derived from Equations (16)–(18) of the CA engine. In this sense, the extended Yvon engine and the CA engine are equivalent.

Herein, we would like to elucidate why the optimization of the extended Yvon engine exhibits greater succinctness than that of the original CA engine. In the optimization process from Equation (11) to Equation (12), the ratio of θ_c/θ_h emerges as a unified quantity, decoupling its role in maximizing P from the heat transfer coefficients $\Gamma_{h,c}$. By contrast, in Ref. [1], the authors individually calculated the derivatives of P [Equation (16)] with respect to θ_c and θ_h, subsequently determining θ_c^* and θ_h^* via $\partial P/\partial\theta_c=\partial P/\partial\theta_h=0$. Eventually, Curzon and Ahlborn arrived at Equation (13) and ascertained its independence from $\Gamma_{h,c}$ or $\gamma_{h,c}$. Essentially, our derivation is more intuitive and effectively highlights the significant universality that the EMP of the endoreversible Carnot engine is invariant to the heat transfer coefficients. Nevertheless, $\Gamma_{h,c}$ impact the specific values of θ_c^* and θ_h^* via Equations (10) and (13). Furthermore, when correlating heat transfer coefficients in the extended Yvon engine with those of the CA heat engine, Equation (20) indicates that the time ratio t_c/t_h of different thermodynamic processes in the cyclic heat engine affects the output power as a unified quantity rather than depending on the specific durations of each individual process. As a final point, similarly to the power–efficiency relation Equation (15) for the extended Yvon engine, the trade-off relation between power and efficiency for the CA engine can be obtained by substituting the upper bound in Equation (20) into Equation (15), which was first derived by Chen and Yan [15] through solving $(\partial P/\partial\theta_h)_\eta = 0$. When expressed in normalized form, as illustrated in Figure 2, the power–efficiency trade-offs for both engines are identical.

5. Conclusions and Discussion

To summarize, we have extended the original model of the Yvon engine with one-sided heat flux to the general endoreversible Carnot engine with two-sided heat fluxes. The extended Yvon engine and the CA engine are essentially two sides of the same coin, namely the steady-state heat engine form and the cyclic heat engine form of the endoreversible Carnot heat engine. Our derivations emphasize the predominance of the temperature ratio over specific temperature values in optimizing this type of engine, explaining why the EMP of endoreversible engines under Newtonian heat transfer is independent of the heat transfer coefficients. For endoreversible heat engines operating under heat transfer laws other than Newton's law, the EMP generally depends significantly on the heat transfer coefficients [15].

It is highlighted that the temperature ratio θ_c/θ_h (or efficiency $\eta = 1 - \theta_c/\theta_h$) constitutes the sole degree of freedom for steady-state endoreversible engines, independent of the heat transfer coefficients [see Equation (11) or (15)], and decouples from time allocation in cyclic engines [see Equation (20)]. It is worth mentioning that (i) Ref. [46] also recognizes the independent role of efficiency, but only for the special case with symmetric heat transfer coefficients; and (ii) Bejan [31] noticed the fact that different irreversible heat engine models, namely Chambadal's [36], Novikov's [37], and Curzon and Ahlborn's [1] share the same EMP, and he explained this with the theory of entropy generation minimization [47]. Nevertheless, the analyses presented in Ref. [31] did not clarify the strict correspondence between the optimization of cyclic heat engines, which incorporated the process durations as per Curzon and Ahlborn's methodology, and that of steady-state heat engines characterized by energy fluxes [35–37]. We provide such strict correspondence relations in Section 4 to fill this gap.

The optimization approach presented in the current work is intuitive and straightforward, allowing for the direct extension of the optimization criteria, from maximizing power to optimizing efficient power, ecological function, or Omega function [14,48]. The extended Yvon engine has potential applications in areas such as chemical engines [49,50] and thermoelectric generators [51,52]. Future research could further explore the optimal control and geometric optimization [13,53–55] of the extended Yvon engine, its performance between finite-sized heat reservoirs [44,56–59], and its experimental realization [41,45,60]. In addition, the extended Yvon engine can serve as a pedagogical example for teaching thermodynamics and engineering thermodynamics, given that its simple and lucid derivation helps students initiate their understanding of nonequilibrium thermodynamics.

As a final remark, perhaps due to historical, linguistic, or other factors [32–34], as well as the particularity of the model, the Yvon engine [35] and its contemporaneous works [36,37] regrettably did not garner the attention they deserved at that time. Twenty years later, the generality and simplicity of the CA engine [1], along with the systematic research on thermodynamics in finite time by the Chicago school [7,61] during the same period, gave birth to the field of finite-time thermodynamics. We hope that the extended Yvon engine proposed in this paper will encourage more people to pay attention to and appreciate Yvon's ingenious ideas regarding the practical thermodynamic cycle within a finite duration and help disseminate the complete history of finite-time thermodynamics.

Author Contributions: Investigation, writing—original draft preparation, writing—review and editing, X.-H.Z. and Y.-H.M.; conceptualization and funding acquisition, Y.-H.M. All authors have read and agreed to the published version of the manuscript.

Funding: This research was funded by the National Natural Science Foundation of China under grant No. 12305037 and the Fundamental Research Funds for the Central Universities under grant No. 2023NTST017.

Institutional Review Board Statement: Not applicable.

Data Availability Statement: The original contributions presented in this study are included in the article. Further inquiries can be directed to the corresponding author.

Acknowledgments: The authors thank Ruo-Xun Zhai for their valuable comments on a preliminary version of this manuscript. Y. H. Ma sincerely thanks Yves Pomeau for recommending J. Yvon's research work on finite-time heat engines [35] in their kind email in November 2019. The authors comprehensively learned the historical background and details of the Yvon engine from the excellent paper of Moreau and Pomeau [33].

Conflicts of Interest: The authors declare no conflicts of interest.

References

1. Curzon, F.L.; Ahlborn, B. Efficiency of a Carnot Engine at Maximum Power Output. *Am. J. Phys.* **1975**, *43*, 22–24. [CrossRef]
2. Andresen, B. Current trends in finite-time thermodynamics. *Angew. Chem. Int. Ed.* **2011**, *50*, 2690–2704. [CrossRef]
3. Callen, H.B. *Thermodynamics and an Introduction to Thermostatistics*, 2nd ed.; Wiley: New York, NY, USA, 1991.
4. Andresen, B.; Salamon, P.; Berry, R.S. Thermodynamics in finite time. *Phys. Today* **1984**, *37*, 62–70. [CrossRef]
5. Tu, Z.C. Recent advance on the efficiency at maximum power of heat engines. *Chin. Phys. B* **2012**, *21*, 020513. [CrossRef]
6. Leff, H.S. Reversible and Irreversible Heat Engine and Refrigerator Cycles. *Am. J. Phys.* **2018**, *86*, 344–353. [CrossRef]
7. Berry, R.S.; Salamon, P.; Andresen, B. (Eds.) *Finite-Time Thermodynamics*; MDPI—Multidisciplinary Digital Publishing Institute: Basel, Switzerland, 2022. [CrossRef]
8. Ma, Y.H.; Zhao, X.H. Finite-Time Thermodynamics: A Journey Beginning with Optimizing Heat Engines. *arXiv* **2024**, arXiv:2411.03853. [CrossRef]
9. Chen, J.; Yan, Z.; Lin, G.; Andresen, B. On the Curzon–Ahlborn Efficiency and Its Connection with the Efficiencies of Real Heat Engines. *Energ. Convers. Manag.* **2001**, *42*, 173–181. [CrossRef]
10. Van den Broeck, C. Thermodynamic Efficiency at Maximum Power. *Phys. Rev. Lett.* **2005**, *95*, 190602. [CrossRef]
11. Apertet, Y.; Ouerdane, H.; Goupil, C.; Lecoeur, P. True Nature of the Curzon-Ahlborn Efficiency. *Phys. Rev. E* **2017**, *96*, 022119. [CrossRef]
12. Dechant, A.; Kiesel, N.; Lutz, E. Underdamped stochastic heat engine at maximum efficiency. *Europhys. Lett.* **2017**, *119*, 50003. [CrossRef]
13. Chen, Y.H.; Chen, J.F.; Fei, Z.; Quan, H.T. Microscopic Theory of the Curzon–Ahlborn Heat Engine Based on a Brownian Particle. *Phys. Rev. E* **2022**, *106*, 024105. [CrossRef] [PubMed]
14. Salamon, P.; Nitzan, A. Finite Time Optimizations of a Newton's Law Carnot Cycle. *J. Chem. Phys.* **1981**, *74*, 3546–3560. [CrossRef]
15. Chen, L.; Yan, Z. The Effect of Heat-transfer Law on Performance of a Two-heat-source Endoreversible Cycle. *J. Chem. Phys.* **1989**, *90*, 3740–3743. [CrossRef]

16. Feldmann, T.; Geva, E.; Kosloff, R.; Salamon, P. Heat Engines in Finite Time Governed by Master Equations. *Am. J. Phys.* **1996**, *64*, 485–492. [CrossRef]
17. Kaushik, S.C.; Kumar, S. Finite Time Thermodynamic Analysis of Endoreversible Stirling Heat Engine with Regenerative Losses. *Energy* **2000**, *25*, 989–1003. [CrossRef]
18. Kosloff, R. A Quantum Mechanical Open System as a Model of a Heat Engine. *J. Chem. Phys.* **1984**, *80*, 1625–1631. [CrossRef]
19. Schmiedl, T.; Seifert, U. Efficiency at Maximum Power: An Analytically Solvable Model for Stochastic Heat Engines. *Europhys. Lett.* **2007**, *81*, 20003. [CrossRef]
20. Tu, Z.C. Efficiency at Maximum Power of Feynman's Ratchet as a Heat Engine. *J. Phys. A Math. Theor.* **2008**, *41*, 312003. [CrossRef]
21. Esposito, M.; Kawai, R.; Lindenberg, K.; Van den Broeck, C. Efficiency at Maximum Power of Low-Dissipation Carnot Engines. *Phys. Rev. Lett.* **2010**, *105*, 150603. [CrossRef]
22. Izumida, Y.; Okuda, K. Efficiency at Maximum Power of Minimally Nonlinear Irreversible Heat Engines. *Europhys. Lett.* **2012**, *97*, 10004. [CrossRef]
23. Ryabov, A.; Holubec, V. Maximum Efficiency of Steady-State Heat Engines at Arbitrary Power. *Phys. Rev. E* **2016**, *93*, 050101. [CrossRef]
24. Shiraishi, N.; Saito, K.; Tasaki, H. Universal trade-off relation between power and efficiency for heat engines. *Phys. Rev. Lett.* **2016**, *117*, 190601. [CrossRef]
25. Ma, Y.H.; Xu, D.; Dong, H.; Sun, C.P. Universal Constraint for Efficiency and Power of a Low-Dissipation Heat Engine. *Phys. Rev. E* **2018**, *98*, 042112. [CrossRef]
26. Pietzonka, P.; Seifert, U. Universal Trade-Off between Power, Efficiency, and Constancy in Steady-State Heat Engines. *Phys. Rev. Lett.* **2018**, *120*, 190602. [CrossRef] [PubMed]
27. Chen, J.F.; Sun, C.P.; Dong, H. Achieve higher efficiency at maximum power with finite-time quantum Otto cycle. *Phys. Rev. E* **2019**, *100*, 062140. [CrossRef]
28. Sheng, S.; Tu, Z.C. Constitutive Relation for Nonlinear Response and Universality of Efficiency at Maximum Power for Tight-Coupling Heat Engines. *Phys. Rev. E* **2015**, *91*, 022136. [CrossRef]
29. Johal, R.S. Heat Engines at Optimal Power: Low-Dissipation versus Endoreversible Model. *Phys. Rev. E* **2017**, *96*, 012151. [CrossRef]
30. Ma, Y.H.; Fu, C. Unified Approach to Power-Efficiency Trade-Off of Generic Thermal Machines. *arXiv* **2024**, arXiv:2411.03849. [CrossRef]
31. Bejan, A. Models of Power Plants That Generate Minimum Entropy While Operating at Maximum Power. *Am. J. Phys.* **1996**, *64*, 1054–1059. [CrossRef]
32. Vaudrey, A.; Lanzetta, F.; Feidt, M. H. B. Reitlinger and the Origins of the Efficiency at Maximum Power Formula for Heat Engines. *J. Non-Equilib. Thermodyn.* **2014**, *39*, 199–203. [CrossRef]
33. Moreau, M.; Pomeau, Y. Carnot Principle and Its Generalizations: A Very Short Story of a Long Journey. *Eur. Phys. J. Spec. Top.* **2015**, *224*, 769–780. [CrossRef]
34. Ouerdane, H.; Apertet, Y.; Goupil, C.; Lecoeur, Ph. Continuity and Boundary Conditions in Thermodynamics: From Carnot's Efficiency to Efficiencies at Maximum Power. *Eur. Phys. J. Spec. Top.* **2015**, *224*, 839–864. [CrossRef]
35. Yvon, J. Saclay Reactor: Acquired Knowledge by Two Years Experience in Heat Transfer Using Compressed Gas. In Proceedings of the 1955 United Nations Conference on the Peaceful Uses of Atomic Energy Geneva Conference, Geneva, Switzerland, August 1955.
36. Chambadal, P. *Les Centrales Nucléaires*; Armand Colin: Paris, France, 1957.
37. Novikov, I.I. Efficiency of an Atomic Power Generating Installation. *Sov. J. At. Energy* **1957**, *3*, 1269–1272. [CrossRef]
38. Tu, Z.C. Abstract Models for Heat Engines. *Front. Phys.* **2021**, *16*, 33202. [CrossRef]
39. Vollmer, M. Newton's Law of Cooling Revisited. *Eur. J. Phys.* **2009**, *30*, 1063. [CrossRef]
40. Izumida, Y.; Okuda, K. Onsager Coefficients of a Finite-Time Carnot Cycle. *Phys. Rev. E* **2009**, *80*, 021121. [CrossRef] [PubMed]
41. Ma, Y.H.; Zhai, R.X.; Chen, J.; Sun, C.P.; Dong, H. Experimental Test of the $1/\tau$-Scaling Entropy Generation in Finite-Time Thermodynamics. *Phys. Rev. Lett.* **2020**, *125*, 210601. [CrossRef]
42. Rubin, M.H. Optimal Configuration of a Class of Irreversible Heat Engines. I. *Phys. Rev. A* **1979**, *19*, 1272–1276. [CrossRef]
43. Demirel, Y. *Nonequilibrium Thermodynamics: Transport and Rate Processes in Physical, Chemical and Biological Systems*, 2nd ed.; Elsevier Science: Amsterdam, The Netherlands, 2007.
44. Yuan, H.; Ma, Y.H.; Sun, C.P. Optimizing Thermodynamic Cycles with Two Finite-Sized Reservoirs. *Phys. Rev. E* **2022**, *105*, L022101. [CrossRef] [PubMed]
45. Zhai, R.X.; Cui, F.M.; Ma, Y.H.; Sun, C.P.; Dong, H. Experimental Test of Power-Efficiency Trade-off in a Finite-Time Carnot Cycle. *Phys. Rev. E* **2023**, *107*, L042101. [CrossRef]

46. Reyes-Ramírez, I.; Barranco-Jiménez, M.A.; Rojas-Pacheco, A.; Guzmán-Vargas, L. Global Stability Analysis of a Curzon–Ahlborn Heat Engine under Different Regimes of Performance. *Entropy* **2014**, *16*, 5796–5809. [CrossRef]
47. Bejan, A. *Entropy Generation Minimization*; CRC Press: Boca Raton, FL, USA, 1996.
48. Hernández, A.C.; Medina, A.; Roco, J.M.M.; White, J.A.; Velasco, S. Unified Optimization Criterion for Energy Converters. *Phys. Rev. E* **2001**, *63*, 037102. [CrossRef]
49. Sieniutycz, S. Analysis of Power and Entropy Generation in a Chemical Engine. *Int. J. Heat Mass Transf.* **2008**, *51*, 5859–5871. [CrossRef]
50. de Parga-Regalado, A.M.A. Optimal Efficiencies of an Isothermal Endoreversible Chemical Engine. *Eur. J. Phys.* **2023**, *44*, 035101. [CrossRef]
51. Kaur, J.; Johal, R.S.; Feidt, M. Thermoelectric Generator in Endoreversible Approximation: The Effect of Heat-Transfer Law under Finite Physical Dimensions Constraint. *Phys. Rev. E* **2022**, *105*, 034122. [CrossRef]
52. Juárez-Huerta, V.H.; Sánchez-Salas, N.; Chimal-Eguía, J.C. Optimization Criteria and Efficiency of a Thermoelectric Generator. *Entropy* **2022**, *24*, 1812. [CrossRef] [PubMed]
53. Salamon, P.; Berry, R.S. Thermodynamic Length and Dissipated Availability. *Phys. Rev. Lett.* **1983**, *51*, 1127–1130. [CrossRef]
54. Chen, J.F. Optimizing Brownian Heat Engine with Shortcut Strategy. *Phys. Rev. E* **2022**, *106*, 054108. [CrossRef] [PubMed]
55. Zhao, X.H.; Gong, Z.N.; Tu, Z.C. Low-Dissipation Engines: Microscopic Construction via Shortcuts to Adiabaticity and Isothermality, the Optimal Relation between Power and Efficiency. *Phys. Rev. E* **2022**, *106*, 064117. [CrossRef]
56. Ondrechen, M.J.; Andresen, B.; Mozurkewich, M.; Berry, R.S. Maximum Work from a Finite Reservoir by Sequential Carnot Cycles. *Am. J. Phys.* **1981**, *49*, 681–685. [CrossRef]
57. Izumida, Y.; Okuda, K. Work Output and Efficiency at Maximum Power of Linear Irreversible Heat Engines Operating with a Finite-Sized Heat Source. *Phys. Rev. Lett.* **2014**, *112*, 180603. [CrossRef] [PubMed]
58. Ma, Y.H. Effect of Finite-Size Heat Source's Heat Capacity on the Efficiency of Heat Engine. *Entropy* **2020**, *22*, 1002. [CrossRef] [PubMed]
59. Ma, Y.H. Simple realization of the polytropic process with a finite-sized reservoir. *Am. J. Phys.* **2023**, *91*, 555–555. [CrossRef]
60. Martínez, I.A.; Roldán, É.; Dinis, L.; Petrov, D.; Parrondo, J.M.R.; Rica, R.A. Brownian Carnot Engine. *Nat. Phys.* **2016**, *12*, 67–70. [CrossRef]
61. Andresen, B.; Salamon, P.; Berry, R.S. Thermodynamics in Finite Time: Extremals for Imperfect Heat Engines. *J. Chem. Phys.* **1977**, *66*, 1571–1577. [CrossRef]

entropy

MDPI

Article

Compressor Power and Efficiency Optimization: A Finite-Time Thermodynamics Approach

François Lanzetta

CNRS, Institut FEMTO-ST, Université Marie et Louis Pasteur, F-90000 Belfort, France; francois.lanzetta@univ-fcomte.fr; Tel.: +33-384-578-224

Abstract: This paper presents a theoretical optimization of an endoreversible compressor under steady-state conditions. A parametric study using finite-time thermodynamic principles highlights the effect of external irreversibilities on compressor performance. A compressor efficiency metric is established based on heat pump theory's analogous performance coefficient concept. The external irreversibilities are characterized as functions of the conductance coefficients between the compressor and the low- and high-pressure reservoirs. In particular, the influence of suction and discharge tube diameters and gas pressures is investigated to determine the optimum compressor operating performance for a given gas mass flow rate. The results highlight the importance of selecting optimal suction and discharge tube diameters to improve compressor power efficiency and minimize energy consumption during gas compression.

Keywords: finite time thermodynamics; compressor; irreversibility; optimization; efficiency

1. Introduction

The optimization of thermal machines has been the subject of numerous engineering studies and has accompanied their industrial development since the late XIXth century. In the 1950s, Chambadal and Novikov optimized thermal machine cycles by considering irreversibilities at the heat exchangers [1,2], leading to an efficiency depending only on the temperatures of the hot source and cold sink. A few years later, Curzon and Alborn laid the groundwork for finite-time thermodynamics [3]. Their article obtained the same results as those of Chambadal and Novikov. Their general approach was to treat time as a finite quantity during each cycle transformation and to incorporate internal and external irreversibilities in the fluid and heat exchangers. Since then, FTT has gone from strength to strength [4–10]. The introduction of finite-time thermodynamics has made it possible to model and optimize thermal machine cycles, starting with Carnot engines [11,12]. Optimal performance metrics, such as efficiencies and power outputs, are determined as a function of heat source and heat sink temperatures. This methodology has been extended to numerous thermal machines, such as Carnot heat machines [13–20], internal combustion engines [21], external heat input machines [11,12,22–25], turbines [21,26,27], co-generation systems [27–29], fuel cells [25,28], refrigeration machines [30–33], heat pumps [34–36], and combined cycles [29,32,37]. The notion of finite time was then subsumed under a more general terminology: finite dimension. This dimension can also be the speed of a piston [38–41] or the surface area of an exchanger [32]. Generally speaking, finite-time thermodynamics can be applied to any energy conversion process that generates work from hydraulic, pneumatic, chemical, thermal, or electrical potential differences, and vice versa. Examples include thermoelectrical systems [42,43] and pneumatic cylinders [10,44–47]. The study of fluid flow in microchannels is relevant to both industry and research, with applications spanning biology, chemistry, microelectronics, space, and micromechanics [48,49]. These microsystems involve creating geometric objects with dimensions as small as a few microns. Under these challenging flow conditions, actuators like valves, pumps, and distributors have operational characteristics that depend heavily on the dimensions and physical properties of the fluids, whether liquid or gas.

In this article, we explore the theoretical study of the optimal performance of a compressor used in high-pressure micro gas flow scenarios. Such issues arise particularly in the cooling of electronic components using thermal machines like pulsed gas tubes, Stirling refrigerators, or Joule Thomson refrigerators [50,51]. We explore the theoretical analysis, drawing a parallel with the finite-time thermodynamics of a heat pump to determine the optimal operating conditions of a compressor functioning under stringent geometric constraints, necessitating suction and discharge flows through tubes with internal diameters ranging from 70 to 200 μm.

2. Mathematical Model

Bejan [10] presented a study on the maximum power generated by fluid flow, using a pneumatic cylinder and drawing an analogy between the conditions for maximum power in fluid and thermal power conversion. Based on this work, we extend the idea to an endoreversible compressor and, by analogy with the study of finite-time optimization of an endoreversible heat pump, we consider that a compressor is a system that provides mechanical work (or mechanical power) to a fluid to transfer it from a source with low energy potential to a sink with high energy potential (Figure 1). In our case, the flow is a gas flow rate, while the potential differences are pressure differences. Thus, the gas flow $\dot{V}$ is the product of a conductance C by a pressure difference ΔP. This notion of a tube as an impedance or resistance in the electrical analogy sense was introduced by Knudsen [52], and Dushman [53] defined the concept of conductance as the ratio between a fluid flow rate and a pressure difference. Conductance measures how easily flow occurs in response to a pressure differential: the higher the conductance for a given pressure differential, the greater the flow rate. In the classical endoreversible heat pump machine, the mechanical power $\dot{W}_c$ pumps the heat from the cold source at temperature T_C to the hot sink at temperature T_H. The present endoreversible model may differ from the behavior of a real compressor due to irreversible mechanical losses in the moving parts of the compressor, rarefaction effects (Knudsen number > 0.001), pressure losses due to the surface condition, and roughness of the compressor wall materials. However, as a preliminary study, we will assume that these losses do not exist, focusing on an endoreversible model with irreversibilities only at the source and sink, and which are represented by the thermal conductances C_m and C_M, respectively [34–36].

A compressor draws a gas through suction tube 1, of length L_1 and internal diameter d_1, from reservoir 1 (equivalent to the cold source of a heat pump) at the average pressure P_1, temperature T_1, and discharges it into reservoir 2 (equivalent to the hot sink of a heat pump) at the average pressure P_2 and temperature T_2 through discharge tube 2 of length L_2 and internal diameter d_2. The pressure at the compressor inlet is P_i while the outlet pressure is P_o. The pressure losses $(P_i - P_1)$ and $(P_2 - P_o)$ are created by the fluid frictions in the suction and discharge tubes, respectively. The fluid is supposed to be incompressible and fully developed, with a constant velocity profile throughout the tube length, and laminar. We impose a constant average mass flow rate of gas $\dot{m}$, without leaks, during the compression phase. We note $\dot{W}_c$, the mechanical compression power (Figure 2).

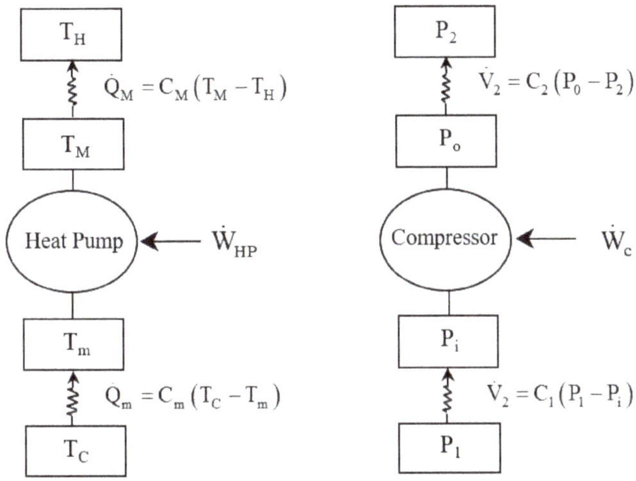

Figure 1. Analogical representation of endoreversible heat pump and an analog endoreversible compressor.

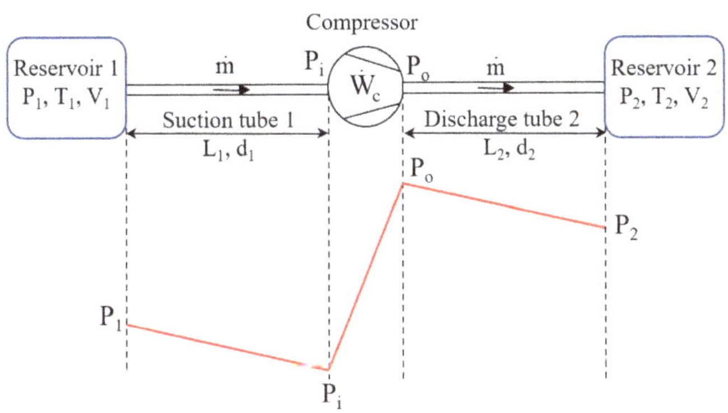

Figure 2. Schematics of a compressor. The compressor generates the mass flow $\dot{m}$ under the $(P_o - P_i)$ pressure difference. The pressure losses $(P_i - P_1)$ and $(P_2 - P_o)$ are created by the fluid frictions in the suction and discharge tubes (L_1, d_1) and (L_2, d_2), respectively.

2.1. Optimal Mechanical Power of the Compressor

The Bernoulli equation written for suction tube 1 makes it possible to link the two pressures P_1 and P_i by the following classical expression:

$$P_1 - P_i = \frac{8}{\pi^2} \frac{\dot{m}^2}{\rho_1 \, d_1^4} \left[1 + \lambda_1 \frac{L_1}{d_1} \right]. \tag{1}$$

With the mass flow inside pipe 1,

$$\dot{m} = \rho_1 \, S_1 \, V_1. \tag{2}$$

Equations (1) and (2) are based on the assumption that the fluid can be treated as a continuum. However, for micro-sized tubes or under certain low-pressure conditions (where multiple characteristic length scales must be considered), the Knudsen number Kn is typically used to assess the validity of this assumption. Defined as the ratio between the mean free path ℓ (the average distance a molecule travels between collisions) and the characteristic length L (such as the hydraulic diameter of the tube), the Knudsen number helps determine the continuum nature of the flow. When Kn < 0.001, the continuum hypothesis holds, and Equations (1) and (2) remain applicable [54].

The tubes under consideration are characterized by small diameters ranging from 50 µm to 200 µm, and relatively long lengths between 50 mm and 100 mm. Although these dimensions suggest that rarefaction effects may be non-negligible—particularly given that the Knudsen number Kn exceeds 0.001—we will not introduce correction terms to account for velocity slip at the wall. This simplification is justified by the assumption that continuum flow remains a valid approximation within the operating conditions of the system. To limit friction losses in the suction and discharge tubes, a laminar flow condition is imposed; then, the pressure loss coefficient λ is expressed by the Blasius relation [55]:

$$\lambda_1 = \frac{64}{\mathrm{Re}_1} = 16\,\pi \frac{\mu_1\, d_1}{\dot{m}}. \tag{3}$$

Equation (3) into (1) gives

$$P_1 - P_i = \frac{8}{\pi^2}\frac{\dot{m}^2}{\rho_1\, d_1^4}\left[1 + 16\,\pi\,\frac{\mu_1\, L_1}{\dot{m}}\right]. \tag{4}$$

Whatever the gas used (see Table 1), taking into account the condition relating to the viscous laminar flow regime, as well as the very low mass flow rates ($\dot{m} \leq 10^{-6}$ kg s^{-1}) and the dimensions of the suction and discharge tubes, we verify the relationship:

$$16\,\pi\,\frac{\mu_1\, L_1}{\dot{m}} >> 1. \tag{5}$$

The pressure difference $(P_1 - P_i)$ is proportional to the volumetric flow rate $\dot{V}_1$ (Figure 1) or mass flow rate $\dot{m} = \rho_1\,\dot{V}_1$. In these conditions, Equation (4) becomes

$$\Delta P_1 = P_1 - P_i = \frac{128}{\pi}\frac{\mu_1\, L_1}{\rho_1\, d_1^4}\dot{m} = \frac{128}{\pi}\frac{\mu_1\, L_1}{d_1^4}\dot{V}_1 = \frac{\dot{V}_1}{C_1} \tag{6}$$

with

$$\frac{1}{C_1} = \frac{128}{\pi}\frac{\mu_1\, L_1}{d_1^4}. \tag{7}$$

We define the factor $K_1 = \frac{1}{\rho 1\, C_1}$ as an analog of a fluidic resistance, and the pressure difference ΔP_1 becomes

$$\Delta P_1 = K_1\,\dot{m} \tag{8}$$

with

$$K_1 = \frac{128}{\pi}\frac{\mu_1\, L_1}{\rho_1\, d_1^4}. \tag{9}$$

Identical reasoning applied to the flow on the discharge side leads to the following relations:

$$\Delta P_2 = P_o - P_2 = \frac{128}{\pi}\frac{\mu_2\, L_2}{\rho_2\, d_2^4}\dot{m} = \frac{128}{\pi}\frac{\mu_2\, L_2}{d_2^4}\dot{V}_2 = \frac{\dot{V}_2}{C_2} \tag{10}$$

with

$$\frac{1}{C_2} = \frac{128}{\pi} \frac{\mu_2 L_2}{d_2^4}. \tag{11}$$

The factor $K_2 = \frac{1}{\rho 2 C_2}$ is defined as an analog of fluidic resistance, and the pressure difference ΔP_2 becomes

$$\Delta P_2 = K_2 \dot{m} \tag{12}$$

with

$$K_2 = \frac{128}{\pi} \frac{\mu_2 L_2}{\rho_2 d_2^4}. \tag{13}$$

The diameter of each tube significantly affects the calculation of pressure losses. In the laminar regime, the pressure loss ΔP varies linearly with the mass flow $\dot{m}$.

Table 1. Thermophysical properties of different gases at 20°C and standard atmospheric pressure 101,325 Pa [56–59].

Gas	Dynamic Viscosity μ (10^{-5}Pa s)	Density ρ (kgm^{-3})	Specific Heat c_p (J kg^{-1} K^{-1})	Thermal Conductivity λ (W m^{-1} K^{-1})	Adiabatic Index γ	Molar Mass $\mathcal{M}$ (kg kmol^{-1})
Air	1.817	1.203	1015	0.02565	1.400	28.97
Argon (Ar)	2.240	1.661	520	0.01737	1.670	39.948
Carbon dioxide (CO_2)	1.493	1.871	851	0.01626	1.294	44.009
Helium (He)	1.973	0.166	5196	0.14929	1.666	4.003
Hydrogen (H_2)	0.867	0.084	14285	0.17690	1.410	2.016
Nitrogen (N_2)	1.757	1.164	1040	0.02543	1.400	28.013

The mechanical power of a compressor being proportional to the variation in enthalpy of the gas, the power supplied by this compressor, noted $\dot{W}_c$, related to the mass flow q_m, is written as follows:

$$\dot{W}_c = \frac{\dot{m}\, c_p\, T_1}{\eta_c} \left[\left(\frac{P_o}{P_i} \right)^{\frac{\gamma-1}{\gamma}} - 1 \right], \tag{14}$$

where η_c is the isentropic compression efficiency. The endoreversible compressor assumes adiabatic efficiency $\eta_c = 1$, to maintain the analogy with an endoreversible heat pump. It means there are no energy losses to compress the gas from the pressure P_i to P_o inside the compressor. Otherwise, we would have an irreversible compressor. Equations (6) and (10) are reported into Equation (14), and

$$\dot{W}_c = \dot{m}\, c_p\, T_1 \left[\left(\frac{P_2 + \Delta P_2}{P_1 - \Delta P_1} \right)^{\frac{\gamma-1}{\gamma}} - 1 \right]. \tag{15}$$

Let us introduce the pressure ratio τ by

$$\tau = \left(\frac{P_2 + \Delta P_2}{P_1 - \Delta P_1} \right)^{\frac{\gamma-1}{\gamma}}. \tag{16}$$

The expression (16) is simplified if we assume that the pressure losses are negligible compared to the average pressure values and $\Delta P_1 < P_1$ and $\Delta P_2 < P_2$. By an expansion limited to the first order of the numerator and the denominator of the relation (16), we finally obtain

$$\tau \approx \left(1 + \frac{\gamma - 1}{\gamma} \frac{\Delta P_1}{P_1} \right) \left(1 + \frac{\gamma - 1}{\gamma} \frac{\Delta P_2}{P_2} \right) \left(\frac{P_2}{P_1} \right)^{\frac{\gamma-1}{\gamma}} \tag{17}$$

and, neglecting the second order terms ($\frac{\Delta P_1}{\Delta P_1}\frac{\Delta P_2}{\Delta P_2}$), the pressure ratio is

$$\tau \approx 1 + \frac{\gamma - 1}{\gamma}\frac{\Delta P_1}{P_1}\left(1 + \frac{P_1}{P_2}\frac{\Delta P_2}{\Delta P_1}\right). \tag{18}$$

The expression of the compressor power $\dot{W}_c$ is written as follows:

$$\dot{W}_c = \dot{m}\, c_p\, T_1 \left\{ \left[1 + \frac{\gamma - 1}{\gamma}\frac{\Delta P_1}{P_1}\left(1 + \frac{P_1}{P_2}\frac{\Delta P_2}{\Delta P_1}\right)\right]\left(\frac{P_2}{P_1}\right)^{\frac{\gamma-1}{\gamma}} - 1 \right\}, \tag{19}$$

and

$$\frac{\Delta P_2}{\Delta P_1} = \frac{\mu_2\, L_2\, T_2}{\mu_1\, L_1\, T_1}\left(\frac{d_1}{d_2}\right)^4\left(\frac{P_1}{P_2}\right)^2. \tag{20}$$

The volumes of the two reservoirs are sufficiently large compared to those of the tubes to consider that the temperatures T_1 and T_2 are equal but that only the pressures P_1 and P_2 are different. And if $T_1 = T_2$ then $\mu_1 = \mu_2$ and $\rho_1 = \rho_2$, and the pressure ratio $\frac{\Delta P_2}{\Delta P_1}$ yields

$$\frac{\Delta P_2}{\Delta P_1} \approx \frac{L_2}{L_1}\left(\frac{d_1}{d_2}\right)^4\left(\frac{P_1}{P_2}\right)^2 \approx \frac{K_2}{K_1}\left(\frac{P_1}{P_2}\right)^2, \tag{21}$$

where the ratio K_2/K_1 is a function of geometric parameters (lengths and diameters) only:

$$\frac{K_2}{K_1} = \frac{L_2}{L_1}\left(\frac{d_1}{d_2}\right)^4. \tag{22}$$

Finally, the compressor power $\dot{W}_c$ is written as follows:

$$\dot{W}_c \approx \dot{m}\, c_p\, T_1 \left\{ \left[1 + \frac{\gamma - 1}{\gamma}\left(1 + \frac{P_1}{P_2}\frac{K_2}{K_1}\right)\frac{K_1\, \dot{m}}{P_1}\right]\left(\frac{P_2}{P_1}\right)^{\frac{\gamma-1}{\gamma}} - 1 \right\}. \tag{23}$$

$$\dot{W}_c \approx \dot{m}\, c_p\, T_1 \left\{ \left[1 + \frac{128}{\pi}\frac{\gamma - 1}{\gamma}\frac{\mu_1\, r\, T_1}{P_1^2}\frac{L_1}{d_1^4}\left(1 + \frac{L_2}{L_1}\left(\frac{d_1}{d_2}\right)^4\left(\frac{P_1}{P_2}\right)^3\right)\dot{m}\right]\left(\frac{P_2}{P_1}\right)^{\frac{\gamma-1}{\gamma}} - 1 \right\}. \tag{24}$$

We define the normalized compression power $\dot{W}_c^*$ as a normalized power from Equation (24):

$$\dot{W}_c^* = \frac{\dot{W}_c}{\dot{m}\, c_p\, T_1}. \tag{25}$$

The normalized power $\dot{W}_c^*$ presents a minimal value for an optimal reservoir pressure P_{2opt} that must be determined (Figure 3). To do this, let us cancel the first derivative of Equation (23), $\frac{\partial \dot{W}_c^*}{\partial P_2} = 0$, and it becomes

$$P_{2opt} = \frac{K_2\, \dot{m}}{\gamma\, P_1 + (\gamma - 1)\, K_1\, \dot{m}}\, P_1. \tag{26}$$

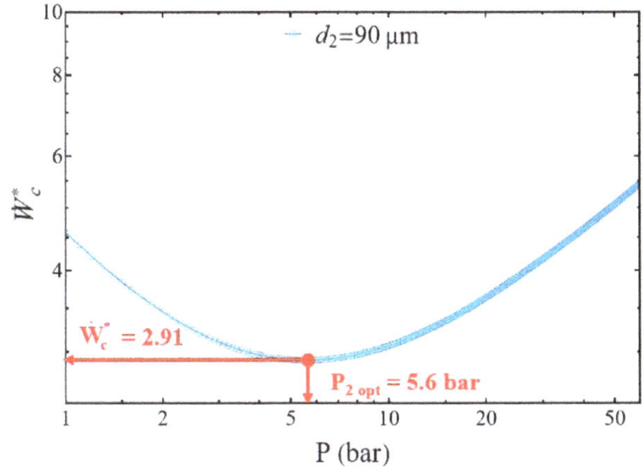

Figure 3. Evolution of the normalized compressor power as a function of the discharge pressure for $d_1/d_2 > 1$. Conditions: helium (Table 1), discharge tube diameter d_2 = 90 μm, d_1 = 200 μm, $L_1 = L_2 = 0.05$ m.

The optimal normalized power is finally written:

$$\dot{W}_c^*\big|_{opt} = \frac{\dot{W}_c\big|_{opt}}{\dot{m}\ c_p\ T_1}. \tag{27}$$

$$\dot{W}_c^*\big|_{opt} = \left[\frac{\gamma-1}{\gamma}\ \frac{1}{P_1^2}\left(\frac{P_2}{P_1}\right)^{-\frac{1}{\gamma}}\right]^{\frac{\gamma}{\gamma-1}}\left[1+\frac{\gamma-1}{\gamma}\ \frac{K_1\dot{m}}{P_1}+K_2\ \dot{m}P_1\left(\frac{P_2}{P_1}\right)^{\frac{1}{\gamma}}\right]-1. \tag{28}$$

Indeed, we show that the second derivative of the power $\frac{\partial^2\dot{W}_c^*}{\partial P_2^2}$ is positive, confirming a minimal point corresponding to P_{2opt}, and it has the following expression:

$$\frac{\partial^2\dot{W}_c^*}{\partial P_2{}^2} = \frac{1}{(P_1P_2\gamma)^2}\left\{\left(\frac{P_2}{P_1}\right)^{-1/\gamma}\frac{\gamma-1}{\gamma}[K_2P_1\dot{m}(\gamma+1)] - P_2\left[K_1\dot{m}\frac{\gamma-1}{\gamma}+P_1\right]\right\}. \tag{29}$$

The compressor's power curve evolves in two different ways. In the first phase, the power decreases with the discharge pressure P_2 up to a minimum corresponding to P_{2opt}, for which $\frac{\partial^2\dot{W}_c^*}{\partial P_2^2}$ is positive. When $d_1 \geq d_2 = 1$, to maintain the mass flow $\dot{m}$ constant, the power of the compressor decreases and reaches a minimum point. Then, from this point, the power increases with pressure P_2 due to the logical increase in pressure losses in the discharge tube.

2.2. Energy Conversion Efficiency of the Compressor

By analogy with the coefficient of performance of a heat pump, the energy conversion efficiency ϵ of the compressor is written as the ratio between the power required to compress the gas from P_o to P_2 corresponding to the viscous power $\dot{W}_{visc\,2}$ expended along the discharge tube, and the total power $\dot{W}_c$ supplied to the compressor to compress the gas from P_1 to P_2 (Figure 1):

$$\varepsilon = \frac{\dot{W}_{visc\,2}}{\dot{W}_c}. \tag{30}$$

Considering the perfect gas law and Equation (12), the viscous power is

$$\dot{W}_{visc\,2} = \dot{V}_2\,\Delta P_2 = \frac{K_2}{\rho_2}\,\dot{m}^2 = c_p\frac{\gamma-1}{\gamma}\,K_2\,\frac{T_2}{P_2}\,\dot{m}^2. \tag{31}$$

From relations (24) and (31), for a given type of gas (adiabatic index γ) and its mass flow rate $\dot{m}$ the energy conversion efficiency ϵ of the compressor (30) is written as a function of the main dimensional parameters, the pressures of the reservoirs, P_1 and P_2, and the conductances of the tubes, K_1 and K_2:

$$\varepsilon = \frac{K_2\,\frac{\gamma-1}{\gamma}\,\dot{m}}{P_2\left\{\left[1+\frac{\gamma-1}{\gamma}\left(1+\frac{P_1}{P_2}\frac{K_2}{K_1}\right)\frac{K_1\,\dot{m}}{P_1}\right]\left(\frac{P_2}{P_1}\right)^{\frac{\gamma-1}{\gamma}}-1\right\}}. \tag{32}$$

3. Results and Discussion

In this section, we address the parametric study of compressor performance for different pressure levels and different tube diameters. We will assume that the gas mass flow rate remains constant ($\dot{m} = 0.3$ µg s^{-1}) and that the tube lengths also remain constant ($L_1 = L_2 = 50$ mm). All simulations were carried out with helium, and the influence of the nature of the gases (Table 1) on the performance of the endoreversible compressor will only be considered in Section 3.2.3.

3.1. Compressor Power

The study of the evolution of the compressor power $\dot{W}_c^*$ as a function of the discharge pressure presents two different behaviors. The normalized power $\dot{W}_c^*$ increases with the pressure P_2, but also when the diameter d_2 of the discharge tube decreases (Figure 4). However, for each diameter d_2 there is a specific pressure at which compression power reaches its minimum.

When the diameter ratio $d_1/d_2 \geq 1$ the power of the compressor presents a minimum for each discharge tube diameter d_2 (Figure 4). This minimum slides with the pressure and increases when the tube diameter d_2 decreases, corresponding to the impact of the pressure losses (Figure 5). The normalized power $\dot{W}^*_{c\,min}$ decreases when the diameter d_2 of the discharge tube increases logically because for the same mass flow $\dot{m}$ the fluid velocity of the gas decreases, generating a corresponding decrease of the pressure losses.

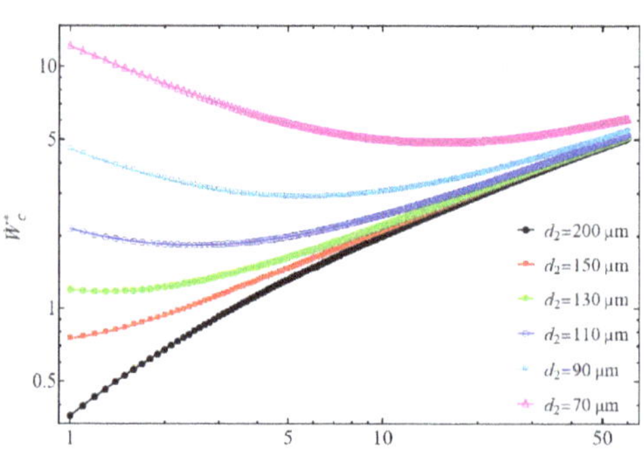

Figure 4. Evolution of the normalized compressor power as a function of the discharge pressure. Influence of the discharge tube diameter d_2 for $d_1/d_2 \geq 1$. Conditions: helium (Table 1), d_1 = 200 µm, $L_1 = L_2$ = 50 mm, and $\dot{m}$ = 0.3 µg s^{-1}.

We show $\frac{\partial^2 \dot{W}_c^*}{\partial P_2^2} > 0$, and the normalized power $\dot{W}_c^*$ presents a minimal value $\dot{W}_{c\,min}^*$ for the pressure P_{2opt} (Equation (26)), and this for all delivery diameter values d_2 from 70 to 200 µm (with d_1 = 200 µm), and this for all discharge diameter values. Under these conditions, for a given tank pressure P_2 and a given mass flow rate $\dot{m}$ we can choose the suction tube diameter allowing the gas to be compressed with a minimum of energy. Figure 5 shows the monotonic decrease of the optimal pressure P_{2opt} as a function of the diameter d_2. It can be observed that the optimal power collapses, starting from a diameter of d_2 = 150 µm.

When the diameter ratio $d_1/d_2 \leq 1$, the power of the compressor $\dot{W}_c^*$ does not present minimum values but increases continuously with the pressure and also when the diameter d_1 of the suction tube decreases due to the increase in pressure drops (Figure 6). The compressor's power does not reveal an optimal value that can be used for sizing the compressor. The normalized power $\dot{W}_{c\,min}^*$ (Figure 7) decreases when the diameter d_1 of the suction tube increases logically because for the same value of the mass flow $\dot{m}$ the tube section presents a bigger area and then a lower velocity. In this configuration, the optimal pressure P_{2opt} increases with the rise in diameter d_1 and tends toward a value of $P_{2optlim}$ = 0.22 bar. This demonstrates a behavior very different from the case where $d_1/d_2 \geq 1$ (Figure 5).

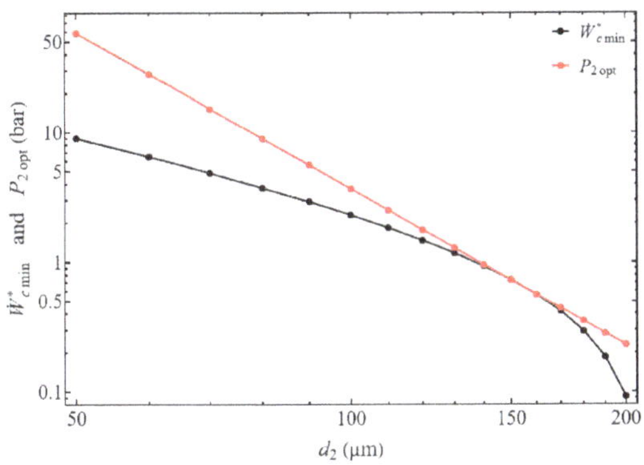

Figure 5. Evolution of normalized power and optimal compressor outlet pressure. Influence of the discharge tube diameter d_2 for $d_1/d2 \geq 1$. Conditions: helium (Table 1), $L_1 = L_2$ = 50 mm, and $\dot{m}$ = 0.3 µg s^{-1}.

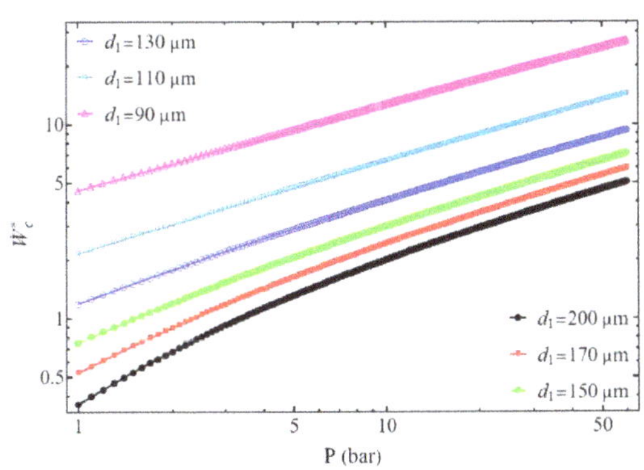

Figure 6. Evolution of the normalized compressor power as a function of the discharge pressure. Influence of the suction tube diameter d_1 for $d_1/d_2 \leq 1$. Conditions: helium Table 1), d_2 = 200 µm, $L_1 = L_2$ = 50 mm, and $\dot{m}$ = 0.3 µg s^{-1}.

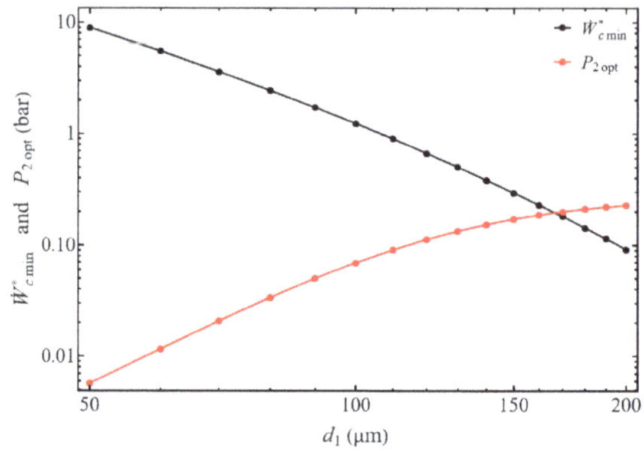

Figure 7. Evolution of normalized power and optimal compressor outlet pressure. Influence of the suction tube diameter d_1 for $d_1/d_2 \leq 1$. Conditions: helium (Table 1),$L_1 = L_2 = 50$ mm, and $\dot{m} = 0.3$ μg s^{-1}.

3.2. *Compressor Efficiency*

The compressor efficiency represents the ability of the compressor to convert the power input $\dot{W}^*_c$ to the power $\dot{W}_{visc\,2}$ to compress the fluid from the pressure P_1 to P_2. The influences of the suction diameter tube d_1 and the discharge diameter tube d_2 are analyzed as a function of the pressure. First, we will analyze the compression efficiency as a function of discharge pressure and then as a function of compression power.

3.2.1. Effect of the Discharged Pressure P_2

In Figures 8 and 9, the efficiency ε decreases with pressure. Overall, for the same mass flow rate $\dot{m}$ this can be explained by the increase in pressure losses as the diameter of the suction and discharge tubes decreases. However, the compressor behaves very differently, depending on the ratio d_1/d_2. The compression efficiency for the same pressure value, P_2, is higher for the case $d_1/d_2 \geq 1$ than for the case $d_1/d_2 \leq 1$. For the case $d_1/d_2 \geq 1$, it is shown that at the same pressure, P_2, the efficiency decreases as the diameter d_2 of the discharge tube increases. Conversely, for the case $d_1/d_2 \leq 1$ the efficiency decreases as the diameter d_1 of the suction tube decreases. This phenomenon can be attributed to the compressor's constant mass flow rate $\dot{m}$ during the compression process. In general, for the case $d_1/d_2 \geq 1$ the compressor demonstrates the highest efficiency values.

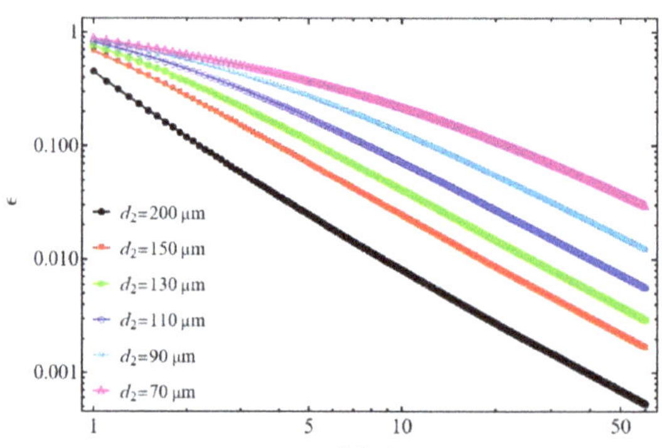

Figure 8. Compressor efficiency. Influence of the discharge tube diameter d_2. Conditions: helium (Table 1), $d_1 = 200$ μm for $d_1/d_2 \geq 1$, $L_1 = L_2 = 50$ mm, and $\dot{m} = 0.3$ μg s^{-1}.

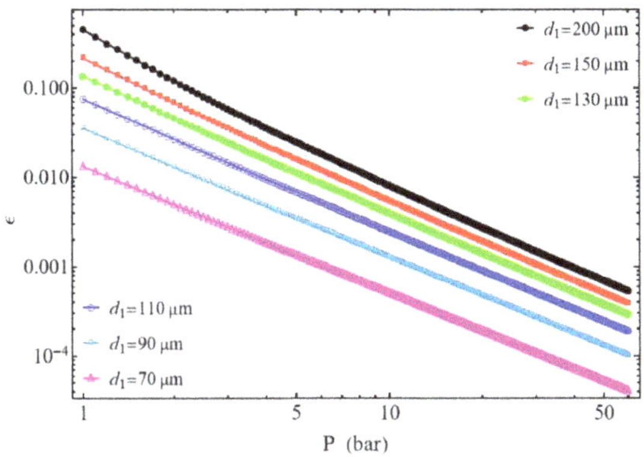

Figure 9. Compressor efficiency. Influence of the suction tube diameter d_1. Conditions: helium (Table 1), d_2 = 200 µm for $d_1/d_2 \leq 1$, $L_1 = L_2$ = 50 mm, and $\dot{m}$ = 0.3 µg s^{-1}.

3.2.2. Bounds of Efficiency and Compression Power

The compressor's performance can be characterized by analyzing the normalized compression power as a function of normalized efficiency. The effects of the ratios d_1/d_2 were assessed, and the results are displayed in Figures 10 and 11. Both figures exhibit the same remarkable point. For the ratio ϵ/ϵ_{max} close to zero, the compressor delivers maximum power to compensate for its very low efficiency to maintain the desired mass flow rate $\dot{m}$ and compression ratio P_2/P_1. On the other hand, at a given normalized efficiency the normalized power of the compressor increases when the diameter of the tubes decreases. This is explained by the fact that the compressor must compensate for the pressure losses, which increase at constant mass flow rate when the tube diameters decrease.

For the diameter ratio $d_1/d_2 \leq 1$ (Figure 10), the normalized power shows a minimum for some tube diameter values $d_2 \leq 110$ µm corresponding to a range of normalized efficiency ϵ/ϵ_{max} between 0.2 and 0.4. For diameters $d_2 > 110$ µm, the normalized power of the compressor decreases with normalized efficiency. For the diameter ratio $d_1/d_2 \geq 1$ (Figure 11), the normalized power constantly decreases with normalized efficiency but does not present a minimum.

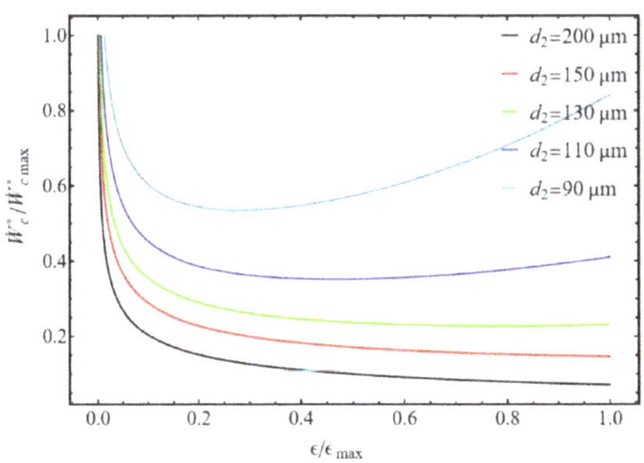

Figure 10. Normalized power versus normalized compressor efficiency. Influence of the discharge tube diameter d_2. Conditions: helium (Table 1), d_1 = 200 µm for $d_1/d_2 \leq 1$, $L_1 = L_2$ = 50 mm, and $\dot{m}$ = 0.3 µg s^{-1}.

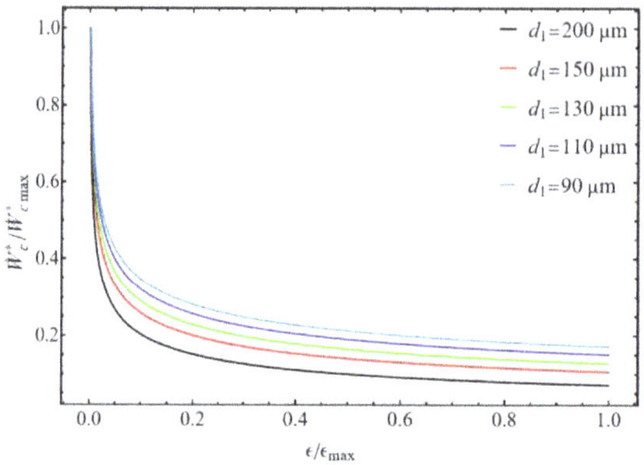

Figure 11. Normalized power versus normalized compressor efficiency. Influence of the suction tube diameter d_1. Conditions: helium (Table 1), d_2 =200 μm for $d_1/d_2 \geq 1$, $L_1 = L_2$ = 50 mm, and $\dot{m}$ = 0.3 μg s^{-1}.

3.2.3. Influence of Gas Type

The nature of the gas significantly influences compressor performance (Table 1). Figure 12 illustrates two major points. The efficiency decreases with the increase of the pressure for all gases, and hydrogen and helium yield the highest compressor efficiency. These gases are lightweight, possess high thermal capacities, and have exceptionally low molar masses, which facilitates their flow through the tubing. At 1 bar, gas type has minimal impact on compressor efficiency (ϵ = 0.45). However, as pressure increases, reaching 5 bar for instance, compressor efficiency for hydrogen and helium is already twice that of other gases.

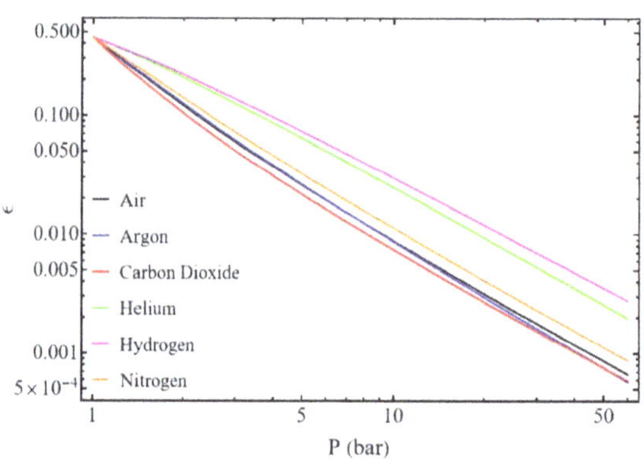

Figure 12. Compressor efficiency. Influence of the thermophysical properties of the gas (Table 1) with d_1 = 200 μm, d_2 = 200 μm, L_1 = 50 mm, L_2 = 50 mm, and $\dot{m}$ = 0.3 μg s^{-1}.

4. Conclusions

We have used finite-time thermodynamics to carry out a parametric study of the performance of an endoreversible compressor by analogy with an endoreversible heat pump. This model cleanly separates internal reversible processes from external irreversibilities (like finite heat transfer rates for the heat pump, finite pressure losses for the compressor). This lets us pinpoint sources of performance loss (power, efficiency) more effectively due to the diameter ratios, tube lengths, mass flows, pressures, and type of gases. This model yields analytical solutions that are easier to interpret and optimize in preliminary design. The compressor compresses a gas reversibly at steady-state conditions, and external irreversibilities occur between the compressor and the low- and high-pressure reservoirs, along the suction and discharge tubes, respectively. Compressor performance is addressed through the evolution of normalized power and normalized compression efficiency. The compressor operates at a constant mass flow rate in the laminar regime. We have shown that the power consumed by the compressor increases with the discharge pressure, due to the increase in pressure losses in the suction and discharge tubes, and that the compression efficiency logically decreases as this pressure increases. The influence of the conductances K_1 and K_2, considered in terms of the diameters of the suction tube d_1 and discharge tube d_2, is important, particularly through the ratio d_1/d_2, which can be greater or less than unity. Although the compressor consumes less power when $d_1/d_2 \leq 1$, the compression efficiency improves when $d_1/d_2 \geq 1$, provided that $d_1 \leq 110$ µm and $d_2 = 200$ µm. Finally, the analysis showed that performance (power and efficiency) is best for low-molar mass gases such as helium and hydrogen. Finite-time thermodynamics is a powerful tool for analyzing the behavior and sizing of an energy conversion system, such as a compressor, and defining its performance limits. This study holds significant technological relevance, as miniaturizing mechanical and pneumatic components presents a major industrial and scientific challenge, particularly within the domain of micro-electromechanical systems (MEMSs).

Funding: This work was supported by the EIPHI Graduate School (contract ANR-17-EURE-0002) and the Region Bourgogne Franche-Comté.

Institutional Review Board Statement: Not applicable.

Data Availability Statement: Data is contained within the article.

Conflicts of Interest: The authors declare no conflicts of interest.

References

1. Chambadal, P. *Les Centrales Nucléaires*; Cambridge University Press: Cambridge, UK, 1957.
2. Novikov, I.I. The efficiency of atomic power stations (a review). *J. Nucl. Energy (1954)* **1958**, *7*, 125–128. [CrossRef]
3. Curzon, F.L.; Ahlborn, B. Efficiency of a Carnot engine at maximum power output. *Am. J. Phys.* **1975**, *43*, 22–24. [CrossRef]
4. Rubin, M.H. Optimal configuration of a class of irreversible heat engines. I. *Phys. Rev. A* **1979**, *19*, 1272. [CrossRef]
5. Rubin, M.H. Optimal configuration of a class of irreversible heat engines. II. *Phys. Rev. A* **1979**, *19*, 1277. [CrossRef]
6. Andresen, B.; Rubin, M.H.; Berry, R.S. Availability for finite-time processes. General theory and a model. *J. Phys. Chem.* **1983**, *87*, 2704–2713. [CrossRef]
7. Berry, R.S.; Salamon, P.; Andresen, B. How it all began. *Entropy* **2020**, *22*, 908. [CrossRef] [PubMed]
8. Andresen, B. Finite-time thermodynamics and thermodynamic length. *Rev. Générale Therm.* **1996**, *35*, 647–650. [CrossRef]
9. Andresen, B.; Salamon, P. Future perspectives of finite-time thermodynamics. *Entropy* **2022**, *24*, 690. [CrossRef] [PubMed]
10. Bejan, A. *Entropy Generation Minimization: The Method of Thermodynamic Optimization of Finite-Size Systems and Finite-Time Processes*; CRC Press: Boca Raton, FL, USA, 2013.
11. Blank, D.A.; Davis, G.W.; Wu, C. Power optimization of an endoreversible Stirling cycle with regeneration. *Energy* **1994**, *19*, 125–133. [CrossRef]
12. Blank, D.A.; Bhattacharyya, S. Chapter Heating rate limit and cooling rate limit of a reversed reciprocating Stirling cycle. In *Recent Advances in Finite Time Thermodynamics*; Nova Science Publishers: New York, NY, USA, 1999; pp. 87–103.
13. Andresen, B.; Salamon, P.; Berry, R.S. Thermodynamics in finite time: Extremals for imperfect heat engines. *J. Chem. Phys.* **1977**, *66*, 1571–1577. [CrossRef]
14. Chen, J. The maximum power output and maximum efficiency of an irreversible Carnot heat engine. *J. Phys. D Appl. Phys.* **1994**, *27*, 1144. [CrossRef]
15. Chen, J. A universal model of an irreversible combined Carnot cycle system and its general performance characteristics. *J. Phys. A Math. Gen.* **1998**, *31*, 3383. [CrossRef]

16. Kodal, A.; Sahin, B.; Yilmaz, T. A comparative performance analysis of irreversible Carnot heat engines under maximum power density and maximum power conditions. *Energy Convers. Manag.* **2000**, *41*, 235–248. [CrossRef]
17. Ozkaynak, S.; Gokun, S.; Yavuz, H. Finite-time thermodynamic analysis of a radiative heat engine with internal irreversibility. *J. Phys. D Appl. Phys.* **1994**, *27*, 1139. [CrossRef]
18. Wu, C.; Kiang, R.L. Finite-time thermodynamic analysis of a Carnot engine with internal irreversibility. *Energy* **1992**, *17*, 1173–1178. [CrossRef]
19. Vaudrey, A.; Lanzetta, F.; Feidt, M. HB Reitlinger and the origins of the efficiency at maximum power formula for heat engines. *J. Non-Equilib. Thermodyn.* **2014**, *39*, 199–203. [CrossRef]
20. Muschik, W.; Hoffmann, K.H. Modeling, simulation, and reconstruction of 2-reservoir heat-to-power processes in finite-time thermodynamics. *Entropy* **2020**, *22*, 997. [CrossRef] [PubMed]
21. Ge, Y.; Chen, L.; Sun, F. Progress in finite time thermodynamic studies for internal combustion engine cycles. *Entropy* **2016**, *18*, 139. [CrossRef]
22. Popescu, G.; Radcenco, V.; Costea, M.; Feidt, M. Optimisation thermodynamique en temps fini du moteur de Stirling endo-et exo-irréversible. *Rev. Générale Therm.* **1996**, *35*, 656–661. [CrossRef]
23. Wu, F.; Chen, L.; Wu, C.; Sun, F. Optimum performance of irreversible Stirling engine with imperfect regeneration. *Energy Convers. Manag.* **1998**, *39*, 727–732. [CrossRef]
24. Lanzetta, F.; Vaudrey, A.; Baucour, P. A new method to optimise finite dimensions thermodynamic models: Application to an irreversible Stirling engine. *Int. J. Ambient Energy* **2018**, *39*, 392–405. [CrossRef]
25. Lai, H.Y.; Li, Y.T.; Chan, Y.H. Efficiency enhancement on hybrid power system composed of irreversible solid oxide fuel cell and Stirling engine by finite time thermodynamics. *Energies* **2021**, *14*, 1037. [CrossRef]
26. Wu, B.; Chen, L.; Ge, Y.; Feng, H.; Liu, X. Finite-time thermodynamic multi-objective optimizations for an irreversible simple Brayton refrigeration cycle based on four objectives, NASG-II algorithm and three decision-making strategies. *J. Therm. Anal. Calorim.* **2025**, *150*, 4653–4668. [CrossRef]
27. Zhang, H.; Wang, L.; Lin, X.; Chen, H. Performance of pumped thermal electricity storage system based on reverse/forward Brayton cycle. *Energy Storage Sci. Technol.* **2021**, *10*, 1796.
28. Qu, J.; Feng, Y.; Wu, B.; Zhu, Y.; Wang, J. Understanding the thermodynamic behaviors of integrated system including solid oxide fuel cell and Carnot battery based on finite time thermodynamics. *Appl. Energy* **2024**, *372*, 123762. [CrossRef]
29. Kaushik, S.C.; Tyagi, S.K.; Kumar, P. *Finite Time Thermodynamics of Power and Refrigeration Cycles*; Springer: Cham, Switzerland, 2017.
30. Dumitrascu, G.; Feidt, M.; Grigorean, S. Closed Irreversible Cycles Analysis Based on Finite Physical Dimensions Thermodynamics. *Proceedings* **2020**, *58*, 37. [CrossRef]
31. Feidt, M. Thermodynamics applied to reverse cycle machines, a review. *Int. J. Refrig.* **2010**, *33*, 1327–1342. [CrossRef]
32. Feidt, M. *Finite Physical Dimensions Optimal Thermodynamics 1: Fundamentals*; Elsevier: Amsterdam, The Netherlands, 2017.
33. Ahmadi, M.H.; Ahmadi, M.A.; Mehrpooya, M.; Hosseinzade, H.; Feidt, M. Thermodynamic and thermo-economic analysis and optimization of performance of irreversible four-temperature-level absorption refrigeration. *Energy Convers. Manag.* **2014**, *88*, 1051–1059. [CrossRef]
34. Sahraie, H.; Mirani, M.R.; Ahmadi, M.H.; Ashouri, M. Thermo-economic and thermodynamic analysis and optimization of a two-stage irreversible heat pump. *Energy Convers. Manag.* **2015**, *99*, 81–91. [CrossRef]
35. Feldmann, T.; Kosloff, R. Performance of discrete heat engines and heat pumps in finite time. *Phys. Rev. E* **2000**, *61*, 4774. [CrossRef]
36. Chen, L.; Wu, C.; Sun, F. Finite time thermodynamic optimization or entropy generation minimization of energy systems. *J. Non-Equilib. Thermodyn.* **1999**, *24*, 327–359. [CrossRef]
37. Wu, C.; Chen, L.; Chen, J. *Recent Advances in Finite-Time Thermodynamics*; Nova Publishers: Hauppauge, NY, USA, 1999.
38. Chen, L.; Feng, H.; Sun, F. Optimal piston speed ratio analyses for irreversible Carnot refrigerator and heat pump using finite time thermodynamics, finite speed thermodynamics and direct method. *J. Energy Inst.* **2011**, *84*, 105–112. [CrossRef]
39. Ge, Y.; Wu, H.; Chen, L.; Feng, H.; Xie, Z. Finite time and finite speed thermodynamic optimization for an irreversible Atkinson cycle. *Energy* **2023**, *270*, 126856. [CrossRef]
40. Grosu, L.; Dobre, C.; Petrescu, S. Study of a Stirling engine used for domestic micro-cogeneration. Thermodynamic analysis and experiment. *Int. J. Energy Res.* **2015**, *39*, 1280–1294. [CrossRef]
41. Petrescu, S.; Feidt, M.; Enache, V.; Costea, M.; Stanciu, C.; Boriaru, N. Unification perspective of finite physical dimensions thermodynamics and finite speed thermodynamics. *Int. J. Energy Environ. Eng.* **2015**, *6*, 245–254. [CrossRef]
42. Agrawal, D.; Menon, V. The thermoelectric generator as an endoreversible Carnot engine. *J. Phys. D Appl. Phys.* **1997**, *30*, 357. [CrossRef]

43. Bejan, A.; Dan, N. Maximum work from an electric battery model. *Energy* **1997**, *22*, 93–102. [CrossRef]
44. Chen, L.; Bi, Y.; Wu, C. The influence of nonlinear flow resistance relations on the power and efficiency from fluid flow. *J. Phys. D Appl. Phys.* **1999**, *32*, 1346. [CrossRef]
45. Lanzetta, F.; Désévaux, P.; Bailly, Y. Optimization performance of a microfluid flow power converter. *Int. J. Fluid Power* **2002**, *3*, 5–12. [CrossRef]
46. Noack, B.R.; Schlegel, M.; Ahlborn, B.; Mutschke, G.; Morzyński, M.; Comte, P.; Tadmor, G. A finite-time thermodynamics of unsteady fluid flows. *J. Non-Equilib. Thermodyn.* **2008**, *33*, 103–148. [CrossRef]
47. Bejan, A. Thermodynamic optimization of geometry in engineering flow systems. *Exergy Int. J.* **2001**, *1*, 269–277. [CrossRef]
48. Madou, M. *From MEMS to Bio-MEMS and Bio-NEMS Manufacturing Techniques and Applications*; Routledge Taylor & Francis Group: London, UK, 2011.
49. Nguyen, N.T.; Huang, X.; Chuan, T.K. MEMS-micropumps: A review. *J. Fluids Eng.* **2002**, *124*, 384–392. [CrossRef]
50. Dang, H.; Tan, J.; Zha, R.; Li, J.; Zhang, T.; Zhao, Y.; Zhao, B.; Tan, H.; Xue, R. Review of recent advances in Stirling-type pulse tube cryocoolers. *IOP Conf. Ser. Mater. Sci. Eng.* **2019**, *502*, 012034. [CrossRef]
51. Pfotenhauer, J.M.; Zhi, X. Pulse Tube Cryocoolers. In *Handbook of Superconductivity*; CRC Press: Boca Raton, FL, USA, 2022; pp. 519–534.
52. Knudsen, M. Eine revision der gleichgewichtsbedingung der gase. Thermische molekularströmung. *Ann. Phys.* **1909**, *336*, 205–229. [CrossRef]
53. Dushman, S. Flow of gases through tubes and orifices. In *Scientific Foundations of Vacuum Technique*; Wiley: Hoboken, NJ, USA, 1949; pp. 84–110.
54. Ivchenko, I.N.; Loyalka, S.K.; Tompson Jr, R. *Analytical Methods for Problems of Molecular Transport*; Springer: Berlin/Heidelberg, Germany, 2007.
55. Cengel, Y.A.; Cimbala, J.M. *Fluid Mechanics: Fundamentals and Applications*, 2nd ed.; Mcgraw and Hill: New York, NY, USA, 2010.
56. Van Wylen, G.J.; Sonntag, R.E.; Borgnakke, C. *Fundamentals of Classical Thermodynamics*; Wiley: Hoboken, NJ, USA, 1994.
57. Kestin, J.; Knierim, K.; Mason, E.; Najafi, B.; Ro, S.; Waldman, M. Equilibrium and transport properties of the noble gases and their mixtures at low density. *J. Phys. Chem. Ref. Data* **1984**, *13*, 229–303. [CrossRef]
58. Chapman, S.; Cowling, T. *The Mathematical Theory of Non-Uniform Gases*; Cambridge University Press: Cambridge, UK, 2009.
59. Poling, B.E.; Prausnitz, J.M.; O'Connell, J.P. *The Properties of Gases and Liquids*, 4th ed.; McGraw-Hill: New York, NY, USA, 2004.

entropy

MDPI

Article

Linking Optimization Success and Stability of Finite-Time Thermodynamics Heat Engines

Julian Gonzalez-Ayala [1,2,*], David Pérez-Gallego [1,2], Alejandro Medina [1,2], José M. Mateos Roco [1,2], Antonio Calvo Hernández [1,2], Santiago Velasco [1,2] and Fernando Angulo-Brown [3]

1 Department of Applied Physics, Universidad de Salamanca, 37008 Salamanca, Spain; dpgallego@usal.es (D.P.-G.); amd385@usal.es (A.M.); roco@usal.es (J.M.M.R.); anca@usal.es (A.C.H.); santi@usal.es (S.V.)

2 Institute of Physics and Mathematics (IUFFyM), Universidad de Salamanca, 37008 Salamanca, Spain

3 Escuela Superior de Física y Matemáticas, Instituto Politécnico Nacional, Mexico City 07700, Mexico

* Correspondence: jgonzalezayala@usal.es

Abstract: In celebration of 50 years of the endoreversible Carnot-like heat engine, this work aims to link the thermodynamic success of the irreversible Carnot-like heat engine with the stability dynamics of the engine. This region of success is defined by two extreme configurations in the interaction between heat reservoirs and the working fluid. The first corresponds to a fully reversible limit, and the second one is the fully dissipative limit; in between both limits, the heat exchange between reservoirs and working fluid produces irreversibilities and entropy generation. The distance between these two extremal configurations is minimized, independently of the chosen metric, in the state where the efficiency is half the Carnot efficiency. This boundary encloses the region where irreversibilities dominate or the reversible behavior dominates (region of success). A general stability dynamics is proposed based on the endoreversible nature of the model and the operation parameter in charge of defining the operation regime. For this purpose, the maximum ecological and maximum Omega regimes are considered. The results show that for single perturbations, the dynamics rapidly directs the system towards the success region, and under random perturbations producing stochastic trajectories, the system remains always in this region. The results are contrasted with the case in which no restitution dynamics exist. It is shown that stability allows the system to depart from the original steady state to other states that enhance the system's performance, which could favor the evolution and specialization of systems in nature and in artificial devices.

Keywords: finite-time thermodynamics; endoreversible hypothesis; optimization; heat engine stability; thermodynamic success; stochastic perturbations; relative entropy

1. Introduction

This work analyzes the role played by the stability of the auxiliary reservoirs appearing in the endoreversible model, under the key premise that the local equilibrium in the working fluid relaxation can be used as a stability mechanism leading to the steady state of the operation regime. Previous works on the framework of low-dissipation systems have shown that in multiobjective optimization, the Pareto front (the best compromise among a variety of thermodynamic objective functions) coincides with the behavior of the endoreversible model. Thus, at first glance, there is a link between optimization and the endoreversible model even before an actual optimization is carried out. It is known that thermodynamic systems evolve to equilibrium states whenever relaxation times are shorter than the operation or interaction times between the system and its surroundings (reservoirs). In this sense, local equilibrium, responsible for the endoreversible effective behavior of the working fluid (out of equilibrium with external reservoirs), acts as an attractor. In this way, the local equilibrium can define stability dynamics, and it is possible to analyze if the stability, under random perturbations, could point to a self-optimization feature. This point will be tested in the context of *thermodynamic success*.

1.1. An Optimization Quest

Since its beginnings, optimization has been a cornerstone of thermodynamics. The role of efficiency in establishing a theoretical upper bound for cyclic energy converters was given in 1824 by Carnot efficiency, 201 years ago [1]. These results represented the first paradigm for heat engine optimization that paved the way in the race for efficiency maximization, where improvements involved the proposal of different geometries in the work and heat (T-S) diagrams aimed at increasing their efficiency and reliability [2–5]. A second paradigmatic model that accounts for the finite time and finite size of real-life devices was proposed in 1975 by Curzon–Ahlborn (CA) [6], 50 years ago. This model is based on an internally reversible Carnot cycle (endoreversible hypothesis), irreversibly coupled with two external thermal reservoirs. The endoreversible hypothesis has been extensively studied [7–11], and beyond its conceptual idealization, studies on molecular dynamics simulation [12,13] have validated its predictions.

In this model, the coupling between the inner reversible component and the external reservoirs is modeled through heat transfer laws and phenomenological conductances that account for the nature of heat fluxes and the properties of materials involved in heat transport. Initially, it allowed the analysis of the maximum power (MP) regime [6,14–21], and the so-called CA efficiency obtained in this regime, without any doubt, is the main result stemming from the new era of finite-time thermodynamics [22–29]. This efficiency has the form

$$\eta = 1 - \sqrt{\tau}, \tag{1}$$

where $\tau = T_c/T_h$ is the ratio of the external cold and hot heat reservoirs and has proven to be linked to a universal feature of maximum power operation regime [30–38]. Over the last five decades, this model and further improvements included the effect of heat leak and internal irreversibilities (affecting the reversible nature of the inner cycle). The model, also denoted as the irreversible Carnot heat engine, has been used to confront a variety of results stemming from macroscopic, mesoscopic, and microscopic scales, allowing the analysis and optimization of several operation regimes and trade-off functions [39–49], leading to similar universal trends for the efficiency at trade-off operation regimes [50–53].

The aim to provide general aspects related to operation regimes beyond specific models and heat transfer mechanisms is still an active field and new ingredients enter into play, such as the operation parameter control. The latter still has some unsolved issues [54,55] that could be exploited in a general approach of optimization, with possible links to the role of constancy (fluctuations in the energetic output records) in power fluctuations with large efficiencies in quasistatic and steady-state HE models [56–59]. In this way, issues such as the Carnot efficiency at finite power and efficiency at maximum power have been widely analyzed by different strategies to account for the control of parameters and engine layouts in macroscopic, mesoscopic, and quantum frameworks [56,60–68]. All this put to test the idea that *you cannot have it all*, the ultimate quest in thermodynamic optimization, although it is generally accepted that it is impossible to optimize power output, efficiency, and entropy production simultaneously.

1.2. Local Equilibrium and Realizations of the Endoreversible Model

The irreversible Carnot-like model has been analyzed analytically through the glasses of mesoscopic heat engines [66], with Otto and Joule–Brayton cycles [4], in the linear irreversible framework [28], as well as in the low dissipation limit [69]. Also, general behaviors typical of the model [70] have been reproduced numerically in simulations of mesoscopic systems [66–68]. From a theoretical point of view, the endoreversible nature of the model and the appearance of the auxiliary thermal reservoirs have been understood from the local equilibrium assumptions [28] applied to the working fluid.

In Ref. [12], a 2D finite-time Carnot-like engine was modeled using molecular dynamics simulations of hard-sphere particles confined in a box with a moving piston. In that work, the connection with the linear irreversible model in [28] was proven by analyzing in detail the endoreversible behavior of the gas. For various operating regimes (e.g., maximum power and maximum ecological efficiency), the numerical results matched well with theoretical predictions, especially when internal irreversibility and heat leaks were properly accounted for. The resulting efficiency–power curves were consistent with the endoreversible engine. Unlike idealized models that assume instantaneous adiabats, the simulations revealed the key role of thermal relaxation in finite-time adiabats and isotherms. The statistical behavior of the gas supported an endoreversible-type performance, with near isotherms that were not in equilibrium with the external reservoirs, which can account for the auxiliary reservoirs of the endoreversible model. For completeness, in Figure 1 the results reported in [12] are depicted. In a similar dynamical simulation model, some results of the linear response regime were recovered [13].

These results point to the fundamental role played by the dynamics between the working fluid and reservoirs in finite-size and finite-time systems [71–76]. This feature provides a first glance at the connection between entropy generation (irreversibilities), working fluid–reservoir correlations, and energetic success, discussed in Section 3. This is also in line with previous efforts to provide a more extended vision of finite-time thermodynamics to include a dynamical structure (involving time or rate constraints) with the theory and add entropy-related requirements to the optimization of operation regimes [77–81].

1.3. Optimization Due to Stability

A series of papers [82–85] explored how a stability mechanism that maintains a heat engine operating under steady-state conditions could induce an energetic optimization. The main idea is to consider that the constancy of the heat fluxes or the energetic function that defines the operation regime is responsible for maintaining the operation variables in the steady state. First, this was analyzed for heat engines and refrigerators in the low-dissipation model [83,84], consistent with the irreversible Carnot-like and the endoreversible models. Later, this feature was analyzed for the endoreversible model [85]. In [83,84] it was shown that the relaxation trajectories to the steady state exhibit an optimization process in power output, efficiency, and entropy generation. This was also proven for consecutive random perturbations on the system, simulating external perturbations that could displace the operation state from the steady state. Statistically, it was shown that the deviation of the operation state is displaced in the direction of the Pareto front of the system, which coincides with the endoreversible limit and in the direction of larger efficiencies and lower entropy generation, with a slight sacrifice in power output. When this behavior was explored in the endoreversible model, the self-optimization was not as evident as in the low-dissipation case.

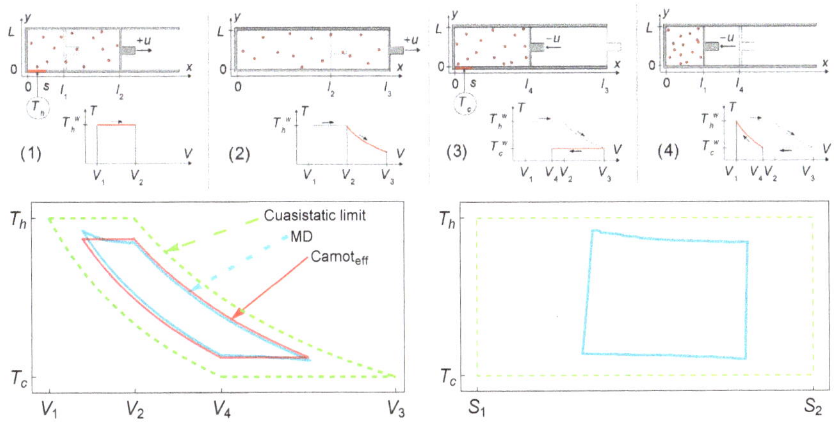

Figure 1. Upper row: Schematic diagram of the finite-time Carnot engine (2D) extracted from Figure 1 of Ref. [12]. **Lower row**: Results extracted from Figures 5 and 6 of Ref. [12] stemming from the molecular dynamics (MD) simulations in the Temperature–Volume (T-V) and Temperature–Entropy (T-S) diagrams. The green dashed lines correspond to the reversible quasistatic cycle, and cyan dotted cycles correspond to the MD results and are the average of 2000 cycles under steady state and maximum power conditions. The red cycle represents the effective Carnot cycle, equivalent to the MD results, with $\tau = 0.5$.

An interesting subject that has not yet been analyzed is what happens when the stability mechanism that keeps the system in a steady state is linked to the local equilibrium that can effectively reproduce the behavior of the endoreversible model. One major goal of studying the lack of control, is to shed some light on "natural optimization preferences" [86–91] in evolution and adaptability to the environment [92,93].

The present paper is organized as follows. In Section 2, the irreversible Carnot-like heat engine (an extension of the endoreversible model) and the operation regimes of maximum power, efficiency, and ecological and Ω functions are addressed. Section 3 addresses a metric to evaluate the success of energy converters. In Section 4, the stability of heat reservoirs (linked to the internal irreversibility) is discussed to justify using the irreversibility of the internal cycle in the definition of stability dynamics. Section 6 presents the energetic trajectories of the system in the relaxation to the steady state. In Section 7 the effect of stochastic perturbations and the evolution of the operation regime induced by the stability is analyzed. Finally, some concluding remarks are discussed.

2. Irreversible Carnot-like Heat Engine

For self-contained purposes, the mathematical model of an irreversible Carnot-like heat engine is summarized. This model is an extension of the original Curzon–Ahlborn endoreversible heat engine proposed in [6] and the simplified versions introduced in [17,19,20,76]. It represent a standard model widely used in finite-time thermodynamics [94]. The endoreversible hypothesis assumes an internal Carnot cycle operating between the temperatures T_h' and $T_c' \leq T_h'$. This Carnot cycle is coupled irreversibly to the external heat baths at temperatures T_h and $T_c \leq T_c' \leq T_h' \leq T_h$, and an external heat leak between these two external heat reservoirs is accounted for. The input and output heat fluxes, $\dot{Q}_h$ and $\dot{Q}_c$, as well as the heat leak $\dot{Q}_L$, obey the Newton heat transfer law

$$|\dot{Q}_h| = \sigma_h(T_h - T_h') \equiv \sigma_h T_h\left(1 - \frac{1}{a_h}\right), \tag{2}$$

$$|\dot{Q}_c| = \sigma_c(T_c' - T_c) \equiv \sigma_c T_c(a_c - 1), \tag{3}$$

$$|\dot{Q}_L| = \sigma_i(T_h - T_c) \equiv \sigma_i T_h(1 - \tau), \tag{4}$$

where $a_h = T_h/T_h' \in [1, \tau^{-1}]$, and $a_c = T_c'/T_c \in \left[1, \tau^{-1}a_h^{-1}\right]$; the σ_c, σ_h and σ_i are the corresponding thermal conductances, which are considered as constant, and t_c and t_h are the operation times. In the endoreversible engine, the entropy generation due to the heat enchanged in the isotherms lead to the Clausius equality, $|\dot{Q}_c|/T_c' - |\dot{Q}_h|/T_h' = 0$. However, to allow for internal irreversibilities, the irreversibility parameter $I \leq 1$ is introduced. This parameter compensates the larger entropy produced in the colder isotherm (or a deficit in the hot reservoir) so the engine fulfills the Clausius equality,

$$I\frac{|\dot{Q}_c|}{T_c'} - \frac{|\dot{Q}_h|}{T_h'} = 0, \tag{5}$$

Thus, it provides a measure of the deviation from the reversible hypothesis for the working fluid. According to this definition, the inner cycle is reversible in the case $I = 1$ and irreversible if $I < 1$. The internal temperatures T_h' and T_c' (i.e., a_h and a_c) are not independent. From Equations (2), (3) and (5),

$$a_c = \frac{I}{I - \sigma_{hc}(a_h - 1)}, \tag{6}$$

where $\sigma_{hc} \equiv \sigma_h/\sigma_c$. From the above equations, the power and the efficiency are given by

$$P = |\dot{Q}_h| - |\dot{Q}_c| = \sigma_h T_h\left(1 - a_h^{-1} - \frac{\tau(a_h - 1)}{I + \sigma_{hc}a_h}\right), \tag{7}$$

$$\eta = 1 - \frac{|\dot{Q}_c| + |\dot{Q}_L|}{|\dot{Q}_h| + |\dot{Q}_L|} = \left(1 - \frac{a_h\tau}{I - \sigma_{hc}(a_h - 1)}\right)\left(\frac{a_h - 1}{a_h - 1 + \sigma_{ih}a_h(1 - \tau)}\right), \tag{8}$$

where $\sigma_{ih} \equiv \sigma_i/\sigma_h$. The Curzon–Ahlborn endoreversible model is recovered if $I = 1$ and $\sigma_i = 0$ [6]. The entropy production is given by

$$\sigma = \frac{|\dot{Q}_c|}{T_c} - \frac{|\dot{Q}_h|}{T_h} + |\dot{Q}_L|\left(\frac{1}{T_c} - \frac{1}{T_h}\right) = \sigma_h\left(\frac{1}{a_h} - 1 + \frac{a_h - 1}{I - \sigma_{hc}(a_h - 1)} + \frac{\sigma_{ih}(1 - \tau)^2}{\tau}\right), \tag{9}$$

Maximizing Power, Efficiency, Ecological and Omega Figures of Merit

The power output is a convex function that exhibits a maximum only in the direction of a_h. The analytical maximum stemming from the condition $\frac{dP}{da_h} = 0$ gives the value

$$a_{h,MP} = \frac{I + \sigma_{hc}}{\sigma_{hc} + \sqrt{I\tau}}, \tag{10}$$

producing an efficiency and power output

$$\eta_{MP} = \frac{\left(\sqrt{I} - \sqrt{\tau}\right)^2}{I - \sqrt{I\tau} + \sigma_{ih}(\sigma_{hc} + I)(1 - \tau)}, \tag{11}$$

$$P_{MP} = \frac{\left(\sqrt{I} - \sqrt{\tau}\right)^2}{I + \sigma_{hc}}, \tag{12}$$

which, in the limit $I \to 1$ and $\sigma_{ih} \to 0$, leads to the well-known Curzon–Ahlborn efficiency, η_{CA} (which is independent of the conductances) and the maximum power given by

$$\eta_{MP} = \eta_{CA} = 1 - \sqrt{\tau}, \tag{13}$$

$$P_{MP} = \frac{\left(1 - \sqrt{\tau}\right)^2}{1 + \sigma_{hc}}. \tag{14}$$

This efficiency is linked with a universal feature of the maximum work and maximum power regimes. The approximation to the first order in the Taylor series in terms of η_C, $\eta_{CA} = \frac{\eta_C}{2} + \frac{\eta_C^2}{8} + ...$ represents a lower bound for the maximum power regime for the low-dissipation regime and for the phenomenological heat transfer law. As it will be shown later, this value, $\frac{\eta_C}{2}$, is meaningful in describing the success of energy conversion.

The corresponding maximization of the efficiency gives the $a_{h,M\eta}$ value

$$a_{h,M\eta} = 1 - \frac{\sigma_{ih}(1-\tau)\left(I\left(\tau - \sqrt{I\tau}\right) + \sigma_{hc}\left(I - \sqrt{I\tau}\right)\right) - \sqrt{I\sigma_{ih}\tau(I-\tau)(1-\tau)(I+\sigma_{hc})}}{\tau(\sigma_{hc}+I) + \sigma_{ih}(I\tau - \sigma_{hc}^2)(1-\tau)}, \tag{15}$$

that for zero heat leak leads to $a_{h,M\eta} = 1$, since the efficiency is a monotonous decreasing function of a_h.

The ecological function, E, and the Omega function, Ω, are between the so-called trade-off figures of merit. It is noteworthy that both functions differ in their interpretation, but for the irreversible Carnot engine, both lead to the same expression [49]. The ecological function was first introduced in Ref. [39] and refers to a compromise between the power output and the entropy generation released to the cold external reservoir

$$E \equiv P - T_c\sigma = (\dot{Q}_h + \dot{Q}_L)(2\eta - \eta_C), \tag{16}$$

while the Ω function is a trade-off between power loss and power gained when compared with a given power output and with fixed heat input, that is,

$$\Omega \equiv P_{gain} - P_{loss} = (P - P_{min}) - (P - P_{max}) = 2P - (\dot{Q}_h + \dot{Q}_L)\eta_C = (\dot{Q}_h + \dot{Q}_L)(2\eta - \eta_C). \tag{17}$$

For the present model, it has the form

$$\Omega = E = \sigma_h T_h \left((1+\tau)\left(1 - a_h^{-1}\right) - \sigma_{ih}(1-\tau)^2 + \frac{2\tau(a_h - 1)}{\sigma_{hc}(a_h - 1) - I} \right). \tag{18}$$

Its maximum value is reached at

$$a_{h,ME} = a_{h,M\Omega} = \frac{I + \sigma_{hc}}{\sigma_{hc} + \sqrt{\frac{2I\tau}{1+\tau}}}, \tag{19}$$

which, in the case of no heat leak, gives the value

$$\eta_{ME} = \eta_{M\Omega} = 1 - \sqrt{\frac{\tau(1+\tau)}{2I}}, \tag{20}$$

and for the endoreversible case, $I = 1$, a well-known ecological efficiency is recovered; this efficiency also has been related to a universal feature of the ecological and the Ω regimes, whose linear term of the Taylor series is in terms of η_C, $\eta_E = \frac{3\eta_C}{4} + \frac{\eta_C^2}{32} + ...$, and also represents a lower bound for the maximum ecological and Omega regimes [49].

Beyond trade-off operation regimes, it is possible to analyze the success of any energy converter operating in cycles. This will be addressed in the following section.

3. Losses and Success of Energy Converters

In [75], stability criteria for the heat reservoirs were defined by the states of minimum internal energy, U_{min}, and maximum entropy, S_{max}, which is analogous to the evolution of closed systems towards equilibrium states. In the stationary operation state, as long as the reservoirs are exchanging heat irreversibly with the working fluid, an inherent entropy generation due to the correlations among them appears. Details on this correlation entropy generation are addressed later in Section 4. These extremal stable states are obtained in the reversible limit, where the changes in internal energy and entropy of a thermal reservoir are only caused by heat exchanges.

$$\Delta U \rightarrow U(t) = U_0 - |Q_h(t)|\eta \leq 0, \tag{21}$$

$$\Delta S \rightarrow S(t) = S_0 + \frac{|Q_h(t)|}{T_h}\frac{\eta_C - \eta}{1-\eta_C} \geq 0, \tag{22}$$

where both are monotonous functions of efficiency. When the efficiency is the highest (η_C), $U_{min}(t)$ is obtained, and at the minimum efficiency, $\eta = W(t) = 0$ (all the heat is dissipated), the value of $S_{max}(t)$ is recovered. Two parameters that provide a measure of the nearness to the extremal situations U_{min} and S_{max} are defined by

$$E_U \equiv \frac{U(t)}{U_{min}(t)} = \frac{\eta}{\eta_C}, \tag{23}$$

$$E_S \equiv \frac{S(t)}{S_{max}(t)} = \frac{\eta_C - \eta}{\eta_C}, \tag{24}$$

fulfilling that $E_U \leq 1$ and $E_S \leq 1$. The Euclidean distance between E_U and E_S can be computed through the $p-$norm

$$N_p \equiv (|E_U|^p + |E_S|^p)^{1/p}, \tag{25}$$

which is a concave function with one minimum at $\eta^* = \frac{\eta_C}{2}$ except for $p = 1$ (Taxicab or Manhattan distance). Thus, at $\eta^* = \frac{\eta_C}{2}$, the two extreme configurations are at the minimum distance. This means that η^* represents a threshold from which the operation state increases the distance between the extremal stable states given by the equilibrium of the reservoir.

In the general analysis presented in [75], an open question remained regarding whether one would expect that the evolution of energy converters will tend to favor optimum energetic performance. In such a case, reversibility (linked to reservoir stability) is another ingredient to consider together with the operation regime. In this context, three quantities are introduced to measure the success of a system, the parameters

$$R \equiv \frac{\eta - \eta_{min}}{\eta_{max} - \eta_{min}}, \tag{26}$$

$$D \equiv \frac{\Delta S_{Univ}}{\Delta S_{Univ}^{max}}, \tag{27}$$

where the reversibility coefficient, $R \in (0,1)$, indicates how close the efficiency is to the maximum efficiency, and the irreversibility coefficient, $D \in (0,1)$, indicates the nearness to the maximum entropy generation state. $\eta_{max} = \eta_C, \eta_{min} = 0$, and the entropy change in the thermodynamic universe is

$$\Delta S_{univ} = -\frac{|Q_h|}{T_h} + \frac{|Q_c|}{T_c} = \frac{|Q_h|}{T_h}\frac{\eta_C - \eta}{1-\eta_C} \geq 0, \tag{28}$$

which has a maximum when no work is delivered, i.e., $\eta = 0$. Thus, $\Delta S_{univ}^{max} = \frac{|Q_h|}{T_h}\frac{\eta_C}{1-\eta_c}$. These definitions of R and D coincide with E_U and E_S (Equations (23) and (24)),

$$R = \frac{\eta}{\eta_C} = E_U, \tag{29}$$

$$D = \frac{\eta_C - \eta}{\eta_C} = 1 - \frac{\eta}{\eta_C} = 1 - R = E_S, \tag{30}$$

respectively, and the difference, which indicates if the system is closer to the state of maximum efficiency or closer to the maximum entropy generation state,

$$R - D = 2R - 1 = \frac{2\eta}{\eta_C} - 1. \tag{31}$$

has a zero at $\eta^* = \frac{\eta_c}{2}$. The case $R > D$ defines the *region of success since the reversibility dominates over irreversibility, while the opposite condition defines the* failure region. Thus, $\eta^* = \frac{\eta_C}{2}$ should be considered as the minimum value of the efficiency of any thermodynamically successful heat engine, which interestingly, is also the lower bound for the efficiency in the maximum power operation regime and where the distance to stable states for reservoirs ($E_U = 1$ and $E_S = 1$) is minimum.

In [76], the parameter D was decomposed into external losses, D_e, and internal losses, D_i. For the irreversible Carnot-like engine, these are defined as

$$D_e = \frac{\eta_c - \eta_{endo}}{\eta_c} = 1 - \frac{\eta_{endo}}{\eta_c}, \tag{32}$$

$$D_i = \frac{\eta_{endo} - \eta}{\eta_{endo}} = 1 - \frac{\eta}{\eta_{endo}}, \tag{33}$$

which gives the deviation of the endoreversible engine from the Carnot cycle and the deviation of the actual engine from the endoreversible one, respectively. These two parameters are linked with D as follows:

$$D \equiv \frac{\eta_C - \eta}{\eta_C} = \frac{\eta_C - \eta_{endo}}{\eta_C} + \frac{\eta_{endo} - \eta}{\eta_{endo}}\left(\frac{\eta_{endo}}{\eta_C}\right) \leq 1, \tag{34}$$

$$D = D_e + \left(\frac{\eta_{endo}}{\eta_C}\right)D_i = D_e + D_i - D_e D_i. \tag{35}$$

The values $I < 1$ and $\sigma_i \neq 0$ will provide information on the internal and external dissipations. Then, both loss factors are computed as

$$D_e = 1 - \frac{\eta(a_h, I = 1, \sigma_{hc}, \sigma_{ih} = 0, \tau)}{1-\tau} = 1 - \frac{\left(1 - \frac{a_h \tau}{1-(a_h-1)\sigma_{hc}}\right)}{1-\tau}, \tag{36}$$

$$D_i = 1 - \frac{\eta(a_h, I, \sigma_{hc}, \sigma_{ih}, \tau)}{\eta(a_h, I = 1, \sigma_{hc}, \sigma_{ih} = 0, \tau)} = 1 - \frac{1 + \frac{a_h \tau}{\sigma_{hc}(a_h-1)-I}}{\left(1 + \frac{a_h \tau}{\sigma_{hc}(a_h-1)-1}\right)\left(1 + \frac{a_h \sigma_{ih}(1-\tau)}{a_h - 1}\right)}. \tag{37}$$

With these parameters and the reversible and irreversible coefficients, R and D, it is possible to evaluate the effect of instabilities and the energetic evolution of the operation regime given by the relaxation mechanism towards stability. In the following, an analysis to connect stability and success will be presented.

4. Heat Reservoirs Stability

Intuition dictates that the natural evolution of cyclic energy converters will tend to favor optimum energetic performance. Thus, if stable points are attraction configurations, those that tend to null performance are doomed to vanish, and those leaning to optimum efficiency will present a trade-off with the actual needs of the energy conversion and the stable conditions of the system acting as a heat source. This point can be explored by analyzing stability dynamics and the effect of perturbations on the system can be studied.

The core of the endoreversible hypothesis lies in the definition of the working fluid isotherms and the non-equilibrium interaction with the external reservoirs, whose stability is compromised by internal irreversibilities. Thus, when departing from reversible exchanges of heat between the reservoirs and the working fluid, the relaxation mechanism in both components is crucial. A good approximation to reversible heat sources requires that relaxation times are sufficiently short compared to operation times. Heat transfers between the working fluid and the reservoirs present unavoidable consequences on their internal modes, and the entropy of the compound system (system+reservoirs) exhibits correlations between them; this is a source of instabilities for the heat reservoirs. This subject has been cleverly addressed in [74] by analyzing the link with an irreversible contribution to the entropy change of a system. From a quantum framework, entropy production can be linked directly to a measure of the correlation/entanglement between the system and reservoirs. Here, the focus is on the macroscopic framework.

The total entropy of the compound system is not only the sum of the entropy of the reservoirs, S_r, and the system, S_s, but there is an additional term due to the correlations between them, S_c, in such a way that

$$S = S_s + \sum_r S_r + S_c. \tag{38}$$

Under initial equilibrium conditions, the reservoirs do not present correlations, and thus $S_c(0) = 0$ (at time $t = 0$). For further times, a coupling is in place, and there is an irreversible contribution to the entropy change of the compound system. This term is measured as the relative entropy between the actual state of the compound system and that when the heat reservoirs are in thermal equilibrium with no correlations between the system and the reservoirs, that is

$$\Delta_i S(t) = D\left[P(t) \| P_s(t) \prod_r P_r^{eq}(t)\right] = P(t) ln\left(\frac{P(t)}{P_s(t) \prod_r P_r^{eq}(t)}\right), \tag{39}$$

where $D[\cdot\|\cdot]$ is the relative entropy, or Kullback–Leibler divergence, P_s is the probability for the system, P_r^{eq} is that for the r-th reservoir at equilibrium conditions and P the actual probability for the system+reservoirs. This total entropy change of the system, $\Delta_i S(t)$, contains at least the irreversible entropy contribution of the correlations, $\Delta_i S(t) \geq -\Delta S_c(t) = -S_c(t) \geq 0$, and fulfills the inequality [74]

$$\Delta_i S(t) \geq -S_c(t) = \Delta S_s(t) + \sum_r \Delta S_r(t) \geq 0. \tag{40}$$

In the case of one reservoir, a version of the maximum work theorem is obtained,

$$T\Delta_i S(t) = W(t) - \Delta F_s(t) \geq 0, \tag{41}$$

where F_s is the free energy of the system, and $T\Delta_i S(t) = 0$ is only for reversible engines.

In the dynamics simulation (Section 1.2), the term associated with the correlations is hidden in the random momentum change in each collision between the particles and the thermalizing wall. However, from Equation (41) for the isothermal process at temperature T_c', and using the Claussius relation given in Equation (5), the irreversible contribution to the entropy change can be linked with the parameter I, as follows:

$$T_c'\Delta_i S = W - \Delta F_s = \Delta U - Q_c - \Delta U + T_c'\Delta S_{T_c'} = -Q_c + T_c'\Delta S_{T_c'} \geq 0, \tag{42}$$

$$\Longrightarrow \Delta_i S = \frac{|Q_c|}{T_c'} + \Delta S_{T_c'} = \frac{|Q_c|}{T_c'} - I\frac{|Q_c|}{T_c'} = (1-I)\frac{|Q_c|}{T_c'} \geq 0. \tag{43}$$

In this way, $\Delta_i S$ is interpreted as the irreversible entropy that is compensated by the parameter I so that the system fulfills the Clausius inequality in Equation (5).

Returning to the expressions of E_U and E_S (Equations (23) and (24)), notice that they coincide with R and D (Equations (29) and (30)), respectively. The configurations where $E_U, R \to 1$ can only be achieved under totally reversible conditions, for which Equations (41) and (43) are zero and reservoirs are in equilibrium with no correlations with the system. The other zero for $\Delta S_i(t)$ is found in the opposite case, where $E_S, D \to 1$ since $W(t) = 0$. Thus, the input heat is directly dissipated into the external cold reservoir at temperature T_c, and no correlations take place between the system and the external reservoirs; in that case,

$$\Delta S = \sum_r \frac{Q_r}{T_r}.$$

These two extreme situations are incompatible with real applications, for one has no use, and the other is linked to zero power. In real situations, there are indeed correlation dynamics between reservoirs and the system that can lead ultimately to instabilities of the heat reservoirs. The previous discussion provides a justification to define stability dynamics where the internal irreversibility of the system enters into play. In the finite-time thermodynamics phenomenological model, it is not possible to properly account for the mentioned correlations. However, the parameter I is linked with $\Delta S_i(t)$, which contains the entropy generated in the correlations (linked with the stability of the heat reservoirs), justifying the use of I as a dynamical quantity involved in the definition of a stability dynamics. This will be addressed in the following section.

5. Stability Dynamics

A stability dynamics that contains information about the operation regime and the stability of the heat reservoirs requires at least two dynamical equations. For this analysis, two functions will be used to define such stability. One will be the Ω function under $M\Omega$ conditions, determined by the control parameter a_h. The other one is the relaxation of the working fluid, responsible for fulfilling the endoreversible hypothesis, and measured by the irreversibility parameter I. By assuming that the $M\Omega$ regime is a steady state, then the most general stability dynamics for small perturbations follows the linear approximation; that is,

$$\begin{pmatrix} \frac{dI}{dt} \\ \frac{da_h}{dt} \end{pmatrix} = -\begin{pmatrix} C_I & 0 \\ 0 & C_{a_h} \end{pmatrix}\begin{pmatrix} I - I^{M\Omega} \\ a_h - a_h^{M\Omega} \end{pmatrix}, \tag{44}$$

which gives the evolution in time, t, of the parameters I and a_h, and C_I and C_{a_h} give the restitution strength in the corresponding variables, which, with the information at hand, cannot be determined, and different scenarios for them will be addressed in the following. These two parameters define the internal working temperature T_c' through $a_c = a_c(a_h, I)$ (Equation (6)) and the Ω function $\Omega(a_h, I)$ (Equations (16) and (17)). The Taylor expansion to linear order of these two quantities gives

$$\begin{pmatrix} a_c - a_c^{M\Omega} \\ \Omega - \Omega^{M\Omega} \end{pmatrix} = J \begin{pmatrix} I - I^{M\Omega} \\ a_h - a_h^{M\Omega} \end{pmatrix} = \begin{pmatrix} \frac{\partial a_c}{\partial I}\Big|_{M\Omega} & \frac{\partial a_c}{\partial a_h}\Big|_{M\Omega} \\ \frac{\partial \Omega}{\partial I}\Big|_{M\Omega} & \frac{\partial \Omega}{\partial a_h}\Big|_{M\Omega} \end{pmatrix} \cdot \begin{pmatrix} I - I^{M\Omega} \\ a_h - a_h^{M\Omega} \end{pmatrix} \tag{45}$$

and from Equations (44) and (45), the final expression for the linear stability dynamics can be written as

$$\begin{pmatrix} \frac{dI}{dt} \\ \frac{da_h}{dt} \end{pmatrix} = -\begin{pmatrix} C_I & 0 \\ 0 & C_{a_h} \end{pmatrix} \begin{pmatrix} \frac{\partial a_c}{\partial I}\Big|_{M\Omega} & \frac{\partial a_c}{\partial a_h}\Big|_{M\Omega} \\ \frac{\partial \Omega}{\partial I}\Big|_{M\Omega} & \frac{\partial \Omega}{\partial a_h}\Big|_{M\Omega} \end{pmatrix}^{-1} \begin{pmatrix} a_c - a_c^{MP} \\ \Omega - \Omega^{M\Omega} \end{pmatrix}, \tag{46}$$

resulting in the dynamics equations

$$\frac{dI}{dt} = g(a_h, I; C_{a_h}, C_I), \tag{47}$$

$$\frac{da_h}{dt} = h(a_h, I; C_{a_h}, C_I). \tag{48}$$

This dynamics in the I-a_h space is depicted in Figure 2. The values $\sigma_{hc} = 1.4$, $\sigma_{ih} = 0.2$, $I_{M\Omega} = 0.9$, and $\tau = 0.5$ are used as a representative configuration as they approximate the behavior of the irreversible Carnot-like heat engine to the one obtained in the simulation shown in Figure 1. The value $a_{h,M\Omega}$ is calculated from Equation (19). Figure 2a represents the case where the dynamics in the a_h direction is stronger, in Figure 2b, where the restitution strength C_I and C_{a_h} are symmetrical, while in Figure 2c the dynamics in C_I is the strongest. The streamlines show a trend of the relaxation trajectories to increase the parameter I, making the engine closer to the endoreversible limit.

After a perturbation, the relaxation trajectories evolve with a variable speed

$$v = \sqrt{(dI/dt)^2 + (da_h/dt)^2}, \tag{49}$$

which is depicted in Figure 3 (left). The influence of the relaxation speed is significant since slow relaxations under many perturbations will effectively modify the operation state. In Figure 3 (right), representative relaxation trajectories starting from conditions near zero power output provide a general idea of how other initial configurations will evolve in time. Also, the states of maximum power (MP, red dashed straight line), the maximum ecological and Omega functions ($M\Omega$, green dashed line), and the maximum efficiency ($M\eta$, blue dashed line) are indicated. All the trajectories will finally arrive at the steady state ($M\Omega$). However, for a given relaxation time, the trajectories that arrive at the steady state are colored in red, and those that require longer times are depicted in light yellow. For this representation, the time defined is a multiple of the characteristic relaxation time, t_{relax}, which is calculated from the eigenvalues of the dynamics matrix appearing in Equation (46). For the representation of Figure 3 (right), a time of $11t_{relax}$ is sufficient to analyze fast and slow relaxation trajectories. Notice in Figure 3 (right) that the states of maximum efficiency are in a region of slow velocity, and, in comparison, the states of maximum power evolve with higher velocities.

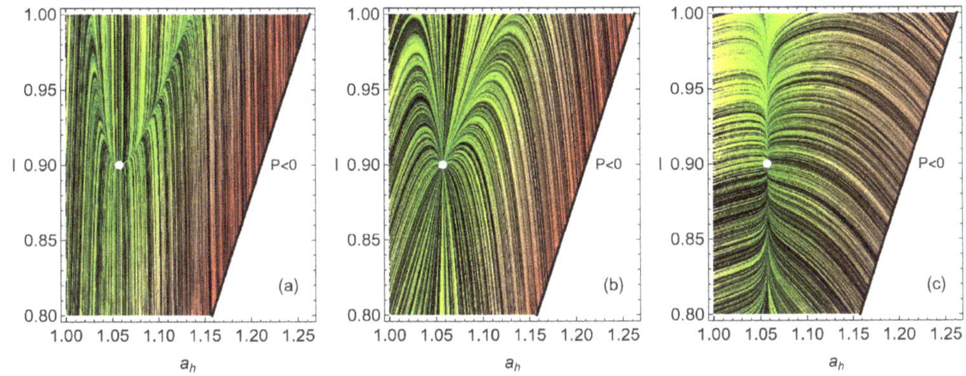

Figure 2. Line integral convolution plot showing streamlines of fixed arc length over a set of random conditions for the dynamics given by solving Equations (47) and (48) for the values $\sigma_{hc} = 1.4$, $\sigma_{ih} = 0.2$, $I_{MP} = 0.9$, $\tau = 0.5$ and for the case (**a**) $C_I = 1$ and $C_{a_h} = 10$, in (**b**) the symmetrical case $C_I = C_{a_h} = 1$ and in (**c**) $C_I = 10$ and $C_{a_h} = 1$. The white dot in the center corresponds to the MP operation state, and the white region on the right side of each figure is the nonphysical region where $P < 0$.

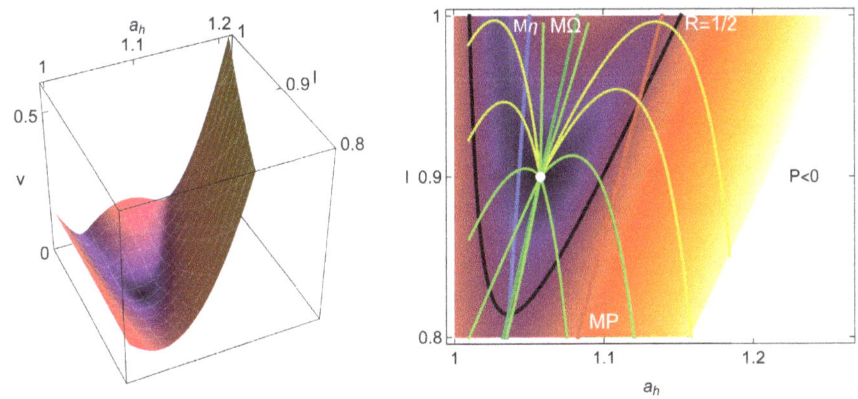

Figure 3. The influence of the velocity of the dynamics (**left**) for the configuration depicted in Figure 2b ($\sigma_{hc} = 1.4$, $\sigma_{ih} = 0.2$, $I_{MP} = 0.9$, $\tau = 0.5$ and the symmetrical case $C_I = C_{a_h} = 1$) and some representative relaxation trajectories (**right**). The dashed lines indicate the possible states of maximum power (*MP*, in red), the maximum ecological and Omega functions ($M\Omega$, in green), and the maximum efficiency ($M\eta$, in blue). The black curve for which $R = 1/2$ indicates the boundary of the success region discussed in Section 3.

In the context of success in the performance of heat devices, it is interesting to keep an eye on the location of the region where the reversible parameter $R \geq 1/2$, which is where $\eta \geq \eta_C/2$. This is also depicted in Figure 3 (right). It is noteworthy that the velocities that take the system out of the unsuccessful area are greater and once the system is in the region of success, the return velocities are slower. Also, the MP states are in the success region only for values of I close to 1, while for $M\eta$ and $M\Omega$ and ME, the region of success covers a more extended area. In all this area, $\Omega > 0$ and its maximum are located between the MP and $M\eta$ states.

6. Energetic Evolution in the Relaxation Towards the Steady State

The relaxation trajectories strongly depend on the restitution strength in the direction of I or a_h, which can be analyzed through the restitution coefficients C_I and C_{a_h}. The velocity is shown in Figure 4 for the three cases of evolution: for the dynamics being dominated more strongly by I, for the symmetrical case $C_I = C_{a_h} = 1$, or where the dynamics are dominated by a_h, this is depicted in Figure 4a, b and c, respectively. Notice that larger velocities are produced when I dominates the dynamics. This behavior is expected if the relaxation time leading to local equilibrium and the effective isothermal processes for the working fluid is considerably smaller than the operation time. On the other hand, when the dynamics are more strongly given by the operation variable a_h, the convergence to $a_{h,M\Omega}$ is fast, but afterward, the evolution is noticeably slow.

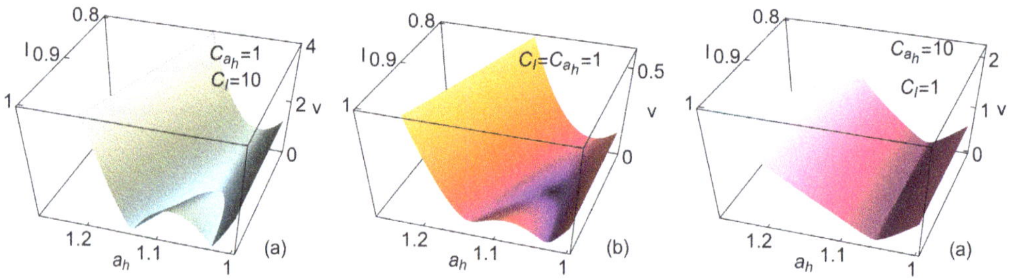

Figure 4. Relaxation velocity for when (**a**) the influence of I is larger, $C_I = 10$ and $C_{a_h} = 1$. In (**b**), the symmetrical case $C_I = C_{a_h} = 1$. In (**c**), a_h dominates the stability dynamics, $C_I = 1$ and $C_{a_h} = 10$. In all the cases, $\sigma_{hc} = 1.4$, $\sigma_{ih} = 0.2$, $I_{MP} = 0.9$, and $\tau = 0.5$.

The relaxation trajectories for the three cases depicted in Figure 4 are shown in Figure 5. In all the cases, it is relevant that the trajectories rapidly enter into the zone where $R > 1/2$, the so-called "success region". In this region, the $M\Omega$ states are located between the MP and $M\eta$ ones and the curves approaching the MP are moving faster than those moving between the $M\Omega$ and $M\eta$ regimes. For illustrative purposes, the trajectories evolving with decreasing values of a_h and are slower and indicated with dashed orange lines, which can be tracked in the forthcoming figures. These trajectories start from points of low or zero power output.

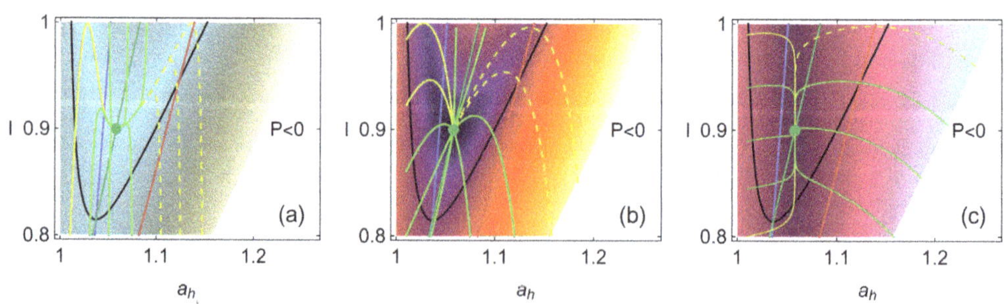

Figure 5. Relaxation trajectories for the cases depicted in Figure 4. The evolution time, t_e, in each case is (**a**) $t_e = 28t_{relax}$, (**b**) $t_e = 11t_{relax}$, and (**c**) $t_e = 22t_{relax}$.

Figure 6 shows the trajectories in the energetic spaces $P - \eta$ and $\sigma - \eta$. The blue-shaded region corresponds to the zone where $R < 1/2$ (or $\eta < \eta_C/2$). All the trajectories that exit the shaded region to enter into the success region in the $P - \eta$ space simultaneously increase power and efficiency, while the trajectories appearing in Figure 5 are evolving from the right exit of the shaded region with trajectories that also decrease entropy generation. Those that evolve by increasing entropy generation tend to do so with small increases in σ.

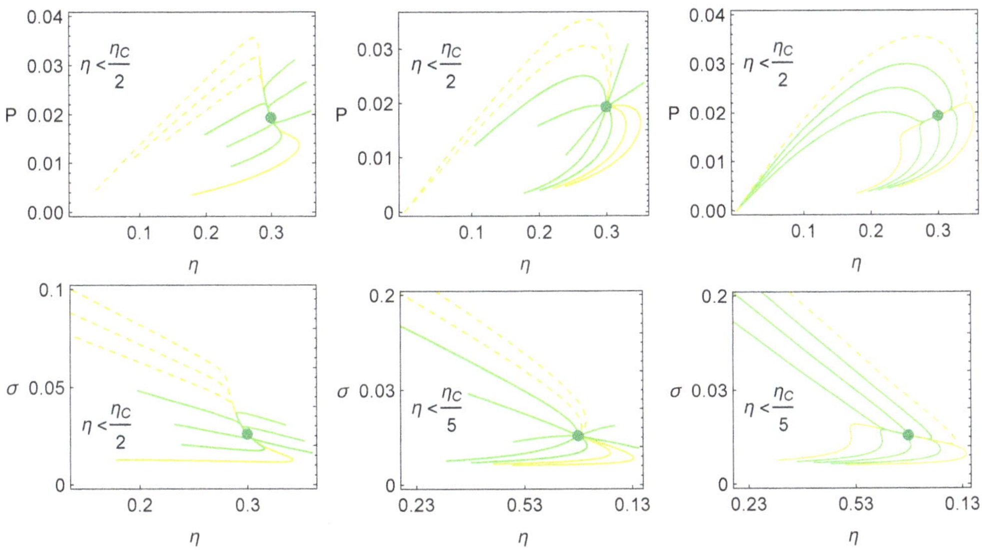

Figure 6. Evolution of σ and η in the relaxation trajectories towards the steady state. The cases correspond to the configurations depicted in Figure 5.

The evolution of the reversible coefficient R with respect to a_h and I is shown in Figure 7 in the first and second rows, respectively. The cases depicted correspond to those presented in Figure 5. Meanwhile, in the first row, the trajectories can exit the non-success region by increasing a_h (short trajectories) or decreasing a_h (with long and rapid trajectories). In the second row, all the trajectories evolve in the shaded region with trajectories that increase the value of I. There are inflection points from which the system cannot continue becoming more reversible, and the system has to converge to the steady state by decreasing R. These trajectories seem to focus on the simultaneous improvement of R and I even though, later, the system has to return to converge to the $M\Omega$ state.

The evolution of the components D_e (Equation (36)) and D_i (Equation (37)) corresponding to the external and internal losses is depicted in Figure 8. The total loss parameter D (Equation (35)) also gives information regarding the success region, which corresponds to the region where $D < 1/2$. Overall, the curves tend to exhibit larger variations in the value of D_e when this parameter has to decrease; otherwise, the increase in D_e is quite small. In any case, the evolution curves always tend to quickly decrease D_i to move to the success region.

All these cases show a clear influence of the stability dynamics in the way the system relaxes to a steady state. Optimum states given by locus near maximum efficiency and maximum trade-off functions such as the ecological and Omega functions produce slower relaxation dynamics, while trajectories passing outside the optimization region of interest exhibit faster relaxation endings to simultaneously improve power, efficiency, and entropy production while improving the reversibility of the system and decreasing the internal irreversibility parameter D_i.

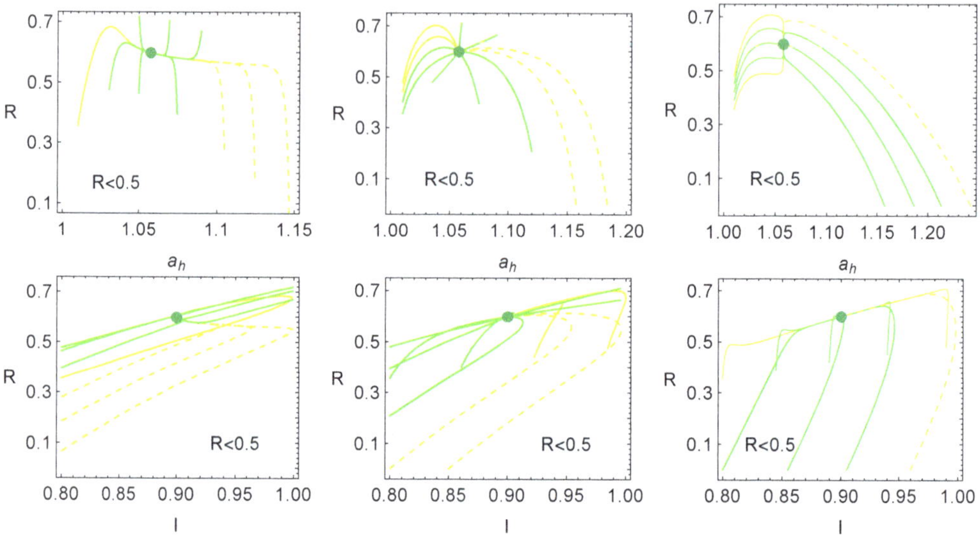

Figure 7. Evolution of the reversible coefficient, R, in relaxation trajectories in terms of a_h (**upper row**) and I (**lower row**) for the cases depicted in Figure 5.

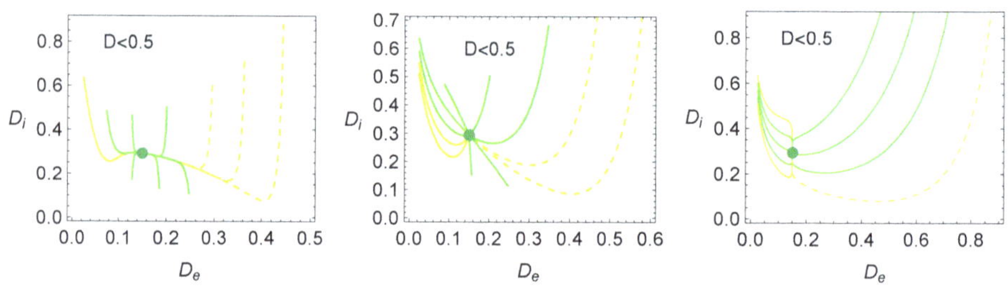

Figure 8. Evolution of internal and external irreversible parameters D_i and D_e in the relaxation of the system towards the steady state. The three cases depicted correspond to those appearing in Figure 4.

7. Consecutive Perturbations

7.1. Stochastic Trajectories

The previous analysis on a single trajectory after a perturbation has indicated the possible role of stability in enhancing the system operation state. However, in realistic conditions, it is expected that heat devices operating under stationary operation conditions experience small fluctuations produced from external factors and from the intrinsic variability due to limited control. Now, it is time to analyze the effect of consecutive perturbations over one cycle. These perturbations will be modeled by stochastic variations in both parameters, a_h and I, and test if the relaxation dynamics, as would be expected, tend to push the operation regime to more optimum states in terms of power, efficiency, and entropy generation. Based on the stability dynamics given by Equations (47) and (48), stochastic perturbations can be incorporated as an additive white noise with two normally distributed random variables. The independent stochastic variables $\{\xi_1, \xi_2\}$ in the I-a_h directions obey a two-dimensional Gaussian distribution

$$f_\xi(I, a_h) = \frac{\beta^2}{\pi\, I_{M\Omega}\, a_{h,M\Omega}}\, e^{-\beta^2\left[\left(\frac{I}{2I_{M\Omega}}\right)^2+\left(\frac{a_h}{2a_{h,M\Omega}}\right)^2\right]}, \tag{50}$$

where $\beta = 40$ is a good compromise to maintain the system around the steady state for long periods. The standard deviations are proportional to the control variables, $\sigma_I = I_{M\Omega}/\sqrt{2}\beta \approx 0.018\, I_{M\Omega}$ and $\sigma_{a_h} = a_{h,M\Omega}/\sqrt{2}\beta \approx 0.018 a_{h,M\Omega}$. Based on Equations (47) and (48), the corresponding stochastic differential equations that numerically describe the system evolution are solved using the Euler–Maruyama method [95].

$$\Delta I = g(a_h, I; C_{a_h}, C_I)\Delta t + \xi_1\sqrt{\Delta t}, \quad (51)$$

$$\Delta a_h = h(a_h, I; C_{a_h}, C_I)\Delta t + \xi_2\sqrt{\Delta t}, \quad (52)$$

and the time evolution is computed by iterating N perturbations equally distributed in time with intervals of length Δt. The evolution time, t_e (see Figure 5), guarantees that for small perturbations the system is always allowed to return to the steady state. Trajectories of 10^4 steps for the $M\Omega$ regime, with $\Delta t = 10^{-4} t_e$, are obtained. Due to the random nature of these external perturbations, one might expect that the system runs away from the fixed point. However (see below), the size of the perturbations is such that even for extended periods, the system will remain close to the stable state.

In Figure 9, a representative case of a stochastic trajectory is depicted. Two trajectories are displayed, one considering the restitution forces (blue) and the other one only the random perturbations without restitution dynamics (pink). As it can be seen, for the particular case depicted, the trajectory with stability effects maintains the system in the zone of thermodynamic success, exhibiting lower values of entropy production and a low loss coefficient (D), with a stable and higher efficiency. Despite the good performance shown in Figure 9, a statistical analysis of many trajectories should be made to guarantee that the induced optimization is an observable quantity.

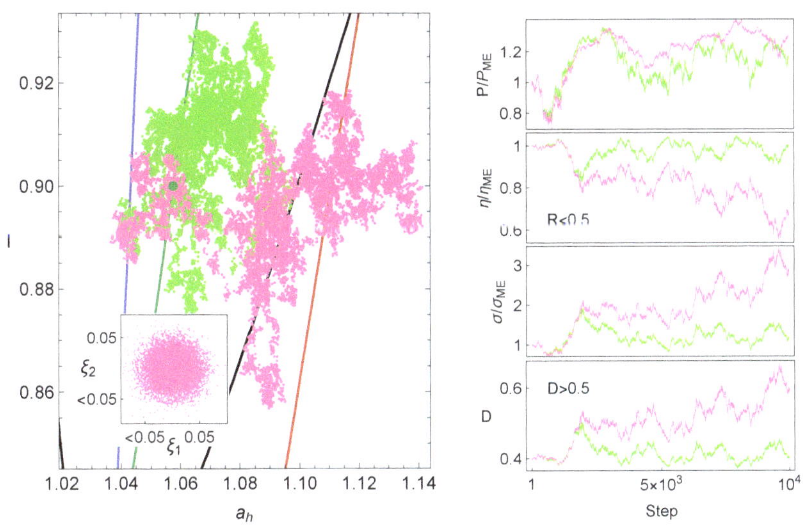

Figure 9. *To the left:* Two representative stochastic trajectories under the influence of the same random perturbations with 10^4 steps. One considers the restitution forces (blue) and the other one does not have restitution dynamics (pink). The random perturbations in the two directions are indicated in the close caption. *To the right:* The time series of P, η, and σ (the three of them normalized to the value of the steady state, $M\Omega$), along with the loss coefficient D. In the time series of the efficiency and in D, the region with $R > 0.5$ (no thermodynamic success) is shaded in blue.

7.2. Statistical Convergence

The previous analysis is repeated for 5×10^4 trajectories. For each one, final and average states for P, η, Ω, and $\tilde{\sigma}$ are computed. To check for statistical convergence of the results, the Kullback–Leibler divergence, D_{KL}, of the distribution of the power output is computed each time that 10^3 trajectories are calculated (the same could be performed for the other thermodynamical functions). This provides a measure of how distant one distribution is compared with the previous one. If $D_{KL} = 0$, the information stemming from both distributions is the same and the results obtained are statistically representative.

The interval between the largest and smallest p values is divided by $\sqrt{N}$ (rounded to the upper next integer) equal intervals or bins; in this way, the same partition is used to compute the discrete probability distributions of the first k-thousand trajectories, ρ_k are obtained, and the D_{KL} is calculated comparing ρ_{k-1} with $\rho_k . D_{KL,k}$.

$$D_{KL,k}(\rho_{k-1} \| \rho_k) = -\sum_i \rho_{k-1,i} \, log\left(\frac{\rho_{k,i}}{\rho_{k-1,i}}\right), \tag{53}$$

giving a measure of how much information is gained by adding more trajectories. In Figure 10a, the resulting relative entropy D_{KL} is shown. In Figure 10b, it is possible to see that from 2×10^4 trajectories the statistical behavior does not vary significantly and adding trajectories will not provide further information. In Figure 10c, the difference between D_{KL} and its previous value is depicted. The difference is small enough to consider that 50×10^5 trajectories are enough to analyze the stability phenomena at hand.

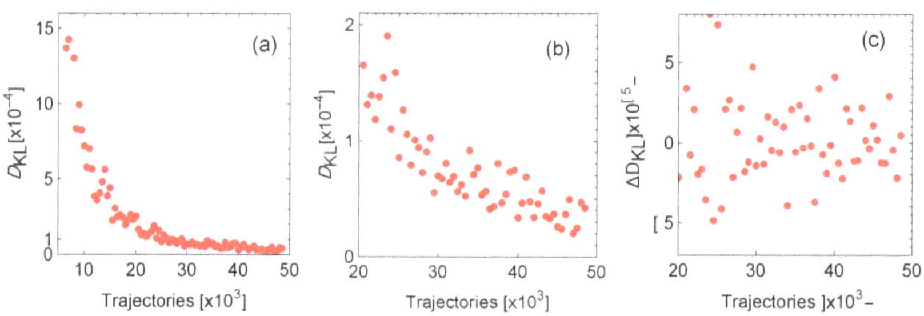

Figure 10. In (**a**) is the Kullback–Leibler divergence, D_{KL}, for the power output distribution of every one thousand trajectories. In (**b**) is a close-up for starting at 20 k trajectories, and in (**c**) is the difference between D_{KL} from one point to the next to show that the series is converging.

7.3. Effect of Stability on Many Trajectories

For each of the 5×10^4 trajectories, the final state of the system is depicted in Figure 11 for the three cases of dynamics: where the influence of I is larger, for the symmetrical case, and when a_h dominates the stability dynamics (blue points), and they are compared with the cases in which there is no stability dynamics (ND, in pink). Each trajectory consists of 10^4 steps. Notice that in every case the system remains inside the region of thermodynamic success ($\Delta > 0$ or equivalently $\eta > \eta_C/2$), and especially for the cases (a) and (c), the system is more concentrated in that region. The mean final states are indicated with a blue $\otimes$ symbol for the ND case and with a black $\odot$ symbol for the case with dynamics.

In the ND case, the average final state is displaced to a slightly lower value of I (less endoreversible) and at a larger a_h (in the direction of the MP state and to the boundary of the success region). However, in the presence of restitution dynamics, the average final state remains at the $M\Omega$ state corresponding to a slightly lower value of I (displaced over the green line but towards a less endoreversible state) if it is strongly influenced by a_h.

Otherwise, the average final state remains located at the $M\Omega$ corresponding to a higher I value (displaced over the green line but towards a more endoreversible configuration).

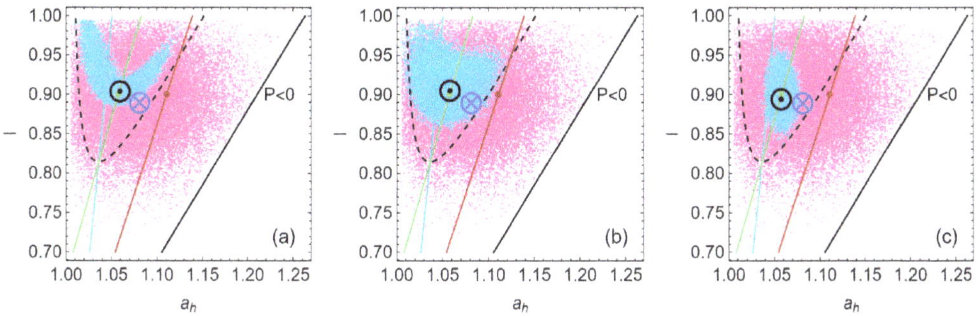

Figure 11. Final states after every stochastic trajectory (5×10^4 trajectories, each one consisting of 10^4 steps). The points in pink correspond to the case where there are no stability dynamics, only random perturbations in the direction of I and a_h. Blue points correspond to the final states in the presence of restitution dynamics. The three cases depicted correspond to those appearing in Figure 4 ($\sigma_{hc} = 1.4$, $\sigma_{ih} = 0.2$, $I_{MP} = 0.9$, and $\tau = 0.5$); in (**a**) the influence of I is larger $C_I = 10$ and $C_{a_h} = 1$. In (**b**), the symmetrical case $C_I = C_{a_h} = 1$. In (**c**), a_h dominates the stability dynamics, $C_I = 1$ and $C_{a_h} = 10$. The black boundary on the right delimits the region where $P > 0$ and the dashed black curve delimits the region of success. The states of maximum power, maximum efficiency, and maximum ecological and Omega functions are indicated as in Figure 3; the green line corresponds to the $M\Omega$ regime, the red line to the MP, and the blue one to $M\eta$, and the points in each line indicate the operation regime with the I value of the steady state.

The average behavior of each trajectory is calculated with the mean values of the energetic functions: power output, efficiency, and entropy production. The average behavior of the loss parameters D_e, D_i, and D are calculated as well. The results for all the trajectories and the three restitution coefficients' cases are shown in Figure 12. The first row corresponds to the case where the dynamics are dominated by I, the second row is the symmetrical case, and the third one is for the case where a_h dominates the dynamics.

In the part of the energetic functions, it is possible to see that the first two rows exhibit dynamics where the system tends to evolve in two directions, first, staying between the maximum efficiency and maximum Omega functions, and second, by increasing the power output without big drops in the efficiency; in some cases the power output even surpasses the MP achievable for the value I of the steady state but with a noticeable lower entropy production (see the larger values of the dynamics compared with the red dot of the MP state). This is only possible by increasing the parameter I in the phase space a_h-I, in which case the system is more endoreversible. Notice that the total losses parameter D varies mostly due to the variation in the external losses (D_e), effectively avoiding getting close to the boundary of the success region.

For the third row, the behavior is different. The averaged states focus mostly on enhancing the efficiency with smaller variations of power output and entropy production, exhibiting a narrow distribution of points with small variations of P and σ. This is achieved while maintaining an almost constant value of D_e, showing a very flat distribution in this direction. Thus, the variations in the total losses, D, come from the variations in the parameter I, which is in agreement with Figure 11c.

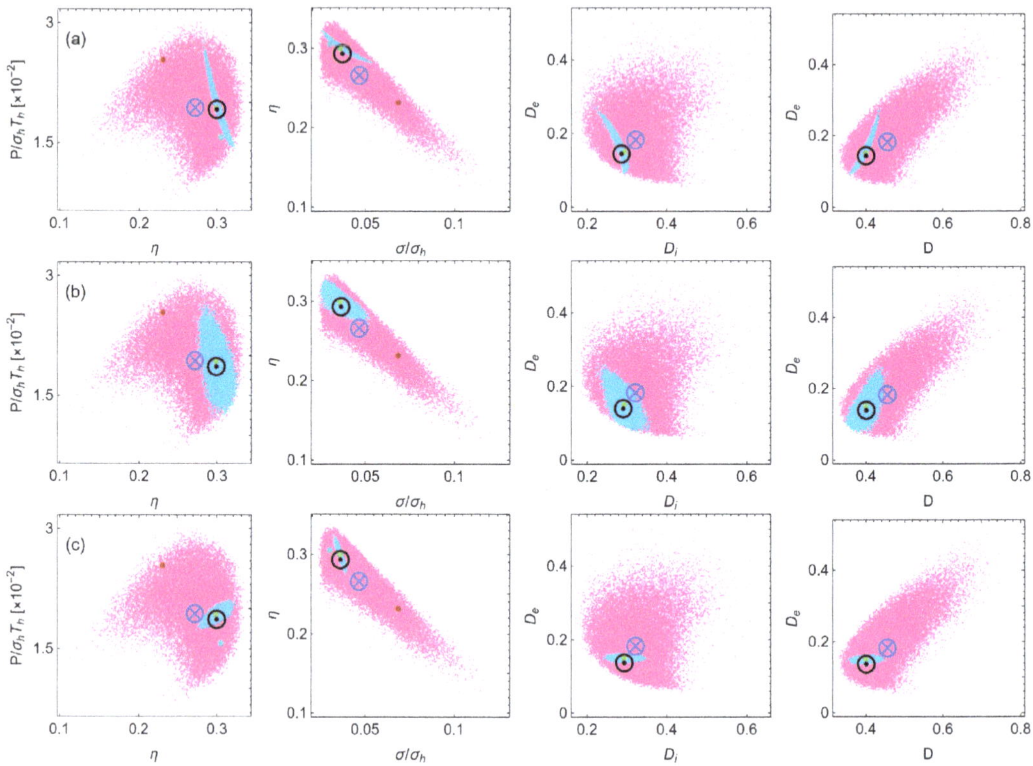

Figure 12. The three cases (**a–c**) depicted correspond to those appearing in Figure 4.

8. Summary and Concluding Remarks

The endoreversible model, proposed 50 years ago, is a phenomenological model that has provided unified and general features related to operation regimes with results relevant in quite different schemes. A noteworthy issue raised in the paper is the analysis of the named thermodynamic region of success for the irreversible Carnot-like heat engine. This region is bounded by considering the distance between the totally reversible and the totally dissipative configurations (both configurations offer the most stable configurations for the heat reservoirs); in between, entropy generation due to the correlations in the exchanges between the reservoirs and the working fluid introduces instabilities. As addressed in the paper, it is possible to define a distance between both extreme states. Regardless of the chosen metric, the inflection point at which both states are at the minimum distance defines the boundary from which the reversibility or irreversibilities dominate. The region where reversibilities dominate (success) is found to be preferred by the stability of the engine.

In the literature, the role of stability as another ingredient in the optimization of heat engines has been discussed, even suggesting that the stability mechanisms can provide a self-optimization mechanism to improve the performance of a heat engine under the influence of external noise and limited control. The proposal for the stability dynamics is justified as follows: the cornerstone of the model is the endoreversible hypothesis, which relies on local equilibrium to produce the effective isotherms which in the model appear as the auxiliary reservoirs. In this work, the stability analysis is made on two components; one is related to the variable a_h that establishes the operation regime, and the other component is the capacity of the system to produce the effective isotherms to fulfill the endoreversible hypothesis. The parameter I provides a measure of how reversible the effective Carnot-like cycle described by the working fluid is (how close the system is to fulfilling the Clausius equality in the inner cycle).

The analysis has been made by selecting the maximum ecological or maximum Omega operation regime, which is a compromise between maximum power and minimum power losses or entropy released to the ambient surroundings. The proposed stability comes solely from recognizing that there exists a steady state and that the stability is well described by the linear approximation (which at least is guaranteed in small perturbations). The resulting dynamics show that restitution velocity favors a slow evolution in the regions between the maximum efficiency and the maximum ecological states. Depending on the direction in which the dynamics is stronger (a_h or I), the slow region shape can vary, but in all cases the system tends to evolve rapidly to enter into the so-called success thermodynamic region and evolve slowly once inside of this success region. The evolution of the system under random perturbations clearly shows that, without the presence of the stability dynamics, the system arbitrarily evolves inside or outside the thermodynamic success region, which significantly decreases the performance of the engine in power output, efficiency, and entropy generation. On the other hand, when the stability dynamics is acting, the system effectively remains inside the success region. The analysis of many stochastic trajectories confirms this fact and shows that if the thermalization dynamics is strong (dynamics mostly dominated by I), the net effect is to produce a slight increase in the reversibility of the system. On the other hand, if the control on the operation parameter a_h is strong, the efficiency is enhanced, the value of I is slightly decreased, and changes in the total losses are linked to internal losses. The results obtained are statistically representative. This is confirmed by the analysis of the Kullback–Leibler divergence, which shows that the system's statistical behavior does not change from around 30 thousand trajectories; here, 50 thousand trajectories are computed.

The stability proposed does not consider specific characteristics of the machinery or technology involved in the engine operation and does not include optimization mechanisms. Nonetheless, a link between optimization and stability is found, strengthening the basic idea that stability is another ingredient to consider in the optimization analysis of heat engines. It would be of interest to incorporate a realistic weak control in the simulations of systems such as the one appearing in [12]. If the role of stability and the tendency of local equilibrium to stabilize the system can be generalized to other models, then the local equilibrium can be seen as a mechanism that optimizes and promotes the evolution of energy converters in natural and artificial heat engines.

Author Contributions: Conceptualization, J.G.-A. and A.C.H.; software, J.G.-A.; writing—original draft preparation, J.G.-A. and A.C.H.; supervision, J.G.-A. and A.C.H.; funding acquisition, A.M. and J.M.M.R. All authors equally contributed to the methodology, formal analysis, and writing—review and editing to provide the final version. All authors have read and agreed to the published version of the manuscript.

Funding: This research was funded by Funds from Universidad de Salamanca; Ministerio de Ciencia, Innovación y Universidades of Spain under grants PID2023-147201OB-I00 and RED2024-153629-T; and from Consejería de Educación de la Junta de Castilla y León under grant SA071G24. D. Pérez-Gallego thanks the financial support from Fondo Social Europeo Plus and Consejería de Educación de la Junta de Castilla y León under their Ph.D. grant program (EDU/1868/2022).

Institutional Review Board Statement: Not applicable.

Data Availability Statement: The raw data supporting the conclusions of this article will be made available by the authors on request.

Conflicts of Interest: The authors declare no conflicts of interest. The funders had no role in the design of the study; in the collection, analyses, or interpretation of data; in the writing of the manuscript; or in the decision to publish the results.

References

1. Carnot, S. *Reflexions sur la Puissance Motrice du Feu et sur les Machines Propres a Developper cette Puissance*; Bachelier: Paris, France, 1824.
2. Ferreira da Silva, M.F. Some considerations about thermodynamic cycles. *Eur. J. Phys.* **2012**, *33*, 13–42. [CrossRef]
3. Gonzalez-Ayala, J.; Angulo-Brown, F. The universality of the Carnot theorem. *Eur. J. Phys.* **2013**, *34*, 273–289. [CrossRef]
4. Gonzalez-Ayala, J.; Arias-Hernandez, L.A.; Angulo-Brown, F. Connection between maximum-work and maximum-power thermal cycles. *Phys. Rev. E* **2013**, *88*, 052142. [CrossRef]
5. Wu, J. Three factors causing the thermal efficiency of a heat engine to be less than unity and their relevance to daily life. *Eur. J. Phys.* **2015**, *36*, 015008. [CrossRef]
6. Curzon, F.L.; Ahlborn, B. Efficiency of a Carnot engine at maximum power output. *Am. J. Phys.* **1975**, *43*, 22. [CrossRef]
7. Sekulic, D.P. A fallacious argument in the finite time thermodynamics concept of endoreversibility. *J. Appl. Phys.* **1998**, *83*, 4561. [CrossRef]
8. Andresen, B. Comment on "A fallacious argument in the finite time thermodynamic concept of endoreversibility" [*J. Appl. Phys.* **1998**, *83*, 4561]. *J. Appl. Phys.* **2001**, *90*, 6557. [CrossRef]
9. Chen, J.; Yan, Z.; Lin, G.; Andresen, B. On the Curzon–Ahlborn efficiency and its connection with the efficiencies of real heat engines. *Energy Convers. Manag.* **2001**, *42*, 173. [CrossRef]
10. Gyftopoulos, E.P. On the Curzon–Ahlborn efficiency and its lack of connection to power producing processes. *Energy Convers. Manag.* **2002**, *43*, 609–615. [CrossRef]
11. Berry, R.S.; Salamon, P.; Andresen, B. How It All Began. *Entropy* **2020**, *22*, 908. [CrossRef]
12. Rojas-Gamboa, D.A.; Rodríguez, J.I.; Gonzalez-Ayala, J.; Angulo-Brown, F. Ecological efficiency of finite-time thermodynamics: A molecular dynamics study. *Phys. Rev. E* **2018**, *98*, 022130. [CrossRef]
13. Izumida, Y.; Okuda, K. Onsager coefficients of a finite-time Carnot cycle. *Phys. Rev. E* **2009**, *80*, 021121. [CrossRef] [PubMed]
14. Leff, H.S. Thermal efficiency at maximum work output: New results for old heat engines. *Am. J. Phys.* **1987**, *55*, 602–610. [CrossRef]
15. Gordon, J.M. Generalized power versus efficiency characteristics of heat engines: The thermoelectric generator as an instructive illustration. *Am. J. Phys.* **1991**, *59*, 551–555. [CrossRef]
16. Gordon, J.M.; Huleihil, M. General performance characteristics of real heat engines. *J. Appl. Phys.* **1992**, *72*, 829–837. [CrossRef]
17. Chen, J. The maximum power output and maximum efficiency of an irreversible Carnot heat engine. *J. Phys. D Appl. Phys.* **1994**, *27*, 1144–1149. [CrossRef]
18. Qin, X.; Chen, L.; Sun, F.; Wu, C. The universal power and efficiency characteristics for irreversible reciprocating heat engines cycles. *Eur. J. Phys.* **2003**, *24*, 359–366. [CrossRef]
19. Agrawal, D.C. A simplified version of the Curzon-Ahlborn engine. *Eur. J. Phys.* **2009**, *30*, 1173–1179. [CrossRef]
20. Pescetti D. On the optimization of endoreversible processes. *Eur. J. Phys.* **2014**, *35*, 025014. [CrossRef]
21. Vaudrey, A.; Lanzetta, F.; Feidt, M.H.B. Reitlinger and the origins of the Efficiency at Maximum Power formula for Heat Engines. *J. Non-Equilib Thermodyn.* **2014**, *39*, 199–203. [CrossRef]
22. Van den Broeck, C.; Kumar, N.; Lindenberg, K. Efficiency of Isothermal Molecular Machines at Maximum Power. *Phys. Rev. Lett.* **2012**, *108*, 210602. [CrossRef]
23. Golubeva, N.; Imparato, A. Efficiency at Maximum Power of Interacting Molecular Machines. *Phys. Rev. Lett.* **2012**, *109*, 190602; Erratum in *Phys. Rev. Lett.* **2013**, *110*, 149902. [CrossRef]
24. Apertet, Y.; Ouerdane, H.; Goupil, C.; Lecoeur, P. Efficiency at maximum power of thermally coupled heat engines. *Phys. Rev. E* **2012**, *85*, 041144. [CrossRef]
25. Wang, R.; Wang, J.; He, J.; Ma, Y. Efficiency at maximum power of a heat engine working with a two-level atomic system. *Phys. Rev. E* **2013**, *87*, 042119. [CrossRef] [PubMed]
26. Wu, F.; He, J.; Ma, Y.; Wang, J. Efficiency at maximum power of a quantum Otto engine: Both within finite-time and irreversible thermodynamics. *Phys. Rev. E* **1990**, *90*, 062134. [CrossRef]
27. Sheng, S.Q.; Tu, Z.C. Hidden symmetries and nonlinear constitutive relations for tight-coupling heat engines. *New J. Phys.* **2015**, *17*, 045013. [CrossRef]
28. Izumida, Y.; Okuda, K. Linear irreversible heat engines based on the local equilibrium assumptions. *New J. Phys.* **2015**, *17*, 085011. [CrossRef]

29. Ouerdane, H.; Apertet, Y. Goupil and Ph. Lecoeur, Continuity and boundary conditions in thermodynamics: From Carnot's efficiency to efficiencies at maximum power. *Eur. Phys. J. Spec. Top.* **2015**, *224*, 839–862. [CrossRef]
30. Landsberg, P.; Leff, H. Thermodynamic cycles with nearly universal maximum-work efficiencies. *J. Phys. A Math. Gen.* **1989**, *22*, 4019. [CrossRef]
31. Esposito, M.; Lindenberg, K.; Van den Broeck, C. Universality of efficiency at maximum power. *Phys. Rev. Lett.* **2009**, *102*, 130602. [CrossRef]
32. Wang, Y.; Tu, Z.C. Efficiency at maximum power output of linear irreversible Carnot-like heat engines. *Phys. Rev. E* **2012**, *85*, 011127. [CrossRef]
33. Guo, J.; Wang, J.; Wang, Y.; Chen, J. Universal efficiency bounds of weak-dissipative thermodynamic cycles at the maximum power output. *Phys. Rev. E* **2013**, *87*, 012133. [CrossRef]
34. Sheng, S.; Tu, Z.C. Universality of energy conversion efficiency for optimal tight-coupling heat engines and refrigerators. *J. Phys. A Math. Theor.* **2013**, *46*, 402001. [CrossRef]
35. Uzdin, R.; Kosloff, R. Universal features in the efficiency at maximum work of hot quantum Otto engines. *Europhys. Lett.* **2014**, *108*, 40001. [CrossRef]
36. Sheng, S.Q.; Tu, Z.C. Constitutive relation for nonlinear response and universality of efficiency at maximum power for tight-coupling heat engines. *Phys. Rev. E* **2015**, *91*, 022136. [CrossRef] [PubMed]
37. Cleuren, B.; Rutten, B.; Van den Broeck, C. Universality of efficiency at maximum power: Macroscopic manifestation of microscopic constraints. *Eur. Phys. J. Spec. Top.* **2015**, *224*, 879. [CrossRef]
38. Ye, Z.; Hu, Y.; He, J.; Wang, J. Universality of maximum-work efficiency of a cyclic heat engine based on a finite system of ultracold atoms. *Sci. Rep.* **2017**, *7*, 6289. [CrossRef] [PubMed]
39. Angulo-Brown, F. An ecological optimization criterion for finite-time heat engines. *J. Appl. Phys.* **1991**, *69*, 7465–7469. [CrossRef]
40. Arias-Hernández, L.A.; Angulo-Brown, F. A general property of endoreversible thermal engines. *J. Appl. Phys.* **1997**, *81*, 2973. [CrossRef]
41. Hern, A.C.; Medina, A.; Roco, J.M.M.; White, J.A.; Velasco, S. Unified optimization criterion for energy converters. *Phys. Rev. E* **2001**, *63*, 037102. [CrossRef]
42. Stucki, J.W. The Optimal Efficiency and the Economic Degrees of Coupling of Oxidative Phosphorylation. *Eur. J. Biochem.* **1980**, *109*, 269–283. [CrossRef] [PubMed]
43. Arias-Hernandez, L.A.; Angulo-Brown, F.; Paez-Hernandez, R.T. First-order irreversible thermodynamic approach to a simple energy converter. *Phys. Rev. E* **2008**, *77*, 011123. [CrossRef]
44. Long, R.; Liu, W. Unified trade-off optimization for general heat devices with nonisothermal processes. *Phys. Rev. E* **2015**, *91*, 042127. [CrossRef] [PubMed]
45. Long, R.; Liu, W. Ecological optimization for general heat engines. *Phys. A Stat. Mech. Appl.* **2015**, *434*, 232–239. [CrossRef]
46. Singh, V.; Johal, R.S. Feynman's Ratchet and Pawl with Ecological Criterion: Optimal Performance versus Estimation with Prior Information. *Entropy* **2017**, *19*, 576. [CrossRef]
47. Iyyappan, I.; Ponmurugan, M. Thermoelectric energy converters under a trade-off figure of merit with broken time reversal symmetry. *J. Stat. Mech.* **2017**, *2017*, 093207. [CrossRef]
48. Lu, C.; Bai, L. Nonlinear Dissipation Heat Devices in Finite-Time Thermodynamics: An Analysis of the Trade-Off Optimization. *J. Non-Equilib. Thermodyn.* **2017**, *42*, 277–286. [CrossRef]
49. Gonzalez-Ayala, J.; Calvo Hernández, A.; Roco, J.M.M. From maximum power to a trade-off optimization of low-dissipation heat engines: Influence of control parameters and the role of entropy generation. *Phys. Rev. E* **2017**, *95*, 022131. [CrossRef]
50. Zhang, Y.; Huang, C.; Lin, G.; Chen, J. Universality of efficiency at unified trade-off optimization. *Phys. Rev. E* **2016**, *93*, 032152. [CrossRef]
51. Holubec, V.; Ryabov, A. Maximum efficiency of low-dissipation heat engines at arbitrary power. *J. Stat. Mech.* **2016**, *2016*, 073204. [CrossRef]
52. Shiraishi, N.; Saito, K.; Tasaki, H. Universal Trade-Off Relation between Power and Efficiency for Heat Engines. *Phys. Rev. Lett.* **2016**, *117*, 190601. [CrossRef]
53. Iyyappan, I.; Ponmurugan, M. General relations between the power, efficiency, and dissipation for the irreversible heat engines in the nonlinear response regime. *Phys. Rev. E* **2018**, *97*, 012141. [CrossRef]
54. Schmiedl, T.; Seifert, U. Optimal Finite-Time Processes In Stochastic Thermodynamics. *Phys. Rev. Lett.* **2007**, *98*, 108301. [CrossRef]
55. Bauer, M.; Brandner, K.; Seifert, U. Optimal performance of periodically driven, stochastic heat engines under limited control. *Phys. Rev. E* **2016**, *93*, 042112. [CrossRef] [PubMed]
56. Funo, K.; Ueda, M. Work Fluctuation-Dissipation Trade-Off in Heat Engines. *Phys. Rev. Lett.* **2015**, *115*, 260601. [CrossRef] [PubMed]

57. Long, R.; Liu, W. Efficiency and its bounds of minimally nonlinear irreversible heat engines at arbitrary power. *Phys. Rev. E* **2016**, *94*, 052114. [CrossRef] [PubMed]
58. Pietzonka, P.; Seifert, U. Universal Trade-Off between Power, Efficiency, and Constancy in Steady-State Heat Engines. *Phys. Rev. Lett.* **2018**, *120*, 190602. [CrossRef] [PubMed]
59. Holubec, V.; Ryabov, A. Cycling tames power fluctuations near optimum efficiency. *Phys. Rev. Lett.* **2018**, *121*, 120601. [CrossRef]
60. Scully, M.O.; Zubairy, M.S.; Agarwal, G.S.; Walther, H. Extracting work from a single heat bath via vanishing quantum coherence. *Science* **2003**, *299*, 862–864. [CrossRef]
61. Dillenschneider, R.; Lutz, E. Energetics of quantum correlations. *Europhys. Lett.* **2009**, *88*, 50003. [CrossRef]
62. Abah, O.; Lutz, E. Efficiency of heat engines coupled to nonequilibrium reservoirs. *Europhys. Lett.* **2014**, *106*, 20001. [CrossRef]
63. Polettini, M.; Verley, G.; Esposito, M. Efficiency Statistics at All Times: Carnot Limit at Finite Power. *Phys. Rev. Lett.* **2015**, *114*, 050601.[CrossRef] [PubMed]
64. Lee, J.S.; Park, H. Carnot efficiency is reachable in an irreversible process. *Sci. Rep.* **2017**, *7*, 10725. [CrossRef]
65. Niedenzu, W.; Mukherjee, V.; Ghosh, A.; Kofman, A.G.; Kurizki, G. Quantum engine efficiency bound beyond the second law of thermodynamics. *Nat. Commun.* **2018**, *9*, 165. [CrossRef]
66. Proesmans, K.; Dreher, Y.; Gavrilov, M.; Bechhoefer, J.; Van den Broeck, C. Brownian Duet: A Novel Tale of Thermodynamic Efficiency. *Phys. Rev. X* **2016**, *6*, 041010. [CrossRef]
67. Uzdin, R.; Levy, A.; Kosloff, R. Equivalence of quantum heat machines, and quantum-thermodynamic signatures. *Phys. Rev. X* **2015**, *5*, 031044. [CrossRef]
68. Kosloff, R.; Rezek, Y. The quantum harmonic Otto cycle. *Entropy* **2017**, *19*, 136. [CrossRef]
69. Gonzalez-Ayala, J.; Roco, J.M.M.; Medina, A.; Hernández, A.C. Carnot-Like Heat Engines Versus Low-Dissipation Models. *Entropy* **2017**, *19*, 182. [CrossRef]
70. Ye, Z.; Holubec, V. Optimizing low-dissipation Carnot-like thermal devices with heat leak. *arXiv* **2025**, arXiv:2504.12655. [CrossRef]
71. Anacleto, J.; Ferreira, J.M. Minimizing the generation of entropy: Which sequence of reservoirs to choose? *Eur. J. Phys.* **2010**, *31*, L1. [CrossRef]
72. Aneja, P.; Katyayan, H.; Johal, R.S. Optimal engine performance using inference for non-identical finite source and sink. *Mod. Phys. Lett. B* **2015**, *29*, 1550217. [CrossRef]
73. Wang, Y. Optimizing work output for finite-sized heat reservoirs: Beyond linear response. *Phys. Rev. E* **2016**, *93*, 012120. [CrossRef]
74. Esposito, M.; Lindenberg, K.; Van den Broeck, C. Entropy production as correlation between system and reservoir. *New J. Phys.* **2010**, *12*, 013013. [CrossRef]
75. Gonzalez-Ayala, J.; Calvo Hernández, A.; White, J.A.; Medina, A.; Roco, J.M.M.; Velasco, S. Success versus failure: Efficient heat devices in thermodynamics. *Phys. Rev. E* **2022**, *105*, 014115. [CrossRef] [PubMed]
76. Gonzalez-Ayala, J.; Angulo-Brown, F.; Calvo Hernández, A.; Velasco, S. On reversible, endoreversible, and irreversible heat device cycles versus the Carnot cycle: A pedagogical approach to account for losses. *Eur. J. Phys.* **2016**, *37*, 045103. [CrossRef]
77. Andresen, B.; Rubin, M.H.; Berry, R.S. Availability for finite-time processes. General theory and a model. *J. Phys. Chem.* **1983**, *87*, 2704–2713. . [CrossRef]
78. Berry, R.S.; Kazakov, V.A.; Sieniutycz, S.; Szwast, Z.; Tsirlin, A.M. *Thermodynamic Optimization of Finite Time Processes*; Wiley: Chichester, UK, 1999; ISBN 0-471-967521.
79. Feidt, M. The History and Perspectives of Efficiency at Maximum Power of the Carnot Engine. *Entropy* **2017**, *19*, 369. [CrossRef]
80. Costea, M.; Petrescu, S.; Feidt, M.; Dobre, C.; Borcila, B. Optimization Modeling of Irreversible Carnot Engine from the Perspective of Combining Finite Speed and Finite Time Analysis. *Entropy* **2021**, *23*, 504. [CrossRef] [PubMed]
81. Tsirlin, A.M.; Balunov, A.I.; Sukin, I.A.; Vasilyev, A.M. Methods of Optimization Thermodynamics in Distillation Processes. *Theor Found Chem. Eng.* **2023**, *57*, 537–548. [CrossRef]
82. Gonzalez-Ayala, J.; Santillán, M.; Reyes-Ramírez, I.; Calvo Hernández, A. Link between optimization and local stability of a low dissipation heat engine: Dynamic and energetic behaviors. *Phys. Rev. E* **2018**, *98*, 032142. [CrossRef]
83. Gonzalez-Ayala, J.; Guo, J.; Medina, A.; Roco, J.M.M.; Calvo Hernández, A. Optimization induced by stability and the role of limited control near a steady state. *Phys. Rev. E* **2019**, *100*, 062128. [CrossRef]
84. Gonzalez-Ayala, J.; Guo, J.; Medina, A.; Roco, J.M.M.; Hernández, A.C. Energetic Self-Optimization Induced by Stability in Low-Dissipation Heat Engines. *Phys. Rev. Lett.* **2020**, *124*, 050603. [CrossRef]
85. Gonzalez-Ayala, J.; Roco, J.M.M.; Medina, A.; Calvo-Hernández, A. Optimization, Stability, and Entropy in Endoreversible Heat Engines. *Entropy* **2020**, *22*, 1323. [CrossRef]
86. England, J.L. Dissipative adaptation in driven self-assembly. *Nat. Nanotechnol.* **2015**, *10*, 919–923. [CrossRef] [PubMed]

87. Belete, M.K.; Bal, G. Optimality and adaptation of phenotypically switching cells in fluctuating environments. *Phys. Rev. E* **2015**, *92*, 062716. [CrossRef]
88. Perunov, N.; Marsland, R.A.; England, J.L. Statistical Physics of Adaptation. *Phys. Rev. X* **2016**, *6*, 021036. [CrossRef]
89. Rao, R.; Esposito, M. Nonequilibrium Thermodynamics of Chemical Reaction Networks: Wisdom from Stochastic Thermodynamics. *Phys. Rev. X* **2016**, *6*, 041064. [CrossRef]
90. Kobayashi, T.J.; Sughiyama, Y. Stochastic and information-thermodynamic structures of population dynamics in a fluctuating environment. *Phys. Rev. E* **2017**, *96*, 012402. [CrossRef] [PubMed]
91. Ouldridge, T.E.; Govern, C.C.; Wolde, P.R.T. Thermodynamics of Computational Copying in Biochemical Systems. *Phys. Rev. X* **2017**, *7*, 021004. [CrossRef]
92. Helbing, D.; Vicsek, T. Optimal self-organization. *New J. Phys.* **1999**, *1*, 13.1–13.17. [CrossRef]
93. Wächtler, C.W.; Strasberg, P.; Klapp, S.H.L.; Schaller, G.; Jarzynski, C. Stochastic thermodynamics of self-oscillations: The electron shuttle. *New J. Phys.* **2019**, *21*, 073009. [CrossRef]
94. Durmayaz, A.; Sogut, O.S.; Sahin, B.; Yavuz, H. Optimization of thermal systems based on finite-time thermodynamics and thermoeconomics. *Prog. Energy Combust. Sci.* **2004**, *30*, 175–217. [CrossRef]
95. Kloeden, P.E.; Platen, E. *Numerical Solution of Stochastic Differential Equations*; Springer: Berlin/Heidelberg, Germany, 1992; ISBN 3-540-54062-8.

entropy

MDPI

Article

On the Equivalence of Gibbs, Boltzmann, and Thermodynamic Entropies in Equilibrium and Nonequilibrium Scenarios

Anil A. Bhalekar *,† and Vijay M. Tangde *

Department of Chemistry, Rashtrasant Tukadoji Maharaj Nagpur University, Nagpur 440 033, India
* Correspondence: anabha@hotmail.com (A.A.B.); vijaytn6@gmail.com (V.M.T.)
† Current address: 803, Wing-A, Shrinivas Crossover County, Lagadmala, Sinhgad Road, Near Lokmat Press, Pune 411 041, India.

Abstract: In this presentation, we have identified the domain of equivalence amongst the Boltzmann, Gibbs, and thermodynamic entropies. In this domain, ergodicity is followed even for (i) all nonequilibrium steady states and (ii) those time-dependent nonequilibrium states belonging to it. The condition of this domain is either that the rate of entropy change is zero or its magnitude is exceedingly small. Its implication is that, in this domain, Jaynes' principle of maximum entropy estimate also holds. Outside this domain, the said equivalence among three entropies is not feasible, and the operation of the Jaynes' principle of maximum entropy estimate does not remain of practical utility.

Keywords: Boltzmann entropy; Gibbs entropy; entropy; CSTR; statistical mechanics; entropy extremization; nonequilibrium thermodynamics

1. Introduction

Boltzmann and Gibbs entropies play a central role in forming a bridge between microscopic and macroscopic properties of matter. The well-established branch named statistical thermodynamics for equilibrium states is a testimony of the soundness of this bridge [1–4]. However, there is a basic difference between the ways these two entropies are computed. In computing the Boltzmann entropy, the number of microstates of a given system is used, and for this purpose, the systems with non-interacting molecules are appropriate. Meanwhile, for computing the Gibbs entropy, the microstates of an ensemble are employed. An ensemble is constructed with a large number of replicas of the system but are in different quantum states commensurate with the thermodynamic conditions operating on it. Thus, the restriction of non-interacting molecules of the former gets removed in the latter. In 1965, Jaynes [5] elucidated this fact and clearly brought forward the distinction between these two entropies with a proof that the difference between the two vanishes when the system is composed of non-interacting molecules. Otherwise, the computation of Boltzmann entropy gives a lower value than that obtained using Gibbs entropy, and the equilibrium thermodynamic entropy matches well with the Gibbs entropy. Thus, this bridging is based on the equivalence of Gibbs and thermodynamic entropies in the equilibrium scenario, whereas it is obvious that, in the case of systems with non-interacting molecules, no distinction exists among these three entropies. We also recall that in the above stated equivalence, one uses the maximized values of Boltzmann and Gibbs entropies. The said maximization is carried out in the statistical mechanical premises and uses the mathematics of chance. That is, this method of maximization has nothing to do with the second law of thermodynamics. De Groot and Mazur [6] have also analyzed the equivalence between these entropies based on the fluctuations about equilibrium states, wherein they use the standard method of maximization of entropy, and their results are in tune with Jaynes' above-stated demonstration for equilibrium states. However, in thermodynamics, we do have maximization of entropy at the final equilibrium state under the adiabatic condition, but this maximization is a macroscopic observation valid only for the end state of corresponding irreversible trajectory. Hence, its premises of operation, the phenomenological, is distinctly different than the above-stated statistical mechanical premises. We also recall that, in 1957, Jaynes propounded the rule of maximum entropy estimates for information-theoretic entropy [7,8]. However, this rule has been in use, since even before it was established in 1957, while computing Boltzmann and Gibbs entropies of an equilibrium state by resorting to the mathematics of chance.

However, with the advent of nonequilibrium thermodynamics [6,9–12], it becomes natural to inquire whether a similar equivalence does exist in the nonequilibrium scenario? Also, what will be the status of Jaynes' maximum Boltzmann and Gibbs entropy estimates when a system is in nonequilibrium?

For answering the above questions, we have adopted an approach that involves the use of a continuously stirred tank reactor (CSTR) with n-number of independent chemical reactions taking place within it. Then, we proceed to construct isentropic hypersurfaces for the nonequilibrium steady states and isentropic planes for equilibrium states appropriate for the considered CSTR. Based on these hypersurfaces and planes, we describe various reversible paths and irreversible trajectories, and then a few relevant thermodynamic characteristic relations are recalled in terms of the variation of thermodynamic, Boltzmann, and Gibbs entropies. With this background, the question of equivalence of three entropies in the nonequilibrium scenario is also discussed. A scope of the validity of the Jaynes' maximum entropy estimates for Boltzmann and Gibbs entropies in the nonequilibrium domain is analyzed. Finally, we present concluding remarks.

This paper has been written on an invitation to contribute in the Special Issue of *entropy* on the occasion of 50 years of the finite time thermodynamics [13–16]. This paper deals with a very basic statistical thermodynamic aspect, the ergodicity, extending to nonequilibrium situations. Hence, this subject is of relevance for all nonequilibrium thermodynamic formulisms, including the finite time thermodynamics.

2. Continuously Stirred Tank Reactor with Several Independent Chemical Reactions

We find that the continuously stirred tank reactor (CSTR) shown in Figure 1 is a simple system, with an advantage that it can be operated in different thermodynamic conditions, such as maintained at an equilibrium or a nonequilibrium steady state, which allows it to evolve irreversibly either to an equilibrium or a nonequilibrium states, etc. It is a spatially uniform and homogeneous system.

In the following discussion, the standard thermodynamic relations suitable for the CSTR of Figure 1 are being used without spelling out their background in details (refer to any standard thermodynamic textbook, for example [17–20]). For example, we recall the following ones:

- When all the flows are stopped, the CSTR depicted in Figure 1 evolves towards an equilibrium state. During this evolution, according to the De Donderian Equation [20], we have the following functional dependence:

$$S(t) = S(U(t), V(t), \{\xi_r(t)\}) \tag{1}$$

where S is the entropy, U is the internal energy, V is the volume of the reactor, $\{\xi_r\}$ are the respective extents of advancement of the chemical reaction identified with the running subscript r, and t is time.

- When this CSTR attains an equilibrium state, the functional dependence of Equation (1) transforms to the following:

$$S = S(U, V) = constant, \tag{2}$$

That is, now we have $\xi_r = \xi_r(U, V)$ for all $\{\xi_r\}$, implying they do not remain independent thermodynamic variables in the case of chemically reacting systems.

- On attainment of a nonequilibrium steady state, the functional dependence of Equation (1) becomes time-independent:

$$S = S(U, V, \{\xi_r\}) = constant. \tag{3}$$

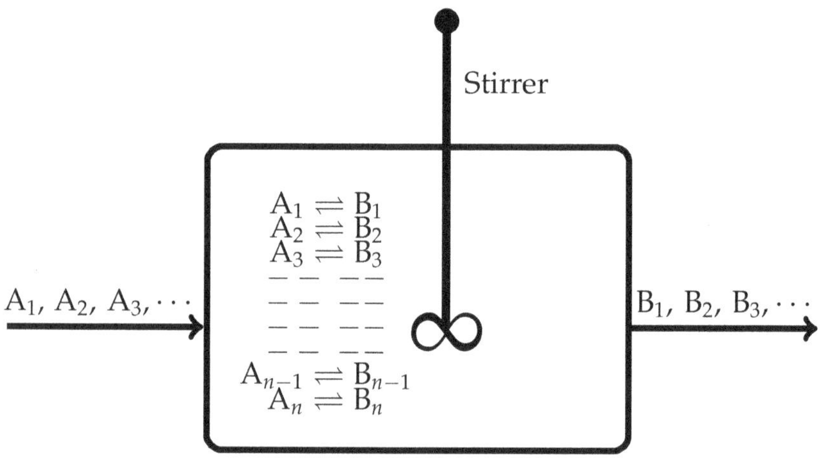

Figure 1. A schematic depiction of a CSTR in which n-number of independent chemical reactions are proceeding at non-vanishing rates. The flow-in of the reactants A_1, A_2, A_3, ... and the flow-out of the products B_1, B_2, B_3, ... can be controlled as per the state of the system that we wish to study. For example, approach towards an equilibrium state or a nonequilibrium steady state or when the system has attained the final equilibrium or nonequilibrium steady state.

3. Geometrical Aspects of Thermodynamic Entropy Based on Isentropic Hypersurfaces Housing Nonequilibrium Steady States

Since the CSTR shown in Figure 1 constitutes a spatially uniform and homogeneous system, when evolving towards an equilibrium state or a nonequilibrium state, the following entropy balance is obeyed [6,9,10]:

$$\frac{dS}{dt} = \frac{d_eS}{dt} + \frac{d_iS}{dt} \tag{4}$$

where $\frac{d_eS}{dt} \gtrless 0$ is the rate of exchange of entropy, and $\frac{d_iS}{dt} \geq 0$ is the positive definite rate of entropy production. The attainment of a nonequilibrium steady state is described by the following:

$$\frac{dS}{dt} = 0 \implies \frac{d_iS}{dt} = -\frac{d_eS}{dt} > 0 \tag{5}$$

That is, at a nonequilibrium steady state, the rate of entropy production equals the rate of entropy driven out of the system. The operative functional dependence for the considered CSTR reads as that given in Equation (3).

Geometrically, the expression in Equation (3) produces an isentropic hypersurface in the thermodynamic space determined by the coordinates $(U,\ V, \{\xi_r\})$. This hypersurface houses time-independent nonequilibrium steady states of identically same values of entropy. Next, by varying the value of the constant in Equation (3), we obtain a family of non-intersecting isentropic hypersurfaces of different values of entropy and those we can stack one over the above in the increasing value of entropy (One of the present authors had earlier identified isentropic hypersurfaces for nonequilibrium steady states by proposing an universal inaccessibility principle. For its details the reader may refer to [21–24]). This we have depicted in Figure 2.

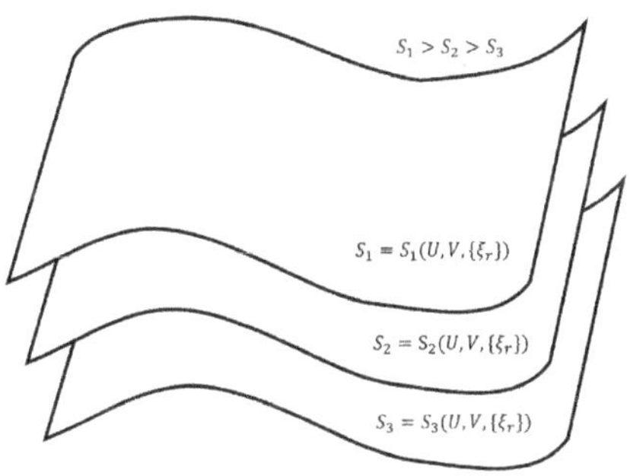

Figure 2. A schematic representation of isentropic hypersurfaces in the $(U,\ V\,\{\xi_r\})$ space with entropy as the vertical axis that is with increasing entropy ($S_1 > S_2 > S_3$). All paths lying exclusively on a given hypersurface are reversible isentropic ones composed of a succession of nonequilibrium steady states, but they are not reversible adiabatic ones because the rate of entropy exchange is non-zero. A reversible non-isentropic path between two end nonequilibrium steady states will be across the isentropic hypersurfaces composed of a succession of nonequilibbrium steady states of varying magnitudes of entropy, which also can be traversed in both the directions. Irreversible non-isentropic paths (trajectories) will also be across the isentropic hypersurfaces but will be accompanied by the variation of entropy as well as both entropy production and entropy exchange.

The transformation of the functional dependence of Equation (3) to that of Equation (2) transforms the non-intersecting isentropic hypersurfaces shown in Figure 2 to the non-intersecting isentropic planes depicted in Figure 3. These isentropic planes house equilibrium states (for a detailed account on the isentropic hypersurfaces for equilibrium states, the reader is directed to refer [18,25,26]).

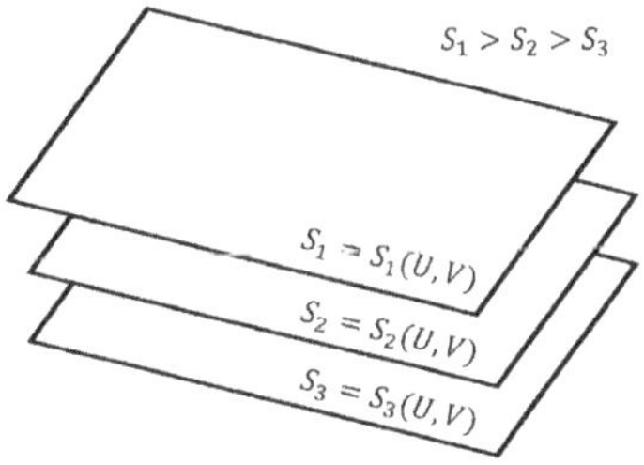

Figure 3. A schematic representation of isentropic planes in the $(U,\ V)$ space with entropy as the vertical axis that is with increasing entropy ($S_1 > S_2 > S_3$). All curves lying exclusively on a given

plane are reversible isentropic ones, which are reversible adiabatic planes too. A non-isentropic (that is non-adiabatic) reversible path will be across the isentropic planes and can be traversed in both the directions, whereas irreversible adiabatic paths will also be across the isentropic planes but will be in the direction of increasing entropy because the change in entropy remains equal to the amount of entropy production. However, the natural direction of non-adiabatic irreversible path (necessarily will be across the isentropic planes) cannot be prescribed in terms of increasing entropy because the sign of dS on such paths also has a contribution from the exchange of entropy with its surrounding but with no definite sign.

Based on the isentropic hypersurfaces in Figure 2 and the isentropic planes in Figure 3, we arrive at the following thermodynamic description of various paths and trajectories:

1. The paths through a succession of nonequilibrium steady states of a given hypersurface of Figure 2 are the reversible isentropic paths. Their thermodynamic description in the De Donderian settings [20] reads as follows:

$$dS = 0 = \frac{1}{T}dU + \frac{p}{T}dV + \sum_r \frac{\mathcal{A}_r}{T}d\xi_r \tag{6}$$

where p is the pressure, T is the temperature and $\{\mathcal{A}_r\}$ are the respective chemical affinities.

The paths through a succession of nonequilibrium steady states across the hypersurfaces in Figure 2 are also reversible ones but are non-isentropic, as the entropy does not remain constant. Its description in the present case is as follows:

$$dS = \frac{1}{T}dU + \frac{p}{T}dV + \sum_r \frac{\mathcal{A}_r}{T}d\xi_r \neq 0 \tag{7}$$

Recall that, by definition, a thermodynamically reversible path is a limiting case but never realized in practice; hence, they are termed as quasi-static paths, which, in principle, can be traversed in either direction. Also, since such paths are prescribed as the ones connecting infinitesimally close successions of time-independent nonequilibrium states, they cannot be termed irreversible paths, though the rate of entropy production remains non-zero. Earlier too, Keizer coined such reversible paths through a succession of nonequilibrium steady states [27–29] in his version of statistical thermodynamics of nonequilibrium processes based on the fluctuations of nonequilibrium steady states. On such paths, the complete compensation of the rate of entropy driven out and the rate of entropy production, as stated in Equation (5), is maintained (In biological systems an example approaching to a reversible path through a succession of nonequilibrium steady states is often presented. For example, biochemical reactions in biological cells, which are usually modelled as undergoing in a CSTR, operate under the nonequilibrium steady state conditions. And often other physico-chemical processes of a biological system are cooperatively associated with them. It has been observed that in biological systems it is not always possible to maintain the existence of the same nonequilibrium steady state continuously, but the system reversibly shifts to a new nonequilibrium steady state which gradually returns to the original one. This transition is considered as very close to a reversible passage through nonequilibrium steady states. That is why, the other biological processes dependent on this nonequilibrium steady state remain practically unhampered).

On the other hand, the isentropic planes depicted in Figure 3 consist of equilibrium states. Therefore, when the thermodynamic reversible paths, through a succession of equilibrium states, are prescribed, then on such paths, no entropy production exists. Hence, in this case, the non-isentropic paths are described by $dS = d_eS \gtrless 0$ and Equation (7), whereas the isentropic paths follow $dS = d_eS = 0$, with the following explicit description:

$$dS = 0 = \frac{1}{T}dU + \frac{p}{T}dV. \tag{8}$$

2. The thermodynamic irreversible trajectories are the ones in which the system leaves an isentropic hypersurface (an isentropic plane), attains a nearby time-dependent nonequilibrium state, and thereafter marches through a succession of time-dependent nonequilibrium states. They can be of two types: (1) in this category, the end state is a time-independent state (a nonequilibrium steady state or an equilibrium state), and (2) in this category, we club all those trajectories not converging to a steady state. In the first category, ultimately, the system attains the state of $\frac{dS}{dt} = 0$. The variation of entropy during the attainment of this final state at constant U and V has two options. Recall that, by definition on an isentropic hypersurface, there is an exact compensation of entropy production and entropy driven out, as described in Equation (5). Therefore, for initiation of an irreversible process, the system has to leave an isentropic hypersurface. This happens only when the balance between the rate of entropy production and the rate of entropy driven out breaks down. Therefore, if the rate of entropy driven out dominates over the rate of entropy production, then the system follows the trajectory of decreasing entropy, whereas when the rate of entropy production dominates over the rate of entropy driven out, then during corresponding evolution of the system, its entropy increases. At this stage of our discussion, we introduce a parameter λ, through which we quantify the said imbalance between the rate of entropy production and the rate of entropy driven out. This parameter is different than the extent of the advancement of reaction ξ. This is so because the said imbalance also contains contributions to the rate of entropy production and the rate of entropy driven out by the non-steady state fluxes in and out of reactants and product molecules. When entropy is decreasing, the operative and simple expression very close to the nonequilibrium steady state is $S = S^{ss}\exp(-k(\lambda^{ss} - \lambda))$, whereas in the case of increasing entropy, we have the simple expressions $S = S^{ss}\exp(k(\lambda^{ss} - \lambda))$. However, when the system is attaining an equilibrium state, implying that the rate of entropy exchange remains zero, we have the expression $S = S^{e}\exp(k(\xi^{e} - \xi))$. In the above three expressions, k is the kinetic constant, and in first two expressions, the superscript ss denotes a nonequilibrium steady state. Moreover, it is easy to comprehend that as the nonequilibrium steady state is approached, the existence time of the time-dependent nonequilibrium states goes on increasing and becomes sufficiently long in the close vicinity of it. The corresponding trajectories near a steady state appear as depicted in Figures 4 and 5. When Figure 4 is used for the evolution to an equilibrium state, the parameters $\xi(t)$ and ξ^{e} need to be used instead of $\lambda(t)$ and λ^{ss}.

Thus, we observe from Figures 4 and 5 that at λ^{ss} we have

$$\left(\frac{\partial S}{\partial \lambda}\right)_{U,V,\lambda^{ss}} = 0, \tag{9}$$

Which is the condition of extremization of entropy at the nonequilibrium steady state. When the final state is an equilibrium state, the variation of entropy will be as depicted in Figure 4, but then it requires the use of $\xi(t)$ and ξ^{e} instead of $\lambda(t)$ and λ^{ss} in this figure, re-expressing Equation (9) as $(\partial S/\partial\xi)_{U,V,\xi^e} = 0$. However, away from the position λ^{ss} on these curves, say at λ', we have

$$\left(\frac{\partial S}{\partial \lambda}\right)_{U,V,\lambda'} \neq 0, \tag{10}$$

And away from an equilibrium state at a position ξ' on the curve of Figure 4, this condition reads as $(\partial S/\partial\xi)_{U,V,\xi'} \neq 0$. It illustrates that there is no possibility at all to have $(\partial S/\partial\xi)_{U,V,\lambda'} = 0$ and $(\partial S/\partial\xi)_{U,V,\xi'} = 0$ for a time-dependent nonequilibrium state even for the situations which are in very close proximity to a steady state. Hence, at the phenomenological level, we can neither claim them as being of maximum entropy nor of minimum entropy with regard to their neighboring states on the trajectory.

Notice that the partial derivatives of the thermodynamic entropy appearing in Equations (9) and (10) belong to the phenomenological premises. We will see in Section 4 that the same equations hold true for Boltzmann and Gibbs entropies in the phenomenological premises.

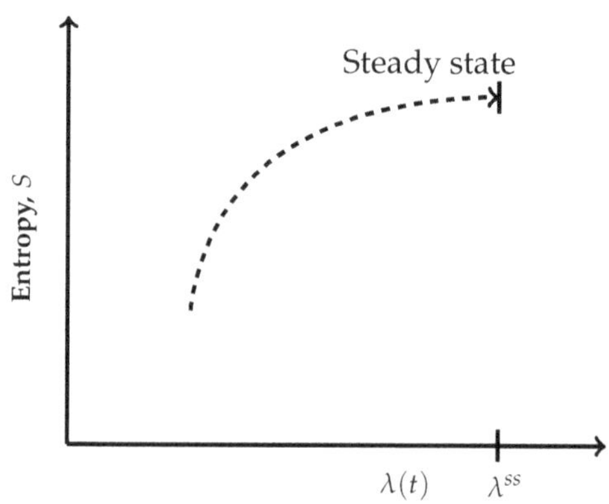

Figure 4. Schematic depiction of the variation of entropy in a CSTR at constant U and V with non-zero rate of entropy driven out, which is dominated by the rate of entropy production near nonequilibrium steady state. Through the parameter λ, we quantify the imbalance between the rate of entropy production and the rate of entropy driven out, whose value is λ^{ss} at the nonequilibrium steady state. A simple expression in the vicinity of the nonequilibrium steady state is $S = S^{ss}\exp(k(\lambda^{ss} - \lambda))$, whereas when the final state is in equilibrium because the rate of entropy exchange remains zero, we have to use the extent of advancement of reaction ξ as the parameter instead of λ. Therefore, in the vicinity of the equilibrium state, we have the expression $S = S^e\exp(k(\xi^e - \xi))$ (hence, during the course of approaching an equilibrium state, the horizontal axis should be interpreted in terms of $\xi(t)$ and ξ^e). In these two expressions, k is the kinetic constant. This trajectory obviously is of increasing entropy, and the final state is a nonequilibrium steady state. However, if we impose the condition of a zero rate of entropy exchange, then the CSTR behaves as if evolving under isolation/adiabatic condition; hence, the final state will be an equilibrium state of highest entropy. Therefore, the same curve represents this situation too. The trajectory has been depicted as a dashed curve to emphasize that it passes through a succession of time-dependent nonequilibrium states.

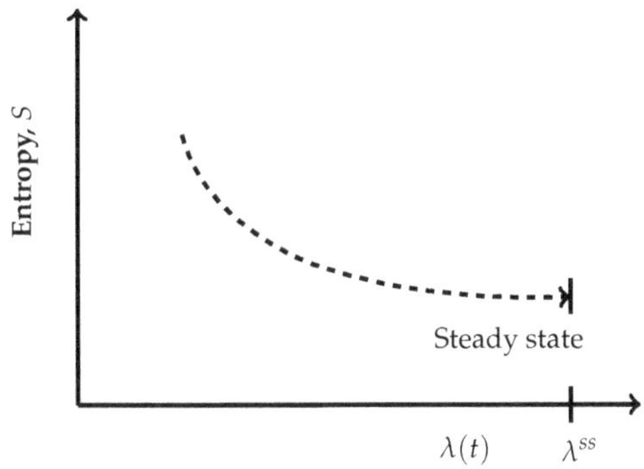

Figure 5. Schematic depiction of the variation in entropy in a CSTR at constant U and V with a non-zero rate of entropy driven out that dominates the rate of entropy production near the nonequilibrium steady state. Through the parameter λ, we quantify the imbalance between the rate of entropy production and the rate of entropy driven out, whose value is λ^{ss} at the nonequilibrium steady state. A simple expression in the vicinity of the nonequilibrium steady state is $S = S^{ss}\exp(-k(\lambda^{ss} - \lambda))$, where k is the kinetic constant. This trajectory obviously is of decreasing entropy, and the final state is a nonequilibrium steady state. The trajectory has been depicted as a dashed curve to emphasize that it passes through a succession of time-dependent nonequilibrium states.

4. Statistical Mechanical Versus Phenomenological Aspects and the Question of Equivalence of Entropies

We recall the following standard expressions: (1) the Boltzmann entropy, S:

$$S = k_B \ln W \tag{11}$$

where k_B is the Boltzmann constant, and W is the thermodynamic probability that measures the number of microstates commensurate with the given macrostate of the system; and (2) the Gibbs entropy, S, which is based on the ensemble averaging, which reads as follows:

$$S = k_B \ln \Omega \tag{12}$$

where Ω is the ensemble based number of microstates. Conventionally, in the case of equilibrium states, the mathematics of chance is employed, which establishes $W \approx W_{max}$ and $\Omega \approx \Omega_{max}$; hence, correspondingly, the Boltzmann and Gibbs entropies calculated using Equations (11) and (12) are the maximized values. It means that while computing Boltzmann and Gibbs entropies, all the microstates are counted. In the case of the ensemble method, it is termed as an ensemble averaged value. However, to grant it a physical status, the concept of the time averaging is used, which means that in sufficiently large time intervals, the system detours all the members of Ω_{max}. The experimentally measured quantities are the results of this detouring. Hence, the time-averaged and ensemble-averaged values are equated, which in statistical mechanics is described as ergodicity [2,3]. With this background, we further assume that the maximum entropy estimates, proposed by Jaynes [7,8] for information-theoretic entropy, can be extended for computing the Boltzmann and Gibbs entropies of a system when it is in (i) a nonequilibrium steady state and (ii) a time-dependent nonequilibrium state, which is very close to a time-independent state (We have not used different symbols for Boltzmann, Gibbs and thermodynamic entropies in the present discussion except for the alternative Gibbs entropy (cf. Equation (14))).

Jaynes has already demonstrated the equivalence of Gibbs and thermodynamic entropies [7] for all systems in equilibrium. The reason is that both take into account all contributions to internal energy, including the existing non-ideality. When a system is in equilibrium and contains monatomic ideal gases, we have equivalence amongst Boltzmann, Gibbs, and thermodynamic entropies. Also, notice that the maximization used in Equations (11) and (12) is with regard to the neighboring Ws and Ωs of lower magnitude. These lower value Ωs (Ws) belong to the same set of say $(U, V, \{\xi_r\})$ and hence cannot be thought as the neighboring points on the considered irreversible trajectory. The use of maximized Ω is physically permitted when an equilibrium state is of the time-independent type described, say by $\frac{dS}{dt} = 0$, and therefore the system has ample time at its disposal to detour all of them. This condition shown in Figure 4 is followed at the maximum irrespective of whether the system is ideal or not.

In the sequel, we recall that the nonequilibrium steady states are also time-independent states described by $\frac{dS}{dt} = 0$ and correspond to the extremum of Figures 4 and 5. In this case too, the system has at its disposal ample time to detour all the microstates counted by Ω_{max} in non-ideal systems and by $W_{max} = \Omega_{max}$ in an ideal system. This means that the ergodicity is obeyed in this case too. Hence, in the case of nonequilibrium steady states too, we have equivalence amongst these three entropies when the system consists of ideal gases and between Gibbs and thermodynamic entropies when the system consists of non-ideal components. Thus, we see that Equation (9) also holds for Gibbs and Boltzmann entropies when the system is at a nonequilibrium steady state.

Next we consider the time-dependent nonequilibrium states. For example, refer to the states lying on the trajectories away from the steady points, λ^e (ξ^e), as shown in Figures 4 and 5. They all, that is the points $\lambda^{'}$ ($\xi^{'}$), follow the expression of Equation (10) for thermodynamic entropy because they are not steady states described by $\frac{dS}{dt} \neq 0$. This raises the question of whether an equivalence amongst these three entropies can be assumed. The answer lies in ascertaining whether the system can detour all the microstates of Ω_{max} and, if applicable, those counted by W_{max} before the next time-independent state is attained. In principle, the answer will be no, but as we get closer to the final time-independent state, we see a decrease in the magnitude of $\frac{dS}{dt}$, and it will be exceedingly small in the close vicinity of the former. In such cases, the system will have sufficient time at its disposal to detour almost all microstates counted by Ω_{max} (W_{max}); that is, the ergodicity remains valid. Thus, for all such time-dependent nonequilibrium states, the said equivalence will come within the physically acceptable domain, of course within the limits of experimental errors or accuracy, even though the expression of Equation (9) is not followed. But a validity of the expression of Equation (10) for Boltzmann and Gibbs entropies does exist. In practice, such situations are numerous, and all of them are not very close to a steady state. However, calculations need to be carried out of transport properties using Gibbs ensemble method to produce suitable examples. However, an encouraging fact is that, in the kinetic theory of gases, these transport properties in a dynamic system have been computed, and the results satisfactorily match with those measured experimentally [30–33].

On the other hand, the time-dependent nonequilibrium states with a significantly large magnitude of $\frac{dS}{dt}$ have short existence times. Since their existence time does not allow the system to detour all the members of the computed Ω_{max}, implying no feasibility of obeying ergodicity, the calculated Gibbs entropy from the physicality point of view cannot be assigned to such short-lived states even though their instantaneous macroscopic conditions are used in the computation. This perspective has been previously emphasized by one of the present authors [34]. Hence, in such situations, not only the said equivalence of entropies but also the validity of Jaynes maximum entropy estimates are of doubtful practical utility.

Of course, there are countless examples of systems in short-lived states. It means that such systems, when in short-lived states, do detour a certain number of microstates, which will be less than the value of Ω_{max} calculated using mathematics of chance for the same thermodynamic conditions. The thermodynamic conditions for a given nonequilibrium state are based on the independent thermodynamic variables appropriate for it. In extended irreversible thermodynamics [12,23,35–44], the physical fluxes are raised to the status of thermodynamic variables. Thus, the entropy dependence therein, for example, reads as follows:

$$s = s(u,\ v,\ \{x_k\},\ \mathbf{q},\ \mathbf{\Pi},\ \{\mathbf{J}_k\}) \tag{13}$$

where s is the per unit mass local entropy, u is the per unit mass local internal energy, v is the local specific volume, $\{x_k\}$ are the local mass fractions of the components, $\mathbf{q}$ is the local heat flux, $\mathbf{\Pi}$ is the local dissipative momentum flux tensor, and $\{\mathbf{J}_k\}$ are the local diffusion fluxes. All these quantities, in general, are position- and time-dependent. Therefore, the ensemble is constructed say for a particular instantaneous set $(u,\ v,\ \{x_k\},\ \mathbf{q},\ \mathbf{\Pi},\ \{\mathbf{J}_k\})$, and then Ω_{max} is computed using the mathematics of chance. Now, suppose the ensemble is constructed for one of the very short-lived nonequilibrium states. In view of the transient nature of the state under consideration, the system will not have adequate time at its disposal to traverse all the members of Ω_{max}—a case of non-compliance of ergodicity. Hence, the calculated maximized value of Gibbs entropy will be higher than the actual thermodynamic entropy. Therefore, instead of maximization, we need to devise a mathematical model to arrive at the appropriate and physically relevant value of Ω. Indeed, one may use some other nonequilibrium thermodynamic framework, say rational thermodynamics, finite time thermodynamics, Keizer's version, thermodynamics with internal variables, or any other; the above assertion does not change because the basic problem being faced herein does not originate in the use of a thermodynamic framework but is due to the very short existence time of a nonequilibrium state; hence, it is the case of the nonfeasibility of traversing all the members of Ω_{max} by the system.

Therefore, in general, the maximized Gibbs entropy given by Equation (12) for short-lived nonequilibrium states will be higher than the actual thermodynamic entropy. This demonstrates why Jaynes' maximum entropy estimate remains inapplicable to such short-lived nonequilibrium states.

Since our discussion involves the traditional definition of Gibbs entropy, as seen in Equation (12), it is pertinent also to consider an alternative Gibbs entropy, defined as follows:

$$S_G(U) = k_B \ln\left(\sum_{U'<U} \Omega(U')\right) \tag{14}$$

This definition can be used also for small systems consisting of tens of molecules. In a recent discussion [45], it is illustrated therein that in the case of equilibrium and for astronomically large values of Ω, S_G coincides with the definition of Equation (12). However, as Ω decreases (corresponding to only tens of molecules), the non-equivalence between the two entropies surfaces out. In this limit, the physical meaning of say hot and cold crumbles down. However, in the case of time-independent states and the time-dependent nonequilibrium states, an additional condition for the said equivalence and nonequivalence will operate in terms of the magnitude of the time rate, $\frac{dS_G}{dt}$.

5. A Brief Note on Natural Fluctuations

The phenomena of fluctuations is common in all experimental measurements carried out on a system in a static condition or evolving with time. In equilibrium statistical thermodynamics, it is quantified in terms of standard deviation. When the Gibbs ensemble is used, the averaging is carried out over all the members of Ω_{max} (refer to any standard statistical mechanics and statistical thermodynamic textbooks, such as those cited in this paper).

Let us examine a system in equilibrium. In view of the above discussion, we envisage that the observed fluctuations will have two types of contributions if the system is chemically reactive. In the traditional description of fluctuations, they are the ones taking place at constant composition at equilibrium. However, when the system is chemically reactive, the extent of advancement of various chemical reactions will also fluctuate. However, this macroscopic fluctuation originates from the corresponding fluctuations in the population of quantum states of reacting partners. In other words, now the origin of fluctuations have contributions from two types of molecular collisions; one is non-chemically reactive type, and the other chemically reactive type. For its illustration, let us consider a canonical ensemble for chemically reactive components. In the present case too, the traditionally computed standard deviation, σ_U, in internal energy, U, will also be a complete description having the following expression [1]:

$$\sigma_U = \sqrt{\overline{(U - \overline{U})^2}} \tag{15}$$

Notice that the two averages involved in Equation (15) are computed by considering population in all the quantum states but the effective contribution to it is from the set of population effectively contributing to Ω_{max}. However, a basic challenge is to devise an appropriate expressions of Ω for chemically reacting systems.

A similar argument also applies to the fluctuations about a nonequilibrium steady state because it is a time-independent state.

However, when we consider fluctuations of a time-dependent nonequilibrium state, similar arguments hold in the case of extremely small magnitude of $\frac{dS}{dt}$ because, in this case, ensemble averaging produces physically realistic results. Only in those cases where the magnitude of $\frac{dS}{dt}$ is significantly large and more, no physically meaningful statistical averaging is feasible, will the statistical computation of the standard deviation in fluctuations be of doubtful physical utility. This result is due to the non-compliance of ergodicity.

6. Concluding Remarks

In the present discussion, we have recalled the demonstration by Jaynes that (1), in general, there exists an equivalence between thermodynamic and Gibbs entropies, and (2) the equivalence between Boltzmann and Gibbs entropies exists only when the systems are composed of monatomic ideal gases. In Section 3, we have clearly illustrated by recalling the thermodynamic descriptions of extremization and maximization of thermodynamic entropy in terms of appropriate derivatives of entropy. For this purpose, we have used a specific and simple CSTR (described in Section 2) and generated corresponding isentropic hypersurfaces for nonequilibrium steady states and isentropic planes for equilibrium states. Thereafter, on combining phenomenological (of Section 3) and microscopic (of Section 4) descriptions, the following conclusions are reached:

1. In determining the equivalence between thermodynamic and statistical mechanically defined entropies, a crucial role is played by the ability of the system to detour all the members of the microstates of the system that is an ability to obey ergodicity.

2. The requirement stated above is meticulously met only when a system is in an equilibrium state but also in nonequilibrium steady state. This is so because they both are time-independent states. Thus, even though a nonequilibrium steady state belongs to the nonequilibrium regime, the equivalence between Gibbs and thermodynamic entropies is guaranteed. Of course, for the equivalence of Boltzmann entropy the absence of non-ideality is demanded.
3. In the case of time-dependent nonequilibrium states with an exceedingly small magnitude of $\frac{dS}{dt}$, the requirement stated in 1 above is practically met; hence, for them, the equilvalence between thermodynamic and Gibbs entropies exists.
4. For all time-dependent nonequilibrium states belonging to considerably large magnitudes of $\frac{dS}{dt}$, because the condition of ergodicity is not met though the Boltzmann and Gibbs entropies, they can be calculated, but their practical utility remains doubtful, and hence their equivalence with the thermodynamic entropy is not possible. However, in practice, we measure thermodynamic parameters to a fair degree of confidence in such systems. It means that the system executes a time averaging within the existence time of the nonequilibrium state. During this period, the members of the Ω or W which get sampled out cannot be identified beforehand. Hence, how to perform the commensurate ensemble averaging is not clear.
5. Thus, except for equilibrium states, nonequilibrium steady states, and the time-dependent non-equilibriun states with exceedingly small magnitudes of $\frac{dS}{dt}$, the Jaynes maximum entropy estimates should not be executed on Boltzmann and Gibbs entropies because of their uncertain practical utilities. And even if it is executed, its practical utility will need to be weighed properly. A similar situation might exist in the other scientific fields wherein the Jaynes maximum entropy estimate is employed; hence, caution needs to be exercised.

Indeed, it will be of interest to carry out corresponding computational investigations to clearly demonstrate what the physicality of the significantly large magnitude of $\frac{dS}{dt}$ is, which will demarcate the region of nonequivalence between statistical and thermodynamic entropies. It is also worth exploring whether the alternative Gibbs entropy of Equation (14) can be coincided with the thermodynamic entropy for very short lived nonequilibrium states. But that is a project in itself, and we plan to direct our future efforts along these lines.

Author Contributions: Both the authors (A.A.B. and V.M.T.) equally contributed in this article. All authors have read and agreed to the published version of the manuscript.

Funding: This research received no external funding.

Institutional Review Board Statement: Not applicable.

Informed Consent Statement: Not applicable.

Data Availability Statement: No data is involved in this presentation.

Acknowledgments: We gratefully acknowledge the invitation to contribute to this special issue. We are also thankful to reviewers for helping us in improving the presentation of the subject matter.

Conflicts of Interest: The authors declare no conflicts of interest.

Abbreviations

The following abbreviation is used in this manuscript:

CSTR Continuously stirred tank reactor

References

1. Hill, T.L. *An Introduction to Statistical Thermodynamics*, 1st ed.; Addison-Wesley Publishing Company, INC.: Reading, MA, USA, 1960.
2. Fowler, R.; Guggenheim, E.A. *Statistical Thermodynamics. A Version of Statistical Mechanics for Students of Physics and Chemistry*; Cambridge University Press: Cambridge, UK, 1956.
3. Tolman, R.C. *The Principles of Statistical Mechanics*, revised ed.; Dover Publications: New York, NY, USA, 2010; ISBN-13978-0486638966.
4. McQuarrie, D.A. *Statistical Mechanics*, 1st ed.; Harper and Row: New York, NY, USA, 1976; ISBN 978-1-891389-15-3.
5. Jaynes, E.T. Gibbs vs. Boltzmann Entropies. *Am. J. Phys.* **1965**, *33*, 391–398. [CrossRef]
6. De Groot, S.R.; Mazur, P. *Non-Equilibrium Thermodynamics*; North Holland: Amsterdam, The Netherlands, 1962.
7. Jaynes, E.T. Information Theory and Statistical Mechanics. *Phys. Rev.* **1957**, *106*, 620–630. [CrossRef]
8. Jaynes, E.T. Information Theory and Statistical Mechanics. II. *Phys. Rev.* **1957**, *108*, 171–190.
9. Prigogine, I. *Introduction to Thermodynamics of Irreversible Processes*; John Wiley-Interscience: New York, NY, USA, 1967.
10. Haase, R. *Thermodynamics of Irreversible Processes*; Addison-Wesley: Reading, MA, USA, 1969.
11. Rastogi, R.P. *Introduction to Non-Equilibrium Physical Chemistry. Towards Complexity and Non-Linear Sciences*; Elsevier: Amsterdam, The Netherlands, 2008.
12. Lebon, G.; Jou, D.; Casas-Vázquez, J. *Understanding Non-Equilibrium Thermodynamics. Foundations, Applications, Frontiers*; Springer: Berlin/Heidelberg, Germany, 2008.
13. Andersen, B. Finite-Time Thermodynamics. Ph.D. Thesis, Physics Laboratory II, University of Copenhagen, Copenhagen, Denmark, 1983.
14. Andresen, B. Tools of Finite-Time Thermodynamics. 2003. Available online: https://www.semanticscholar.org/paper/TOOLS-OF-FINITE-TIME-THERMODYNAMICS-Andresen/c8cf8c2ac0f16ac35e8689e1d4682d40eccc9831 (accessed on 2 February 2007).
15. Andresen, B. Aktuelle Trends in der Thermodynamik in Endlicher Zeit. *Angew. Chem.* **2011**, *123*, 2742–2757. [CrossRef]
16. Salamon, P.; Andresen, B.; Berry, R.S. Thermodynamics in Finite Time. II. Potentials for Finite-Time Processes. *Phys. Rev. A* **1977**, *15*, 2094–2102. [CrossRef]
17. Glasstone, S. *Thermodynamics for Chemists*; Princeton: New Jersey, NJ, USA, 1967.
18. Blinder, S.M. *Advanced Physical Chemistry*; Collier-Macmillan: New York, NY, USA, 1969.
19. Atkins, P.; de Paula, J. *Atkins' Physical Chemistry*, 8th ed.; W. H. Freeman and Company: New York, NY, USA, 2006.
20. Prigogine, I.; Defay, R. *Chemical Thermodynamics*; Everett, D.H., Translator; Longmans Green: London, UK, 1954.
21. Bhalekar, A.A. On the Generalized Phenomenological Irreversible Thermodynamic Theory (GPITT). *J. Math. Chem.* **1990**, *5*, 187–196. [CrossRef]
22. Bhalekar, A.A. Measure of Dissipation in the Framework of Generalized Phenomenological Irreversible Thermodynamic Theory (GPITT). In *Proceedings of the 1992 International Symposium on Efficiency, Costs, Optimization and Simulation of Energy Systems (ECOS'92), Zaragoza, Spain, 15–18 June 1992*; Valero, A., Tsatsaronis, G., Eds.; American Society of Mechanical Engineers: New York, NY, USA, 1992; pp. 121–128.
23. García-Colín, L.S.; Bhalekar, A.A. Recent Trends in Irreversible Thermodynamics. *Proc. Pakistan Acad. Sci.* **1997**, *34*, 35–58.
24. Bhalekar, A.A. Universal Inaccessibility Principle. *Pramana—J. Phys.* **1998**, *50*, 281–294. [CrossRef]
25. Blinder, S. Carathéodory's Formulation of the Second Law. In *Physical Chemistry: An Advanced Treatise*, 1st ed.; Jost, W., Ed.; Academic Press: New York, NY, USA, 1971; Chapter 10, pp. 613–637. [CrossRef]
26. Blinder, S.M. Mathematical methods in elementary thermodynamics. *J. Chem. Educ.* **1996**, *43*, 85–92. [CrossRef]
27. Keizer, J. Fluctuations, Stability, and Generalized State Functions at Nonequilibrium Steady States. *J. Chem. Phys.* **1976**, *65*, 4431–4444. [CrossRef]
28. Keizer, J. Thermodynamics at Nonequilibrium Steady States. *J. Chem. Phys.* **1978**, *69*, 2609–2620. [CrossRef]
29. Keizer, J. *Statistical Thermodynamics of Nonequilibrium Processes*; Springer: Berlin/Heidelberg, Germany, 1987. [CrossRef]
30. Chapman, S.; Cowling, T.G. *Mathematical Theory of Non-Uniform Gases: An Account of the Kinetic Theory of Viscosity, Thermal Conduction and Diffusion in Gases*, 3rd ed.; Cambridge University Press: Cambridge, UK, 1970.
31. Grad, H. On the Kinetic Theory of Rarefied Gases. *Commun. Pure Appl. Math.* **1949**, *2*, 331–407. [CrossRef]
32. Grad, H. Principles of the Kinetic Theory of Gases. In *Thermodynamik der Gase/Thermodynamics of Gases*; Flügge, S., Ed.; Springer: Berlin/Heidelberg, Germany, 1958; Handbuch der Physik/Encyclopedia of Physics; Volume XII, pp. 205–294.
33. Liboff, R.L. *Kinetic Theory: Classical, Quantum, and Relativistic Descriptions*, 3rd ed.; Graduate Texts in Contemporary Physics; Springer: Berlin/Heidelberg, Germany, 2003.
34. Bhalekar, A.A. The Universe of Operations of Thermodynamics vis-à-vis Boltzmann Integro-Differential Equation. *Indian J. Phys.* **2003**, *77*, 391–397.

35. García-Colín, L.S. Extended Nonequilibrium Thermodynamics, Scope and Limitations. *Rev. Mex. Fís.* **1988**, *34*, 344–366.
36. García-Colín, L.S.; Uribe, F.J. Extended Irreversible Thermodynamics Beyond the Linear Regime. A Critical Overview. *J. Non-Equilib. Thermodyn.* **1991**, *16*, 89–128.
37. García-Colín, L.S. Extended Irreversible Thermodynamics: Some Unsolved Questions. In Proceedings of the AIP Conference Proceedings, CAM-94 Physics Meeting, Cancun, Mexíco, 26–30 September 1994; Zepeda, A., Ed.; Centro de Investigacion y de Estudios Avanzados del IPN: Woodbury, NY, USA, 1995; Volume 342, pp. 709–714.
38. García-Colín, L.S. Extended Irreversible Thermodynamics. An Unfinished Task. *Mol. Phys.* **1995**, *86*, 697–706. [CrossRef]
39. Jou, D.; Casas-Vázquez, J.; Lebon, G. *Extended Irreversible Thermodynamics*, 2nd ed.; Springer: Berlin/Heidelberg, Germany, 1996.
40. Nettleton, R.E.; Sobolev, S.L. Applications of Extended Thermodynamics to Chemical, Rheological, and Transport Processes: A Special Survey. Part I. Approaches and Scalar Rate Processes. *J. Non-Equilib. Thermodyn.* **1995**, *20*, 200–229.
41. Nettleton, R.E.; Sobolev, S.L. Applications of Extended Thermodynamics to Chemical, Rheological, and Transport Processes: A Special Survey. Part II. Vector Transport Processes, Shear Relaxation, and Rheology. *J. Non-Equilib. Thermodyn.* **1995**, *20*, 297–331.
42. Nettleton, R.E.; Sobolev, S.L. Applications of Extended Thermodynamics to Chemical, Rheological, and Transport Processes: A Special Survey. Part III. Wave Phenomena. *J. Non-Equilib. Thermodyn.* **1996**, *21*, 1–16.
43. Müller, I.; Weiss, W. Irreversible Thermodynamics—Past and Present. *Eur. Phys. J. H* **2012**, *37*, 139–236.
44. Lebon, G.; Jou, D. Early History of Extended Irreversible Thermodynamics (1953–1983): An Exploration Beyond Local Equilibrium and Classical Transport Theory. *Eur. Phys. J. H* **2015**, *40*, 205–240. [CrossRef]
45. Vilar, J.M.G.; Rubi, J.M. Communication: System-size scaling of Boltzmann and alternate Gibbs entropies. *J. Chem. Phys.* **2014**, *140*, 201101. [CrossRef] [PubMed]

entropy

MDPI

Article

Thermodynamic Theory of Macrosystems: Entropy Production as a Metric

Sergey Amelkin

Institute of System and Software Engineering and Information Technology, National Research University of Electronic Technology, 124498 Moscow, Russia; amelkin@ist.education

Abstract: The article considers the description of a macrosystem in terms that do not depend on the nature of the macrosystem. The results obtained can be used to describe macrosystem models of thermodynamic processes, and to create interdisciplinary models that take into account interactions of various natures. The macrosystem model is based on its representation in the form of a self-similar oriented weighted graph where the equation of state is fulfilled for each node, which connects extensive variables. One of the extensive variables is entropy, the maximum of which corresponds to the state of equilibrium. For processes in which fluxes are linearly dependent on driving forces, Onsager's relations are shown to be true, which makes it possible to prove that in the space of stationary processes, entropy production in a closed macrosystem is a metric similar to the Mahalanobis metric, which determines the distance between processes. Zero in such a space indicates reversible processes, and thus the production of entropy shows the degree of irreversibility as the distance from a researched process to a reversible one.

Keywords: thermodynamic processes; macrosystem; entropy production; extensive variables; intensive variables; measure space

1. Introduction

Thermodynamics has historically been the first and main research focus of complex systems [1]. It is thermodynamic analogies that are used in the description of other systems such as economic, social, informational, and algorithmic systems when control is only possible in an averaged sense. We mean *averaged* because it is impossible to observe or control the state and behaviour of individual elementary particles, whose large aggregate forms a complex system.

In a thermodynamic system, the elementary particles are molecules, even though only statistical mechanics need discrete particles as a concept. The reason is that phenomenological (i.e., experimentally observable) patterns related to the state of a thermodynamic system are carried out through dynamic or statistical averaging of the interactions between individual molecules. However, such averaging is only possible under a priori assumptions about the nature of such interactions as they cannot be verified due to the incredibly large number of molecules in the system and their extremely small size, which makes them neither observable nor controllable [2].

The validity of such a framework is supported by the existence of phenomenological laws. However, in this case, the assumptions about the properties of individual molecular interactions must be derived from the observations of thermodynamic systems as they cannot serve as proof of the truth of the phenomenological laws per se.

The approach suggested in this work is the reverse one, which is based on a structural induction where the basic point is any scale level at which the properties of the system can be experimentally verified. This presupposes the fulfilment of the self-similarity condition, which is generally defined as follows: a set X is self-similar if there exists a finite set K indexing a collection of injective and non-surjective mappings $\{\phi_\kappa\}_{\kappa\in K}$, such that $X = \bigcup_{\kappa\in K} \phi_\kappa(X)$ [3]. The self-similarity condition is actively used in thermodynamics [4], even though it is strictly fulfilled only when the set of subsystems whose union forms the given thermodynamic system is a continuum. This assumption is too strong for real thermodynamic systems but can be considered a good approximation.

The approach considered below is not limited to describing thermodynamic systems alone, but is intentionally extended to a more general representation. This allows for the consideration of a class of macrosystem models that includes thermodynamic systems.

2. Definition of Macrosystems

When analyzing patterns that arise in the exchange processes in systems of various natures (e.g., physical, chemical, economic, informational, and social), it is often advisable to use a macrosystem approach.

Macrosystems are systems in which the following conditions are satisfied:

1. Macrosystems consist of a large number of elementary objects, and this number is so large that any macrosystem can be considered as a continuum and can be divided into any finite number of subsystems, including those sufficient to determine the statistical characteristics of any given accuracy; each of the subsystems can be considered as a macrosystem. Any macrosystem Y is a nonempty collection $\mathfrak{S}$ of pairwise disjoint subsets of Y closed under complement, countable unions, and countable intersections. $\mathfrak{S}$ is a σ-algebra and ordered pair $(Y, \mathfrak{S})$ is a measurable space. The macrosystem Y can be considered as a union of a finite number of a lower level macrosystems Y_i: $Y = \bigcup_{i\in\mathcal{Y}} Y_i$. This condition is called the self-similarity condition [5].
2. The macrosystem state is determined by the vector Q of state variables. It is assumed that the state variables satisfy the conservation law; therefore, we can consider vector Q as a vector of extensive variables. Because of non-negativity and countable additivity, vector Q is a vector measure and $(Y, \mathfrak{S}, Q)$ is a measure space. In thermodynamic macrosystems, the extensive variables are internal energy, mol number, and volume [6], which the macrosystem can exchange with its external environment [7]; we will also consider the external environment as a set of macrosystems (and a higher-level macrosystem is the union of the macrosystem and its external environment). In the process of interaction between the subsystems X and Y, the values of the vectors Q_X and Q_Y change over time. In this way, exchange processes and their corresponding fluxes are formed, which are understood as the rates of change of the extensive quantities. We will denote fluxes between subsystems X and Y as q_{XY}.
3. It is impossible to control each elementary object due to their extremely large number [7]. The macrosystem control can be organized only by impact on the parameters averaged over a set of elementary objects, namely
   - Changes in the parameters of the macrosystem's external environment, the interaction with which determines the change in the values of extensive variables in the macrosystem;
   - Change in the values of extensive variables (for example, their extraction) in the macrosystem due to external interventions;
   - Changes in the characteristics of the exchange infrastructure to accelerate or, conversely, slow down the exchange processes.

3. Representation of the Macrosystem as a Graph

The macrosystem can be represented as a self-similar oriented weighted graph, the nodes of which correspond to the subsystems and the macrosystem's external environment, and the edges correspond to the fluxes between the subsystems and between the subsystems and the macrosystem's external environment. Each graph node (each subsystem) is characterized by a vector $Q = (Q_0, \dots, Q_N)$, and fluxes between subsystems can be functionally unrelated to each other.

Self-similar—each graph node can be represented as a graph that describes the interaction of subsystems that form this node with each other and with their environment. Let us consider two interacting (the vector of the fluxes of extensive variables is q_{XY}) disjoint macrosystems X and Y. Due to the self-similarity condition, we can introduce sets $X = \bigcup_i X_i, Y = \bigcup_j Y_j$ of pairwise disjoint subsystems corresponding to these macrosystems. Each subsystem is characterized by its own values of extensive variables: Q_i for X_i and Q_j for Y_j. Flux q_{ij} is formed between subsystems from different sets X and Y. In this notation, the following equations are valid:

$$\left.\begin{array}{c} \sum\limits_{i\in\mathcal{X}} Q_i = Q_X; \sum\limits_{j\in\mathcal{Y}} Q_j = Q_Y; \\ \sum\limits_{\substack{i\in X \\ j\in Y}} q_{ij} = q_{XY}. \end{array}\right\} \tag{1}$$

Oriented—the direction of each edge determines the positive sign of each flux; thus, if the edges are directed from node X to node Y, then

$$q_{XY} = -\nabla Q_X = \nabla Q_Y. \tag{2}$$

Weighted—the weight of the edge shows the flux intensity: the positive value of the ν-th flux $q_{XY,\nu} > 0$ if the real flux of the extensive variable is directed towards the edge, and $q_{XY,\nu} < 0$ if it is opposite to the edge's direction. In addition, a matrix A of infrastructure coefficients is determined for all fluxes between nodes X and Y (that is, for all fluxes q_{XY}). This matrix is the metadata for the edges connecting nodes X and Y, and determines both the facility of exchange process and complementary or substitutionary features of the extensive variables [8].

Figure 1 shows an example of the self-similar oriented weighted graph of a part of a macrosystem.

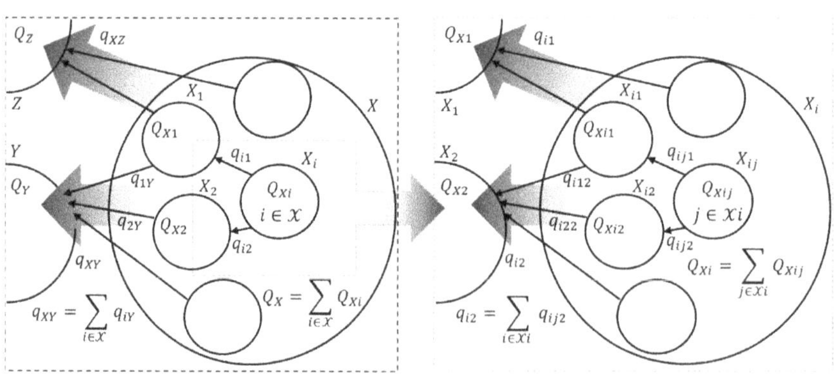

Figure 1. An example of the self-similar oriented weighted graph corresponding to a macrosystem. The nodes (the circles) correspond to the subsystems, and the edges (the arrows) correspond to the fluxes between the subsystems.

4. Equilibrium State of the Macrosystem

Let us assume that two macrosystems X and Y exchange the vector of the extensive variables Q. At each moment of time t, the reserves of the extensive variables describing the state of macrosystems are equal to $Q_X(t),\ Q_Y(t)$. We represent macrosystems as a union of a finite set of subsystems: X and Y, respectively. An equilibrium state shall be a state $(Q_X,\ Q_Y)$ in which the sum of fluxes at each moment of time t for each $\nu = 0, \ldots, N$

$$q_{XY,\nu}(t) = \sum_{\substack{i\in X\\ j\in Y}} q_{ij,\nu}(t) = 0. \quad (3)$$

Thus, the equilibrium in the macrosystem is considered dynamic. The equilibrium condition can be defined as follows: at the any level of sets of subsystems $X, Y : X = \bigcup_{i\in\mathcal{X}} X_i; Y = \bigcup_{j\in\mathcal{Y}} Y_j$, a vector of fluxes $q_{ij}(i \in \mathcal{X}, j \in \mathcal{Y})$ is described by a time-independent distribution $f\left(\tilde{q}\right)$ with expectation $E\left[\tilde{q}\right] = 0$.

Note that the fluxes $q_{ij}(t),\ (i \in \mathcal{X}, j \in \mathcal{Y})$, and the random variable $\tilde{q}$, which describes the subsystem fluxes distribution, are vectors. At this level of subsystems, we can assume that a large number of factors affects the distribution of fluxes, which means that the distribution $f_q\left(\tilde{q}\right)$ can be described by a multivariate normal distribution. The parameters of this distribution are the expectation $E\left[\tilde{q}\right]$, which determines fluxes at the level of subsystems X and Y in accordance with (3), and the covariance matrix $\mathrm{Cov}\left[\tilde{q}\right]$.

Assume that fluxes q_{XY} arise due to the activity of some driving forces, which we can also consider at the subsystem level as a random vector, such that

- Fluxes q_{XY} are linearly dependent on the exchange driving forces φ_{XY} [6]: $q_{XY} = A\varphi_{XY}$—under this assumption and suggesting that the matrix of infrastructure coefficients A is constant, the driving forces distribution is also normal;
- The covariance matrix $\mathrm{Cov}\left[\tilde{q}\right]$ of subsystems of the $X \cup Y$ macrosystem depends on the driving forces intensity $\varphi_{XY} = E\left[\tilde{\varphi}\right]$ causing these fluxes, so that the matrices $\mathrm{Cov}\left[\tilde{q}\right]$ and $\mathrm{Cov}\left[\tilde{\varphi}\right]$ are jointly normalizable (their eigenvectors coincide)—due to the linear relationship between fluxes and driving forces;
- Limits of correlation coefficients corresponding to the covariance matrix $\mathrm{Cov}\left[\tilde{q}\right]$ for any $\nu, \kappa = 0, \ldots, N$: $\lim\limits_{\varphi_{XY}\to 0} \rho_{\nu\kappa} = 0,\ \lim\limits_{\varphi_{XY}\to\infty} \rho_{\nu\kappa} = 1$—due to the redistribution of the extensive variables over a variety of subsystems, depending on the number of intermediate nodes in the graph chain to the contact point.

If the equilibrium condition is satisfied for a macrosystem X when interacting with each macrosystem from its environment, then such a macrosystem is called closed. If, for a macrosystem X, all its subsystems are in equilibrium when interacting with each other (but not necessarily with the environment of the macrosystem X)

$$\forall t, \forall i \in \mathcal{X} : \sum_{j\in\mathcal{X}} q_{ij}(t) = 0, \quad (4)$$

then we can say that this macrosystem is in a state of internal equilibrium. For a macrosystem in internally equilibrium, all fluxes can be observed only at the boundaries of the macrosystem and its environment.

5. Extensive and Intensive Variables

Let us assume that two macrosystems X and Y exchange the vector of the extensive variables Q. At each moment of time t, the values of the extensive variables describe the state of macrosystems.

The macrosystem shall be described by a set of extensive and intensive variables:

- Extensive variables are such that for any two disjoint macrosystems X and Y (not necessarily in equilibrium):

$$Q_{X \cup Y} = Q_X + Q_Y; \tag{5}$$

Due to conservation law, all the values of the extensive variables are extensive variables;
- Intensive variables v are such that for any two systems X and Y that are in equilibrium:

$$v_{X \cup Y} = v_X = v_Y. \tag{6}$$

Extensive variables satisfy the neutral scale effect condition: if all extensive variables in all subsystems of the macrosystem are increased by n times, then the flux intensities in the macrosystem will not change. In particular, such a proportional increase in the extensive variables will not bring the macrosystem out of the internal equilibrium state if before that, the system was in internal equilibrium. The neutral scale effect condition provides the self-similarity of the macrosystem.

As a consequence of (6), intensive variable dependencies on the extensive variable values that determine the macrosystem state should be homogeneous functions of the zero-order.

6. Entropy of the Macrosystem

The set of extensive variables describes the macrosystem state. Among the extensive variables, we single out the value $S = Q_0$, which characterizes the objective function of the system. S and other extensive variables are functionally related: $S = \mathcal{S}(Q)$, $Q = (Q_1, \ldots, Q_N)$. We call this equation the system state equation.

The choice of S as the objective function is determined by the Levitin–Popkov Axioms [9], which impose the following conditions on this variable:

1. For a controlled system, $S = \mathcal{S}(Q)$ given a fixed deterministic control v. Its stochastic state, which is characterized by a vector flux, is transformed into a deterministic vector $Q(v)$, called the steady or stationary state, which belongs to a permissible set $D(v)$.
2. For any fixed vector $v \in D(v)$, there exists a vector $p(v)$ of a priori probabilities for the distribution of fluxes in the system $\mathcal{S}(Q)$, such that the stationary state $Q(v)$ of the macrosystem under that given fixed control v is the optimal solution to the entropy-based optimization problem: $Q(v) = z[v](p(v))$, where $z[v](p(v))$ is the entropy operator defined as

$$z[u](p) = \text{argmax}\{\mathcal{S}(p, Q) : v \in D(v)\}. \tag{7}$$

Thus, the pair $(p(v),\ Q(v))$ simultaneously provides both the required vector of prior probabilities and the corresponding stationary state vector.

3. There exists an inverse mapping $p = \zeta[v](Q)$ such that the desired pair $(p(v),\ Q(v))$ is the unique solution to the system of the relations:

$$\begin{cases} Q(v) = z[v](p(v)), \\ \quad p = \zeta[v](Q). \end{cases} \tag{8}$$

Within the framework of the considered model, these statements can be interpreted as follows:

1. For any fixed deterministic control $v = \text{const}$, there exists a stable steady state $Q^*(v)$.
2. The stable steady state $Q^*(v)$ is a state of an internal equilibrium corresponding to the maximum of $\mathcal{S}(Q)$.
3. The stable steady state $Q^*(v)$ is unique.

For a closed system, the condition $\mathcal{S}(Q) \to \max$ determines the spontaneous direction of exchange of extensive variables. However, a question arises about the homogeneity of the function $\mathcal{S}(Q)$: entropy is a homogeneous function of degree one only in the systems that are in a state of internal equilibrium. In cases where the distribution $p(v)$ is scale-invariant but does not correspond to internal equilibrium, a fractal structure of the macrosystem is observed, in which $\mathcal{S}(Q)$ is a homogeneous function with a degree of homogeneity less than one. Taking this remark into account, entropy can still be considered an extensive quantity. The question to use the degree of homogeneity $\mathcal{S}(Q)$ as a measure of equilibrium of the macrosystem should be given further consideration.

Since S is an extensive variable in the condition of internal equilibrium, then $\mathcal{S}(Q)$ is a homogeneous function of the first-order: when scaling the system by n times or combining n identical macrosystems in the equilibrium state:

$$nS = \mathcal{S}(nQ). \tag{9}$$

In accordance with the Euler relations for homogeneous functions

$$\mathcal{S}(Q) = Q\nabla\mathcal{S} = \sum_{\nu=1}^{N} Q_\nu \frac{\partial\mathcal{S}}{\partial Q_\nu}. \tag{10}$$

Let us denote $v_\nu = \partial\mathcal{S}/\partial Q_\nu$. Since $\mathcal{S}(Q)$ is a homogeneous function of the first-order, then $v_\nu(Q)$ $(\nu = 1,\ldots,N)$ are homogeneous of the zero-order, i.e., they are intensive variables: for any n value $v_\nu(nQ) = v_\nu(Q)$. This means that a proportional increase in the extensive variables will not bring the macrosystem out of the internal equilibrium state if before that, the system was in internal equilibrium.

Under the assumption that the function $\mathcal{S}(Q)$ is differentiable and its partial derivatives are continuous, in accordance with the necessary condition for the function to be differentiable, there exists a total differential

$$d\mathcal{S} = \sum_{\nu-1}^{N} \frac{\partial\mathcal{S}}{\partial Q_\nu} dQ_\nu = \sum_{\nu=1}^{N} v_\nu dQ_\nu. \tag{11}$$

Two consequences of Equation (11) can be formulated.

Consequence 1. By differentiating the Euler relation (10), we obtain

$$d\mathcal{S} = \sum_{\nu=1}^{N} (v_\nu dQ_\nu + Q_\nu dv_\nu). \tag{12}$$

By comparing (11) and (12), we see that the second term in (12) should be equal to zero:

$$\sum_{\nu=1}^{N} Q_\nu dv_\nu = 0. \tag{13}$$

Consequence 2. The exchange process between subsystems X and Y (with a positive direction of fluxes from X to Y) can be described using Equations (2) and (11) as follows:

$$\frac{dS}{dt} = \frac{dS_Y}{dt} + \frac{dS_X}{dt} = \sum_{\nu=1}^{N} (v_{Y\nu} - v_{X\nu}) q_{XY,\nu}. \tag{14}$$

The parameter S is an extensive variable and its dependence on other state variables $\mathcal{S}(Q)$ is an objective function for spontaneous processes in the macrosystem.

From the established relations, the following conclusion can be drawn:

Let Q be a vector measure on a measurable space $(Y, \mathfrak{S}, Q)$. Then, the entropy of a subsystem $X \in Y$ is defined as a function $S_X = \mathcal{S}(Q_X)$, such that

- $S_X \leq \mathcal{S}(Q^*)$, where $\mathcal{S}(Q^*)$ is the entropy of X under internal equilibrium of all subsystems X;
- $\mathcal{S}(Q_X) + \mathcal{S}(0) = \mathcal{S}(Q_X)$, where $\mathcal{S}(0)$ is the entropy of a subsystem with a zero-vector measure;
- For any two subsystems X_1, X_2 such that $Y = X_1 \cup X_2$, $Q_{X_1 \cap X_2} = 0$, it holds that $S_Y = S_{X_1} + S_{X_2}$.

These properties correspond to Shannon–Khinchin Axioms for entropy [10].

The entropy function meets the conditions of Shannon entropy if the number of subsystems is large enough. S statistically corresponds to the entropy of the distribution $p(v(Q))$ of state parameters of subsystems of the macrosystem. Let us explain this statement. Since there are no restrictions on the possible random vector values with a finite variance, the normal distribution corresponds to the maximum entropy value (this can be considered as a justification for the distribution law of subsystems parameters that form a macrosystem).

The entropy of the normal distribution consists of two terms: $S = S_0 + 0.5\ln \det R$, where $R = (\rho_{\nu\kappa},\ \nu, \kappa = 1, \ldots, N)$ is the correlation matrix. The first addend corresponds to the complete independence of the system elements and characterizes the structure of the macrosystem, and the second describes the relationships in the macrosystem. The correlation $\rho_{\nu\kappa}$ $(\nu, \kappa = 1, \ldots, N)$ between fluxes increases, which is observed with the growing magnitude of the exchange driving forces φ_{XY}. Figure 2 illustrates the kind of dependency of $\rho_{\nu\kappa}(\varphi_{XY})$ when components φ_ν and φ_κ of the vector φ_{XY} are changed. The second term of entropy expression, which is always negative at $R \neq E$, (E is the identity matrix), decreases. This explains the statement about S as an objective function that reaches its maximum when the macrosystem reaches an equilibrium state, and a state function that sets the direction of processes in a closed system, increasing the dispersion of low-level subsystem parameters while simplifying the macrosystem structure at a high level.

In the process of spontaneous, not forced exchange, the S value of a closed macrosystem cannot decrease since $\mathcal{S}(Q)$ is the objective function for the system behaviour. In order for the value $\sigma = dS/dt$, called the entropy production, to be positive during the any process of given non-zero intensity, it is sufficient to fulfil the condition

$$\operatorname{sign}(v_{Y\nu} - v_{X\nu}) = \operatorname{sign}(q_{XY,\nu}) \forall \nu = 1, \ldots, N. \tag{15}$$

If the process duration is infinitely long, then the macrosystem's natural evolution leads it to a state of internal equilibrium, which corresponds to the achievement of the maximum $\mathcal{S}(Q)$ value. This is equivalent to the statement that in the internal equilibrium state, $\mathcal{S}(Q)$ is maximum.

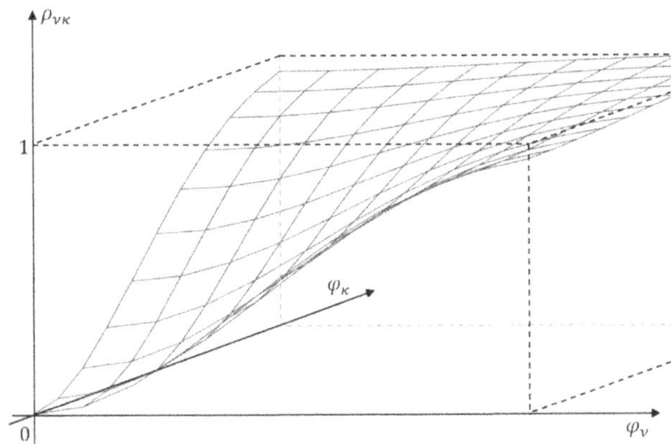

Figure 2. Dependency of correlation between fluxes $\rho_{\nu\kappa}(\varphi_{XY})$ in a macrosystem: φ_ν and φ_κ are the components of the vector φ_{XY}.

Intensive variables $v = \nabla\mathcal{S}$ can be considered as specific potentials. Their difference $\varphi_{XY} = v_Y - v_X$ according to (15) is the driving force of the exchange process. The fluxes are directed from subsystems with lower values of intensive parameters to subsystems with higher values of intensive parameters. Thus, Equation (14) can be rewritten as follows:

$$\sigma(v_X, v_Y) = \sum_{\nu=1}^{N} \varphi_\nu(v_X, v_Y) q_{XY,\nu}(A, \varphi(v_X, v_Y)), \tag{16}$$

where A is the matrix of infrastructure coefficients. Note that the infrastructure of the exchange process involves a wide variety of features of the subsystems' boundaries medium. For example, in thermodynamic macrosystems these features are surface area, roughness, etc., all these parameters can be controls in the corresponding optimization problems, but here, A is assumed to be constant. If the driving force during the process is constant, then such a process is called stationary. Since $\varphi(v_X, v_Y)$ is a linear relation and therefore the superposition principle is being applied, we single out the classes of reversible $(\varphi(v_X, v_Y) \to 0)$ processes and processes of minimal dissipation $\sigma(\varphi) \to \min_{\varphi} \mid q_{XY}(\varphi) = \text{fix}$.

7. Differential Form of the State Equation

The state function can be given in differential form, as is typical for thermodynamic systems:

$$\delta\Phi = \sum_{\nu=1}^{N} F_\nu(Q) dQ_\nu. \tag{17}$$

If we cannot write a state function explicitly, then the problem of integrability of $\Phi(Q)$ should be solved. Equation (17) is the Pfaffian form. A Pfaffian form is said to be holonomic if there exists an integrating multiplier $w(Q)$ such that

$$w(Q)\delta\Phi = \sum_{\nu=1}^{N} \frac{\partial \mathcal{S}}{\partial Q_\nu} dQ_\nu = d\mathcal{S}, \text{ where } \frac{\partial \mathcal{S}}{\partial Q_\nu} = w(Q)F_\nu(Q), \nu = 1, \ldots, N. \quad (18)$$

The Pfaffian form of two independent variables is always holonomic, that is, there is always an integrating multiplier $w(Q)$. However, for $N > 2$, the integrating multiplier exists if the holonomy conditions are satisfied: for any three different κ, μ, ν,

$$F_\kappa(Q)\left(\frac{\partial F_\mu}{\partial Q_\nu} - \frac{\partial F_\nu}{\partial Q_\mu}\right) + F_\mu(Q)\left(\frac{\partial F_\nu}{\partial Q_\kappa} - \frac{\partial F_\kappa}{\partial Q_\nu}\right) + F_\nu(Q)\left(\frac{\partial F_\kappa}{\partial Q_\mu} - \frac{\partial F_\mu}{\partial Q_\kappa}\right) = 0. \quad (19)$$

These conditions are obtained from the equality of the second mixed derivatives with respect to any variable pairs (Maxwell's relations)

$$\frac{\partial^2 \mathcal{S}}{\partial Q_\nu \partial Q_\kappa} = \frac{\partial (w(Q)F_\nu(Q))}{\partial Q_\kappa} \quad (20)$$

and by exclusion of the integrating factor $w(Q)$ from these equalities. In addition to conditions (19), it is necessary that all products $w(Q)F_\nu(Q)$ be homogeneous of zero-order homogeneity.

8. Concavity of Entropy Function

The $\mathcal{S}(Q)$ function gradient determines the intensive variable vector of the macrosystem. As the extensive variables increase, the S value also increases, but at a slower rate (the diminishing returns law), so that for all $= 1, \ldots, N$ $\partial\mathcal{S}/\partial Q_\nu = v_\nu > 0$, $\partial^2\mathcal{S}/\partial Q_\nu^2 = \partial v_\nu/\partial Q_\nu < 0$. Moreover, the Hessian matrix $\mathcal{H}_S = (\partial^2\mathcal{S}/(\partial Q_\nu \partial Q_\kappa))$ for homogeneous functions of the first-order homogeneity is negative semi-definite. Indeed, it follows from the Euler relations (10) that differentiating both parts of this equation per Q_κ,

$$\forall \kappa = 1, \ldots, N : \sum_{\nu=1}^{N} Q_\nu \frac{\partial^2 \mathcal{S}}{\partial Q_\nu \partial Q_\kappa} = 0. \quad (21)$$

For an arbitrary vector x, the necessary conditions for the extremum of a quadratic form in x are

$$x^{\mathrm{T}}\mathcal{H}_S x = \sum_{\nu=1}^{N} x_\nu^2 \frac{\partial^2 \mathcal{S}}{\partial Q_\nu^2} + \sum_{\nu=1}^{N-1} \sum_{\kappa=\nu+1}^{N} 2x_\nu x_\kappa \frac{\partial^2 \mathcal{S}}{\partial Q_\nu \partial Q_\kappa} \to \max_x . \quad (22)$$

$$\frac{\partial (x^{\mathrm{T}}\mathcal{H}_S x)}{\partial x_\nu} = \sum_{\kappa=1}^{N} x_\kappa \frac{\partial^2 \mathcal{S}}{\partial Q_\nu \partial Q_\kappa} = 0, \quad (23)$$

which corresponds to the maximum point (according to the diminishing returns law) of the quadratic form $x_\nu = Q_\nu, \nu = 1, \ldots, N$. Since the product $Q^{\mathrm{T}}\mathcal{H}_S$ (21) is equal to zero, then $x^{\mathrm{T}}\mathcal{H}_S x$ at the maximum point is also equal to zero. Thus, for any value of $x : x^{\mathrm{T}}\mathcal{H}_S x \leq 0$, which has to be proved. Note that since the Hesse matrix is symmetric, then all its eigenvalues are real numbers.

The negative semi-definiteness of the Hessian matrix $\mathcal{H}_S$ corresponds to the upward convexity (concavity) of the $\mathcal{S}(Q)$ function and S unimodality as the objective parameter. As the reserves of the extensive variables in the macrosystem increase, S increases, and the intensive parameters decrease, reducing the magnitude of the resource exchange driving force. In accordance with (14), this behaviour of intensive parameters leads to the fact that when the macrosystem is affected, changing its internal equilibrium conditions; the resource exchange processes are directed towards counteracting changes, and thus the Le Chatelier principle is fulfilled.

Partial derivatives $\partial^2 S/\partial Q_\nu^2$ describe the saturation of the system with the extensive variables, and $\partial^2 S/(\partial Q_\nu \partial Q_\kappa)$ determines the substitution and complementation of the extensive variables in the macrosystem. If the extensive variables are substitutes, then an increase in one of them reduces the intensive variables adjoined with the other extensive variable. If the extensive variables are complements, then an increase in one of them, on the contrary, increases the intensive variables adjoined with the other extensive variable.

9. Metric Features of Entropy

Let us consider a particular case of exchange processes in a macrosystem consisting of two subsystems X and Y, where the fluxes linearly depend on the differences between the intensive variables of the subsystems:

$$q_{XY,\nu}(v_X, v_Y) = \frac{dQ_{Y\nu}}{dt} = \sum_{\kappa=1}^{N} \alpha_{\nu\kappa}\varphi_\kappa(v_{X\kappa}, v_{Y\kappa}), \tag{24}$$

where $\varphi_\kappa(v_{X\kappa}, v_{Y\kappa}) = v_{Y\kappa} - v_{X\kappa}$.

Proposition 1. *Matrix $A = (\alpha_{\nu\kappa})$ is the matrix of infrastructure coefficients describing the exchange possibilities at the boundary between subsystems and it is a symmetric matrix.*

Proof. Given the self-similarity property, the system can be divided into a statistically significant disjoint set of subsystems such that the characteristics of these subsystems form a representative sample of random variables $\tilde{q}, \tilde{\varphi}$, etc. The eigenvectors of the matrices $\mathrm{Cov}\left[\tilde{q}\right]$ and $\mathrm{Cov}\left[\tilde{q}, \tilde{\varphi}\right]$ coincide and form a system of orthonormal vectors. Since the eigenvectors of any matrix and its inverse are the same, $\mathrm{Cov}\left[\tilde{q}\right]$ and $\mathrm{Cov}\left[\tilde{q}, \tilde{\varphi}\right]$ are symmetric and commutative, so their product is a symmetric matrix. Since $\mathrm{M}\left[\tilde{q}\right] = \mathrm{Cov}^{\mathrm{T}}\left[\tilde{q}, \tilde{\varphi}\right]\mathrm{Cov}\left[\tilde{q}\right]\mathrm{M}\left[\tilde{q}\right]$, the matrix of phenomenological coefficients $A = \mathrm{Cov}^{\mathrm{T}}\left[\tilde{q}, \tilde{\varphi}\right]\mathrm{Cov}\left[\tilde{q}\right]$ is symmetric (Onsager conditions). □

Proposition 2. *Matrix $A = (\alpha_{\nu\kappa})$ is a positive definite matrix.*

Proof. The right part of Equation (16) under condition (24) has a quadratic form:

$$\sigma(v_X, v_Y) = \varphi^{\mathrm{T}}(v_X, v_Y) A \varphi(v_X, v_Y). \tag{25}$$

For any positive values of the exchange driving forces $\varphi(v_X, v_Y)$, the entropy production is positive, as shown in (15). Therefore, A is a positive definite matrix. It means that of the dependencies $\sigma(\varphi)$ and $\sigma(q)$ under the linear dependence of the flows on the driving forces are convex. □

The consequence of this proposition is that entropy production is a metric in the space of stationary processes and can be used to determine the distance between processes

$$\delta(a,b) = \sqrt{\Delta\sigma(v_X, v_Y)} = \sqrt{(\varphi_a - \varphi_b)^{\mathrm{T}} A (\varphi_a - \varphi_b)}. \tag{26}$$

Let us present some properties of this metric:

- Zero in the space of stationary processes represents reversible processes for which $\sigma = 0$. According to the third Levitin–Popkov axiom, there is the only reversible process in the macrosystem;
- The distance between two processes a and b is determined as $\delta^2(a,b) = (\varphi_a - \varphi_b)^T A(\varphi_a - \varphi_b)$. It is evident that $\delta(a,a) = 0; \delta(a,b) = \delta(b,a)$;
- The distance $\delta(a,b)$ satisfies the triangle inequality due to A being a positive definite symmetric matrix; all its eigenvalues λ are positive real numbers.

Thus, the entropy production in a macrosystem characterizes the distance of stationary exchange processes occurring in such a system from the corresponding reversible process. For linear systems, it is possible to consider generalized, averaged over probability to measure stationary processes: $\forall s \in \mathbb{R} \exists \mu(s) = P(\sigma(v_X, v_Y) = s) : \overline{\sigma} = \int_{-\infty}^{+\infty} s\, \mu(s) ds$. Such an averaging for cyclic and stochastic processes is illustrated on Figure 3.

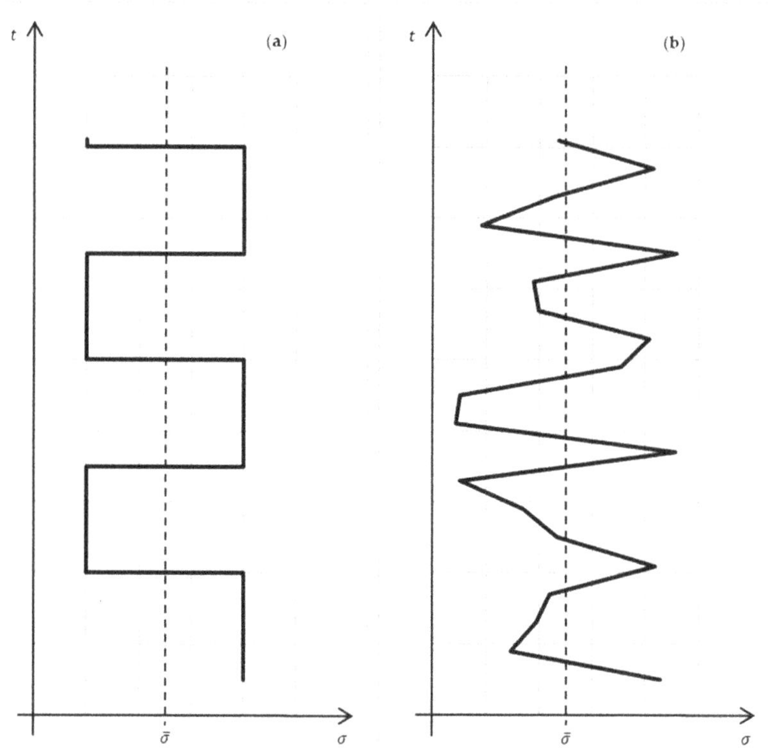

Figure 3. Averaging of entropy production in cyclic (**a**), and stochastic (**b**) stationary processes.

The distance between irreversible processes is an important concept for analyzing complex systems. The class of minimal dissipation processes indicates the limit of performance of macrosystems when the average intensities of the processes in it are restricted. To determine the effectiveness of an arbitrary process, it is necessary to find a distance between this process and the minimum dissipation process. This distance is determined by the production of entropy.

10. Trajectories of the Exchange Process

For non-stationary exchange processes, it is necessary to determine the trajectory of the process, i.e., the change in time of all subsystem parameters when approaching the equilibrium state. For a linear dependence between fluxes and driving forces (24), we derive a differential equation that determines the change in the driving forces of the exchange process in a macrosystem consisting of two subsystems X and Y. Under the given initial conditions $\varphi_{XY}(0) = \varphi_0$, this equation describes all the parameters of the subsystems.

Full differentials of functions $v_{i\kappa}(Q_i), i \in \{X, Y\}, \kappa = 1, \ldots, N$ shall be written as

$$\forall \kappa = 1, \ldots, N : \sum_{\nu=1}^{N} Q_\nu \frac{\partial^2 S}{\partial Q_\nu \partial Q_\kappa} = 0. \quad (27)$$

Subtracting the equations for subsystem X from the equations for subsystem Y and taking into account the linear dependencies of fluxes on driving forces (24) and the fact that the driving forces are the differences between the corresponding intensive variables of the subsystems ($d\varphi_{XY}/dt = dv_Y/dt - dv_X/dt$), we obtain

$$\forall \kappa = 1, \ldots, N : \sum_{\nu=1}^{N} Q_\nu \frac{\partial^2 S}{\partial Q_\nu \partial Q_\kappa} = 0. \quad (28)$$

Equation (28), together with the initial conditions, determines the trajectory of the exchange process.

11. Conclusions

The metric properties of entropy production allow for the determination of both the class of minimally irreversible processes [11] and the quantitative distance between the processes. The stationarity restriction can be removed either by averaging the process parameters over time or due to the superposition principle for processes in linear systems. A generalized, independent-of-the-nature-of-the-processes, macrosystem model is useful for the formalization and investigation of extreme performances of complex, hierarchically related systems. In these macrosystem, there is a vector of entropy functions. It gives the macrosystem as additional degree of freedom at the level of subsystems.

The given proofs of the phenomenological properties of macrosystems are valid for thermodynamic systems but can also be applied to the systems of a different nature, particularly economic systems [12,13], information exchange systems in communication networks [14], and high-performance computers [15], where the number of computational cores is already comparable to the number of molecules in 1 μm^3 of gas. The main difference in the description of macrosystems of different natures lies in the definition of the extensive quantities that describe the state of a system and in the relationships between the fluxes that arise from interactions between subsystems. In thermodynamic macrosystems, the extensive variables are internal energy, mol number, and volume [6]; in economic macrosystems, the extensive variables are resources, goods, and welfare [12]; in information macrosystems like recommendation systems, the extensive variables are database size and the number of exposed marks [16]; and so on. In information macrosystems, we restrict ourselves to describing syntactic information exchange [17]. Social macrosystems with semantic and pragmatic information exchange processes require a special conceptual apparatus, which is markedly different from the one usually used in formal logic [18]. For signal transmission systems, the extensive variables are the number of processors of a given type, the amount of memory, and the entropy properties of the computing power; the intensive variables are determined by the contribution to the increase in computing power that each type of hardware provides. When modelling algorithms as complex systems, the extensive variables can be used as control values, the entropy corresponds to the objective function, and the intensive variables are the values of the Lagrange multipliers. Economic analogies of complex systems are often considered. At the micro level, extensive quantities include the stock of goods while intensive quantities include the prices and the values of the goods. The welfare function has entropy properties in microeconomic systems. At the macro level, it is advisable to choose the gross regional product as the entropy, which is a function of vector of production factors. The intensive variables in this model are prices in real terms.

Entropy as an extensive quantity is introduced into different models of self-similar systems. Integration problems of the welfare function have been observed in [12]. The first-order homogeneity properties for gross regional product have been proven: the Cobb–Douglas production function is most commonly used in approximation. For communication networks, the entropy properties of the indicator of congestion have been proven [14]. These examples show the universality of a macrosystemic approach to modelling complex systems of different natures.

The graph of interaction between subsystems can be different for each type of nature of extensive variables. Such a complex system can be represented as a multigraph. Multigraphs where two graph nodes can be connected both by edges of different colours and by groups of edges of different colours, which correspond to multiple contact points between subsystems. The absence of edges between the multigraph's nodes means that the nodes are isolated from each other. Edges of different colours corresponding to different extensive variables can be either co-directed or opposite.

The specific behaviour of individual elementary entities is especially relevant for social systems, where free and often unmotivated decision-making is possible and can complicate the application of the model. Nevertheless, the model can still be used to describe processes of various natures occurring simultaneously within a system. For example, a high-performance computer can be represented as a macrosystem in which hardware, software, and engineering subsystems exchange energy, signals, and information, and the average intensities of the computational processes and heat transfer processes are considered to be given. A computer does not perform mechanical work; therefore, all consumed electrical energy is converted into heat and must be dissipated into the surrounding environment.

Funding: This research received no external funding.

Data Availability Statement: No new data were created or analyzed in this study. Data sharing is not applicable to this article.

Conflicts of Interest: The author declares no conflict of interest.

References

1. Andresen, B.; Salamon, P. Future Perspectives of Finite-Time Thermodynamics. *Entropy* **2022**, *24*, 690. [CrossRef] [PubMed]
2. Fitzpatrick, R. *Thermodynamics and Statistical Mechanics*; World Scientific Publishing Company: Chennai, India, 2020; 360p, ISBN 978-981-12-2335-8.
3. Prunescu, M. Self-Similar Carpets over Finite Fields. *Eur. J. Comb.* **2009**, *30*, 866–878. [CrossRef]
4. Tropea, C.; Yarin, A.L.; Foss, J.F. *Springer Handbook of Experimental Fluid Mechanics*; Springer: Berlin/Heidelberg, Germany, 2007; 557p, ISBN 978-3-54-030299-5.
5. Popkov, Y.S. Macrosystem Models of Fluxes in Communication-Computing Networks (GRID Technology). *IFAC Proc. Vol.* **2004**, *37*, 47–57. [CrossRef]

6. Rozonoér, L.I. Resource Exchange and Allocation (a Generalized Thermodynamic Approach). *Autom. Remote Control* **1973**, *34*, 915–927.
7. Popkov, Y.S. *Macrosystems Theory and Its Applications. Equilibrium Models*; Ser.: Lecture Notes in Control and Information Sciences; Springer: Berlin, Germany, 1995; Volume 203, 323p, ISBN 3-540-19955-1.
8. Amelkin, S.A.; Tsirlin, A.M. Optimal Choice of Prices and Fluxes in a Complex Open Industrial System. *Open Syst. Inf. Dyn.* **2001**, *8*, 169–181. [CrossRef]
9. Levitin, E.S.; Popkov, Y.S. Axiomatic Approach to Mathematical Macrosystems Theory with Simultaneous Searching for Aprioristic Probabilities and Stochastic Flows Stationary Values. *Proc. Inst. Syst. Anal. RAS* **2014**, *64*, 35–40.
10. Khinchin, A.I. *Mathematical Foundations of Information Theory*; Dover: New York, NY, USA, 1957; 120p, Original publication in Russian is: Khinchin, A.I. The Concept of Entropy in Probability Theory. *Uspekhi Matematicheskikh Nauk* **1953**, *8*, 3–20.
11. Tsirlin, A.M. Minimum Dissipation Processes in Irreversible Thermodynamics. *J. Eng. Phys. Thermophys.* **2016**, *89*, 1067–1078. [CrossRef]
12. Tsirlin, A.M. External Principles and the Limiting Capacity of Open Thermodynamic and Economic Macrosystems. *Autom. Remote Control* **2005**, *66*, 449–464. [CrossRef]
13. Tsirlin, A.M.; Gagarina, L.G. Finite-Time Thermodynamics in Economics. *Entropy* **2020**, *22*, 891. [CrossRef] [PubMed]
14. Dodds, P.; Watts, D.; Sabel, C. Information Exchange and the Robustness of Organizational Networks. *Proc. Natl. Acad. Sci. USA* **2003**, *100*, 12516–12521. [CrossRef] [PubMed]
15. De Vos, A. Endoreversible Models for the Thermodynamics of Computing. *Entropy* **2022**, *22*, 660. [CrossRef] [PubMed]
16. Pebesma, E.; Bivand, R. *Spatial Data Science: With Applications in R*; Chapman and Hall/CRC: Boca Raton, FL, USA; New York, NY, USA, 2023; 314p. [CrossRef]
17. Martinás, K. Entropy and Information. *World Futures* **1997**, *50*, 483–493. [CrossRef]
18. Jøsang, A. *Subjective Logic. A Formalism for Reasoning Under Uncertainty*; Springer: Cham, Switzerland, 2016; 337p, ISBN 978-3-319-42335-7. [CrossRef]

entropy

MDPI

Article

Transfer Irreversibilities in the Lenoir Cycle: FTT Design Criteria with $\varepsilon - NTU$

Ricardo T. Páez-Hernández [1,*], Juan Carlos Pacheco-Paez [2,3,*] Juan Carlos Chimal-Eguía [4], Delfino Ladino-Luna [1] and Javier Contreras-Sánchez [3]

1 Area de Física de Procesos Irreversibles, Departamento de Ciencias Básicas, Universidad Autónoma Metropolitana, U-Azcapotzalco, Av. San Pablo No. 420, Col. Nueva el Rosario, Alcaldía Azcapotzalco, Ciudad de México C.P. 02128, Mexico; dll@azc.uam.mx

2 Departamento de Ciencias Básicas, Universidad Autónoma Metropolitana, U-Azcapotzalco, Av. San Pablo No. 420, Col. Nueva el Rosario, Alcaldía Azcapotzalco, Ciudad de México C.P. 02128, Mexico

3 Departamento de Biofísica, Escuela Nacional de Ciencias Biológicas, Instituto Politécnico Nacional, D.R. Instituto Politécnico Nacional (IPN). Av. Luis Enrique Erro S/N, Unidad Profesional Adolfo López Mateos, Zacatenco, Alcaldía Gustavo A. Madero, Ciudad de México C.P. 07738, Mexico; javcontreras@ipn.mx

4 Laboratorio de Ciencias Matemáticas y Computacionales, Centro de Investigación en Computación, Instituto Politécnico Nacional, Ciudad de México C.P. 07738, Mexico; jchimale@ipn.mx

* Correspondence: phrt@azc.uam.mx (R.T.P.-H.); jcpacheco@azc.uam.mx (J.C.P.-P.)

Abstract: This work extends the steady flow Lenoir cycle within finite-time thermodynamics (FTT) by incorporating heat transfer irreversibilities through the $\varepsilon - NTU$ formalism and a non-isentropic expansion modeled via the expander isentropic efficiency η_E. The total conductance U_T (sum for the two heat exchangers) is partitioned between hot and cold units using $u_L = U_L/U_T$, with $U_T = U_H + U_L$. For each triplet ($\tau = T_H/T_L$, U_L, U_T), we closed the cycle by determining T_1, the working fluid temperature at the cooler outlet and heater inlet, T_2, the heater outlet and expander inlet, and T_3, the expander outlet and cooler inlet. Using these states, we compute the heat rates $\dot{Q}_{12}$, $\dot{Q}_{31}$ and the net power P. In addition to the thermal efficiency η, the following extended objective functions are evaluated: the efficient power E_F, the ecological efficiency ϕ, and the second law efficiency η_{II}. Parametric sweeps on u_L for $\tau \in \{3.25, 3.75\}$ and $U_T \in \{2.5, 5.0, 7.5, 10\}$ kW show unimodal curves for P(u_L) and maxima. A robust result places the optima of P, η, E_F, ϕ, and η_{II} in a distribution band at $u_L \sim 0.6$. This guideline offers clear design guidance for allocating exchange area in heat recovery and microgeneration, maximizing power, high η, and exergetic utilization with contained entropic penalty.

Keywords: finite-time thermodynamics; steady flow Lenoir cycle; heat conductance; efficient power; bounded ecological function; second law efficiency

1. Introduction

Finite-Time Thermodynamics (FTT) [1–5] is a branch of thermodynamics developed to analyze irreversible processes in real systems, particularly in power plants and heat engines, where thermodynamic equilibrium is not reached instantaneously. FTT seeks theoretical models that describe the operating limits of these systems under realistic constraints of dissipation, irreversibility, efficiency, and power.

One of the key motivations of FTT is to improve the design and performance of thermal devices by considering not only the ideal cycle efficiency but also other relevant variables such as entropy production, power output, or even functions that balance multiple performance criteria [6–10]. A historical milestone was the Curzon–Ahlborn heat engine [1], which proposed a Carnot engine model irreversibly coupled to its heat reservoirs by finite thermal resistances. The model predicted an efficiency at maximum power that surprisingly coincides with efficiencies observed in real power plants.

Velasco et al. [11] introduced the so-called saving functions, dimensionless compromise criteria that compare a given operating point with a reference one and quantify the saving in fuel consumption or in entropy generation. By simultaneously weighting power output and dissipation, these functions provide a unified way to assess operating regimes beyond the traditional focus on either maximum efficiency or maximum power. Building on this formalism, Barranco-Jiménez et al. [12] applied the saving functions to the Novikov engine [13,14] and analyzed different levels of participation of irreversible processes and considered two alternative heat transfer laws, which made it possible to evaluate the sensitivity of power and efficiency to variations in the model parameters.

The advancement of the FTT has also incorporated new multi-objective criteria, such as the ecological function and efficient power, and both approaches aimed at simultaneously minimizing energy loss and environmental degradation [15,16]. These criteria are fundamental for the energy transition towards cleaner and more efficient technologies and have been applied to a wide variety of endoreversible and irreversible cycle models, including modified Brayton cycles and thermoelectric generators [16–33].

In the face of increasing global energy demands and environmental constraints, the optimization of thermodynamic cycles has become a cornerstone of modern energy system design. While the Carnot and Otto cycles have traditionally dominated the landscape of thermodynamic analysis, the Lenoir cycle [34], historically associated with early internal combustion engines, is gaining renewed interest due to its relevance in pulsed detonation engines, fast-acting pneumatic systems, and micro-scale energy conversion devices.

In these contexts, the Lenoir cycle often serves as a simplified model of constant-volume combustion followed by expansion, and the conductance–allocation guidelines developed in this work are directly applicable to the design of heat recovery units and micro-generation systems based on Lenoir-type architectures.

Several authors have already analyzed the Lenoir cycle within irreversible and finite-time frameworks. Georgiou investigated the ideal Lenoir cycle with regenerative preheating, clarifying the impact of regeneration on useful work and thermal efficiency [35]. Shen et al. considered an endoreversible Lenoir cycle coupled to constant-temperature reservoirs, identifying the influence of finite heat transfer rates on performance [36]. Ahmadi et al. carried out a thermo-economic, multi-objective optimization of an endoreversible Lenoir engine, exploring trade-offs between power, efficiency, and economic cost [37]. More closely related to the present work, Rubio and Wang [38] studied an irreversible steady flow Lenoir cycle with external irreversibilities modeled by $\varepsilon - NTU$ heat exchangers and internal irreversibilities in the expander and optimized power and thermal efficiency under a fixed (symmetric) allocation of the total thermal conductance between the hot and cold exchangers. Building on this framework, the present paper focuses on a steady flow Lenoir cycle with explicit $\varepsilon - NTU$ heat exchangers and a finite isentropic efficiency of the expander and extends Rubio and Wang's model by allowing the total thermal conductance to be distributed through an allocation parameter φ and by comparing several objective functions (power, efficient power, bounded ecological function, and second law efficiency) within a unified design framework.

The classical Lenoir cycle comprises three fundamental processes: a constant-volume heat addition $(1 \rightarrow 2)$, an adiabatic expansion $(2 \rightarrow 3)$, and a constant-pressure heat rejection $(3 \rightarrow 1)$. In Figure 1a,b, we also present T-s and P-V diagrams of the cycle for better thermodynamic insight. However, real systems inherently deviate from these idealizations due to irreversibilities, such as internal dissipation and finite-rate heat transfer. To better reflect realistic energy conversion performance, irreversible cycle models have been proposed, introducing internal irreversibility parameters and nonlinear heat transfer laws.

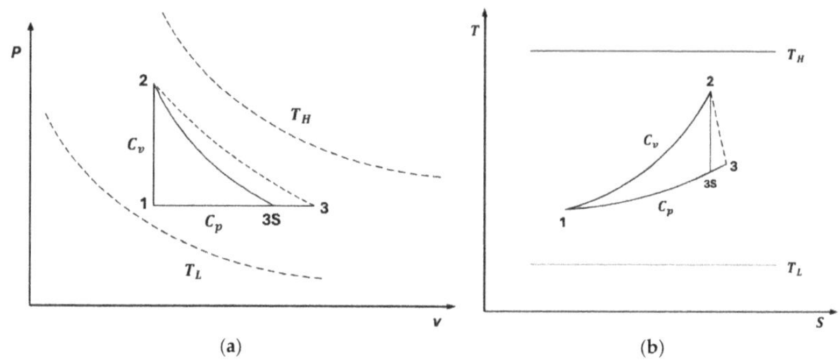

Figure 1. (**a**) $p - v$ diagram for the irreversible Lenoir cycle. (**b**) $T - s$ diagram for the irreversible Lenoir cycle.

As we have mentioned, this work focuses on the optimization of an irreversible steady flow Lenoir cycle following the approach of [34–38] and incorporating the following:

1. Internal irreversibilities represented by an effective isentropic efficiency of the expander.
2. Newtonian heat transfer laws to account for finite thermal gradients.
3. Ecological function (E_F), which balances power output and entropy generation to address environmental considerations.
4. A second law efficiency (η_{II}), defined as the ratio between the net power output and the thermal exergy rate supplied by the hot source, which quantifies how effectively the available exergy is converted into useful work.
5. Efficient power (E_P), which optimizes the product of power and thermal efficiency, representing a compromise between quantity and quality of energy conversion.
6. An extended analytical framework for these optimization regimes, together with a detailed numerical example using realistic parameter values.

2. Thermodynamic Description of the Irreversible Lenoir Cycle

In this section, we construct the irreversible model of the Lenoir cycle under steady flow conditions. Our goal is to consistently calculate the net power output of the cycle and the turbine efficiency based on design parameters, such as the quality of the heat exchangers and the losses in the expansion stage. To arrive at these expressions, we need to rigorously describe the internal thermodynamic states of the cycle and the energy and entropy balances in each process.

2.1. Working Fluid and Thermodynamic Properties

Since the Lenoir cycle we are analyzing operates with a gas, we adopt the ideal gas model with constant specific heats. This means that the specific enthalpy h and the specific internal energy u depend linearly on the temperature such that $h = C_pT$ and $u = C_vT$, where C_p and C_v are the specific heats at constant pressure and volume, respectively, and are taken as constant over the operating range. Furthermore, the relationship between these two heat capacities is given by

$$k \equiv \frac{C_p}{C_v}. \tag{1}$$

This relationship indicates how much work a gas can do when it expands or contracts; it also allows us to easily convert any ΔT jump into an enthalpy or energy jump and allows us to relate temperature and pressure in isentropic processes.

2.2. Physical Structure of the Lenoir Cycle in Steady Flow

The cycle we are considering has three main stages, which correspond directly to the classic stages of the Lenoir cycle but in a steady state and with explicit irreversibilities.

Process $1 \to 2$ (constant-volume heat addition) is modeled as nearly isochoric heating in a combustion chamber or plenum upstream of the expander, where the working gas is heated from T_1 to T_2 at an essentially fixed volume (or fixed vessel geometry).

Process $2 \to 3$ (expansion) is where the fluid then expands in a gas turbine or expansion device, producing shaft work. Ideally, this would be an isentropic expansion, but in practice, losses are present and are captured by an isentropic efficiency η_E.

Process $3 \to 1$ (constant pressure heat rejection) is where, finally, the fluid is cooled at approximately constant pressure in a heat recovery or exhaust heat exchanger, returning from state 3 to state 1 and closing the cycle.

According to [38], stages $1 \to 2$ and $3 \to 1$ are explicitly modeled as processes where heat exchange is limited by actual transfer surfaces and is, therefore, irreversible. To describe this thermal irreversibility in a quantifiable way, the authors use the formulation of $\varepsilon - NTU$ [38] heat exchangers. Our work follows the same philosophy.

2.3. Heat Exchangers with the ε–NTU Model

Let us assume that the cycle receives and rejects heat through two heat exchangers (HX): one "hot" (U_H), which connects the working fluid to a hot heat source at a fixed temperature T_H, and one "cold" (U_L), which connects the working fluid to a cold source at a fixed temperature T_L.

We know that these heat exchangers are not ideal; that is, their heat transfer capacity is limited by their area, convective coefficients, etc. All this complexity is summarized in an overall conductance U. Assuming that the sum of the available conductance is fixed, we obtain the following:

$$U_T = U_H + U_L, \tag{2}$$

where U_H is the part allocated to the hot side (heating $1 \to 2$) and U_L is the part allocated to the cold side (rejection $3 \to 1$). To analyze how to distribute this "total exchange capacity" between the heater and the cooler, we introduce $U_H = (1 - u_L)U_T$ and $U_L = u_L U_T$, where u_L is a fraction between 0 and 1. In this work, the total thermal conductance U_T is defined in the usual way as the ratio between a heat transfer rate and a temperature difference, $U_T \equiv \dot{Q}/\Delta T$, so its physical dimension is that of power divided by temperature. Since all heat rates and the net power of the cycle are expressed in kilowatts, we consistently take U_T, U_H, and U_L in U_T in kW/K. This is entirely equivalent to using W/K and simply corresponds to a change in scale (1 kW/K $= 10^3$ W/K), but it is more convenient here because it keeps the numerical values of conductances of the same order of magnitude as the power output. Thus, throughout the paper, the units of the overall thermal conductances are kW/K, in accordance with the definition $U \equiv \dot{Q}/\Delta T$.

In the case where one of the two streams in the exchanger acts as an almost immutable temperature reservoir (i.e., the hot source is maintained at T_H or the cold source at T_L), the so-called thermal effectiveness E of the exchanger can be written as follows:

$$E = 1 - \exp(-NTU), \tag{3}$$

The ratio of heat capacities between the streams is very small, which is precisely the hypothesis used in [38]. Here, "NTU" (Number of Transfer Units) is

$$NTU = \frac{U}{\dot{m}C_{cap}}, \tag{4}$$

where $\dot{m}C_{cap}$ is the heat capacity rate (or thermal capacity rate) of the stream whose temperature changes. For the heat exchanger (given in $1 \to 2$), the fluid absorbs heat with thermal capacity $\dot{m}C_v$; then

$$N_H = \frac{U_H}{\dot{m}C_v} = \frac{(1-u_L)U_T}{\dot{m}C_v}, \tag{5}$$

$$E_H = 1 - e^{-N_H}, \tag{6}$$

where E_H measures how closely the internal fluid approaches temperature T_H. In the cold exchanger (given in $3 \to 1$), the fluid releases heat at constant pressure, so its relevant capacity is $\dot{m}C_p = \dot{m}kC_v$. We define the following:

$$N_L = \frac{U_L}{\dot{m}C_p} = \frac{u_L U_T}{\dot{m}kC_v}, \tag{7}$$

$$E_L = 1 - e^{-N_L} \tag{8}$$

These quantities are fundamental because they allow us to go directly from design parameters of the exchanger (area, coefficients, etc.) to cycle temperatures.

During process $1 \to 2$, the cycle gas exchanges heat with the hot reservoir at constant temperature T_H. Because the process is essentially modeled as an isochoric process for the working fluid, the pressure rises, and the temperature increases from T_1, which is the working fluid temperature at the cooler outlet and heater inlet, to T_2, which is the heater outlet and expander inlet. The thermal effectiveness E_H of the heat exchanger is then defined as follows:

$$E_H = \frac{T_2 - T_1}{T_H - T_1}. \tag{9}$$

In this equation, when $E_H = 1$, the gas will exit at exactly the temperature of the hot source ($T_2 = T_H$); if $E_H = 0$, the gas is not heated ($T_2 = T_1$). For any intermediate value of E_H, the outlet temperature is a linear interpolation; that is

$$T_2 = T_1 + E_H(T_H - T_1). \tag{10}$$

After heating, the fluid expands to produce work (during process $2 \to 3$). Before introducing the irreversibilities of the machine, we first construct the ideal state, state $3s$, which would be the final state if the expansion were isentropic. For an ideal gas with constant specific heats, a sentropic expansion obeys the relation $Tp^{\frac{1-k}{k}} = constant$. The Lenoir cycle has a useful geometric feature. Process $1 \to 2$ is considered isochoric (constant volume), which implies that the relationship between pressure and temperature satisfies $p \propto T$ in this process. This implies

$$\frac{T_2}{T_{3s}} = \left(\frac{p_2}{p_{3s}}\right)^{\frac{k-1}{k}} = \left(\frac{T_2}{T_1}\right)^{\frac{k-1}{k}}, \tag{11}$$

therefore, we can write

$$T_{3s} = T_1 \left(\frac{T_2}{T_1}\right)^{\frac{1}{k}}. \tag{12}$$

This is the lowest temperature the gas could reach after expansion, if the expansion were perfectly reversible (without internal losses). In practice, the expansion is not ideal; there are irreversibilities, and the machine does not extract the maximum possible work, introducing this loss through an isentropic efficiency of the expansion, η_E, which is defined as

$$\eta_E = \frac{h_2 - h_3}{h_2 - h_{3s}} = \frac{T_2 - T_3}{T_2 - T_{3s}}. \tag{13}$$

This definition is standard for expansions in turbomachines. If $\eta_E = 1$, the expansion is ideal, and the actual outlet temperature T_3 coincides with T_{3s}. If $\eta_E < 1$, the expansion is worse, and, therefore, T_3 will be hotter than T_{3s} (work potential is wasted).

We solve for T_3 from Equation (13) in a direct algebraic way as follows:

$$T_3 = (1 - \eta_E)T_2 + \eta_E T_{3s}. \tag{14}$$

This expression tells us how the mechanical/thermofluid dynamic inefficiency in the expansion stage "raises" the outlet temperature T_3 with respect to the ideal temperature T_{3s}, penalizing the work. After the cycle expansion, the still-hot fluid (state 3 at temperature T_3) must then be cooled back to the initial state 1 at temperature T_1. Using the $\varepsilon - NTU$ formulation, the cooling heat exchanger's effectiveness E_L is defined as

$$E_L = \frac{T_3 - T_1}{T_3 - T_L}. \tag{15}$$

E_L is physically entirely parallel to the hot case. If $E_L = 1$, the fluid would exit at exactly T_L; if $E_L = 0$, it would not cool at all. Solving this relationship for T_3 in terms of T_1 and T_L is performed as follows:

$$T_3 = \frac{T_1 - E_L T_L}{1 - E_L}. \tag{16}$$

This equation provides information that, given the effectiveness of the cold exchanger, the actual outlet temperature of the expansion T_3 cannot be arbitrary; it is strongly controlled by how well one rejects heat to the cold sink. Combining Equations (10), (12), (14) and (16), we obtain

$$\frac{T_1 - E_L T_L}{1 - E_L} = (1 - \eta_E)[T_1 + E_H(T_H - T_1)] + \eta_E T_1 \left(\frac{T_1 + E_H(T_H - T_1)}{T_1}\right)^{\frac{1}{k}}. \tag{17}$$

Now, we can calculate the heat flows; for process $1 \to 2$, the fluid absorbs heat from the hot source, and then we obtain

$$\dot{Q}_{12} = \dot{m} C_v E_H (T_H - T_1) = \dot{m} C_v (T_2 - T_1). \tag{18}$$

Whereas for process $3 \to 1$, the fluid gives up heat to the cold source, so we can write

$$\dot{Q}_{31} = \dot{m}C_p E_L (T_3 - T_L) = \dot{m}C_p (T_3 - T_1). \tag{19}$$

By combining the previous equation with Equation (16), we obtain

$$\dot{Q}_{31} = \dot{m}C_p \frac{E_L}{1 - E_L}(T_1 - T_L) = \dot{m}kC_v \frac{E_L}{1 - E_L}(T_1 - T_L). \tag{20}$$

The net power output of the steady-state cycle can be obtained from the first law of thermodynamics; that is

$$P = \dot{W} = \dot{Q}_{12} - \dot{Q}_{31}. \tag{21}$$

Substituting the heat flow expressions, which are Equations (18)–(20), we have

$$P = \dot{m}C_v \left[E_H (T_H - T_1) - k\frac{E_L}{1 - E_L}(T_1 - T_L) \right]. \tag{22}$$

We can now write the efficiency in terms of temperatures and effectiveness, which is given by

$$\eta = \frac{P}{\dot{Q}_{12}} = 1 - \frac{kE_L (T_1 - T_L)}{E_H (1 - E_L)(T_H - T_1)}. \tag{23}$$

2.4. Entropy Production and the Origin of Thermal Irreversibility

In addition to energy, we need to characterize irreversibilities. In this cycle, we assume that all relevant dissipative behavior comes from two sources: (i) the finite temperature difference in the heat exchangers and (ii) the internal irreversibility of the expansion, already captured by η_E.

If we consider only the heat exchangers, we can estimate the total entropy production $\dot{S}_{gen}$ associated with heat transfer between the fluid and the thermal reservoirs. To perform this, we model that, in the hot heat exchanger, the hot reservoir loses heat $\dot{Q}_{12}$ at a nearly constant temperature T_H, while the fluid gains heat, but at an effective mean temperature $\overline{T}_{12}$, which we approximate as

$$\overline{T}_{12} = \frac{T_1 + T_L}{2}. \tag{24}$$

Similarly, in the cold exchanger, the fluid discharges heat $\dot{Q}_{31}$ at an average temperature.

$$\overline{T}_{31} = \frac{T_1 + T_3}{2}. \tag{25}$$

The cold reservoir receives it at T_L, which we assume to be constant

$$\dot{S}_{gen} = \dot{Q}_{12}\left(\frac{1}{\overline{T}_{12}} - \frac{1}{T_L}\right) + \dot{Q}_{31}\left(\frac{1}{T_L} - \frac{1}{\overline{T}_{31}}\right) \geq 0. \tag{26}$$

This equation shows the actual heat transfer of a reversible transfer and gives us a quantity that we can later combine with power to build "eco-efficiency" metrics that have physical meaning (e.g., how much power we obtain per unit of irreversibility).

3. Objective Functions

So far, we have only discussed power and thermal efficiency, which are classical metrics. But now, we want to incorporate the entropic cost, that is, the irreversibilities generated in the cycle, especially in the heat exchangers. Recall that the total rate of entropy generation that we attribute to the exchangers is given by Equations (24)–(26), which give us information about how far we are from the reversible ideal. If the exchangers were infinitely large and there was no temperature difference between fluids and reservoirs, $\dot{S}_{gen}$ would tend to be zero. On this basis, we define three useful operating regimes.

3.1. Bounded Ecological Function Regime

Some of the literature on irreversible cycles uses the so-called "ecological function" ($E_F = P - T_L\dot{S}_{gen}$), but this function can become negative if losses due to irreversibility are large. To avoid this sign ambiguity, we propose using the positive dimensionless version.

$$\phi = \frac{P}{P + T_L\dot{S}_{gen}}. \tag{27}$$

This quantity is between 0 and 1; ϕ is always positive, always bounded, and always easy to compare. The closer it is to 1, the more of the "total available resource" (which, in this case, we interpret as useful power plus equivalent entropic destruction power measured at T_L) ends up as useful power and not as irreversibility.

3.2. Second Law Efficiency

Another common approach in exergy analysis is to compare the power generated with the exergy available in the absorbed heat, B_{in}. If the hot source is modeled as a reservoir at a fixed temperature T_H, the thermal exergy supplied by $\dot{Q}_{12}$ can be approximated as

$$B_{in} \approx \left(1 - \frac{T_L}{T_H}\right)\dot{Q}_{12}, \tag{28}$$

where T_L is the reference cold temperature. We define the second law efficiency by

$$\eta_{II} = \frac{P}{B_{in}} = \frac{\dot{m}C_v\left[E_H(T_H - T_1) - k\frac{E_L}{1-E_L}(T_1 - T_L)\right]}{\left(1 - \frac{T_L}{T_H}\right)\dot{Q}_{12}}. \tag{29}$$

This quantity compares the actual power generated with the available thermal exergy from the heat absorbed from the hot source, and this quantity ranges between 0 and 1.

In the $\phi(u_L)$ and $\eta_{II}(u_L)$ graphs, we again observe well-defined maxima. Most interestingly, the u_L values that maximize ϕ and η_{II} tend to be close to the value that maximizes P. This is extremely convenient from a design perspective; it means we are not forced to sacrifice almost all our power just to improve our "ecological quality" or our exergy efficiency. Generally, there is a reasonable compromise zone around a given u_L distribution.

3.3. Efficient Power Regime

In addition to P and η, in irreversible cycle thermodynamics, efficient power is frequently used, which is defined as follows:

$$E_P = P\eta. \tag{30}$$

This objective function optimizes power to achieve maximum performance with minimum energy consumption, thus minimizing losses and waste. Substituting Equations (22) and (23) into the previous equation and simplifying, we obtain the following:

$$E_P = \left(\dot{m}C_v\left[E_H(T_H - T_1) - k\frac{E_L}{1 - E_L}(T_1 - T_L)\right]\right)\left(1 - \frac{kE_L(T_1 - T_L)}{E_H(1 - E_L)(T_H - T_1)}\right). \tag{31}$$

When $P_E(u_L)$ is plotted, a maximum also appears. And again, this maximum usually falls in a region very close to the one that maximizes P and maximizes ϕ. This consistency between different criteria (gross power P, efficient power E_P, bounded ecological efficiency ϕ, second law efficiency η_{II}) is exactly what one wants to see from a design perspective. It indicates that there is a physically relevant range of u_L, where we not only obtain a lot of power but we obtain it in a reasonably efficient way and with bounded energy quality destruction.

4. Numerical Results

This section presents and analyzes the numerical results for the irreversible steady flow Lenoir cycle. External heat transfer irreversibilities in the hot and cold exchangers are modeled through the effectivenesses (E_H, E_L) derived from an $\varepsilon - NTU$ framework, while the internal irreversibility of the expansion is represented by a finite isentropic efficiency η_E. To allow a direct comparison with previous studies, we adopt parameter values similar to those used by Rubio and Wang [38] and in related works [26,34–37]. The working fluid is assumed to be air with constant properties, namely, a specific heat at a constant volume $C_v = 0.7165$ kJ/(kg·K) and a specific heat ratio $k = 1.4$. The isentropic efficiency of the gas turbine is taken as $\eta_E = 0.92$. Unless otherwise stated, the mass flow rate and the cold reservoir temperature are fixed at $\dot{m} = 1.1165$ kg/s and $T_L = 320$ K, respectively.

Unless otherwise stated, all performance maps shown in Figures 2–7 are computed using this same baseline data. The overall thermal conductance U_T varies within the range indicated in the captions, and the conductance–allocation parameter φ distributes U_T between the hot and cold exchangers. Figures 2 and 3 display the behaviour of the net power and the pure thermal efficiency under these parametric sweeps, while Figures 4–6 show the corresponding responses of the efficient power, the bounded ecological function, and the second law efficiency, respectively. Figure 7 summarizes and compares the optimal values of φ obtained from these different objective functions.

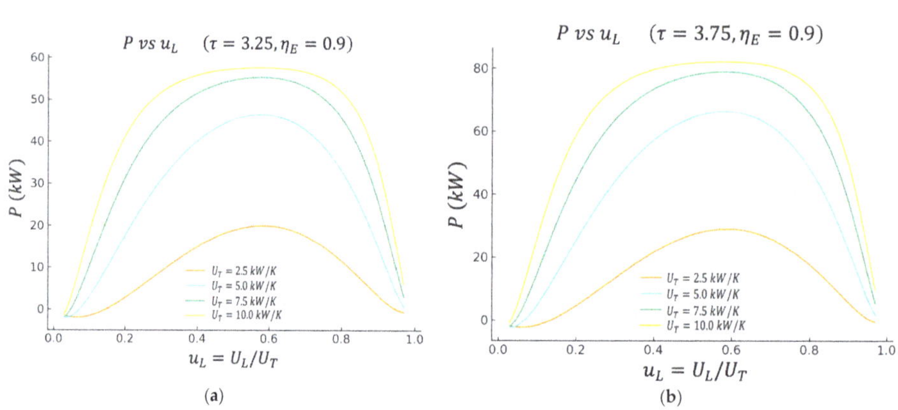

Figure 2. (**a**) Effect of U_T on P vs. u_L characteristics when $\tau = 3.25$. (**b**) Effect of U_T on P vs. u_L characteristics when $\tau = 3.75$.

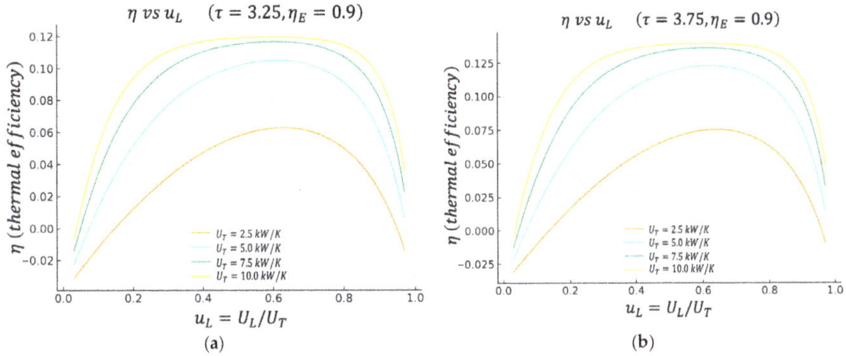

Figure 3. (**a**) Effect of U_T on η vs. u_L characteristics when $\tau = 3.25$. (**b**) Effect of U_T on η vs. u_L characteristics when $\tau = 3.75$.

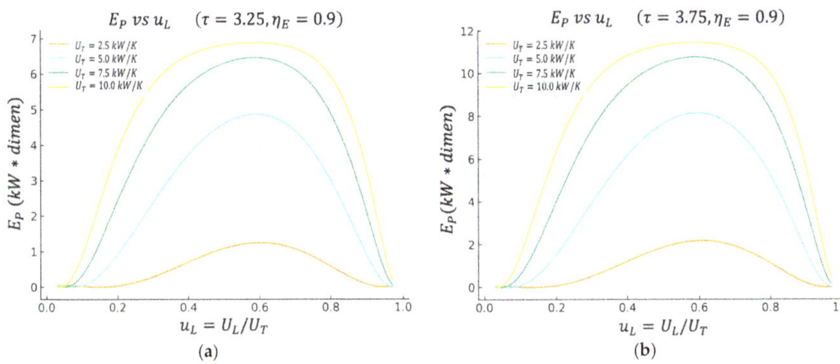

Figure 4. (**a**) Effect of U_T on E_P vs. u_L characteristics when $\tau = 3.25$. (**b**) Effect of U_T on E_P vs. u_L characteristics when $\tau = 3.75$.

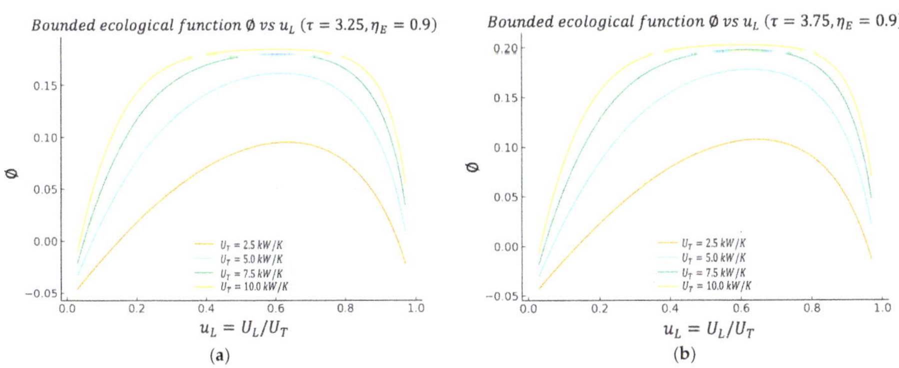

Figure 5. (**a**) Effect of U_T on ϕ vs. u_L characteristics when $\tau = 3.25$. (**b**) Effect of U_T on ϕ vs. u_L characteristics when $\tau = 3.75$.

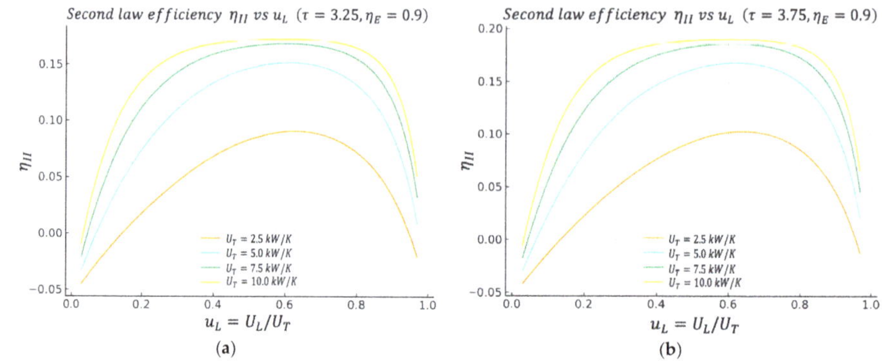

Figure 6. (**a**) Effect of U_T on η_{II} vs. u_L characteristics when $\tau = 3.25$. (**b**) Effect of U_T on η_{II} vs. u_L characteristics when $\tau = 3.75$.

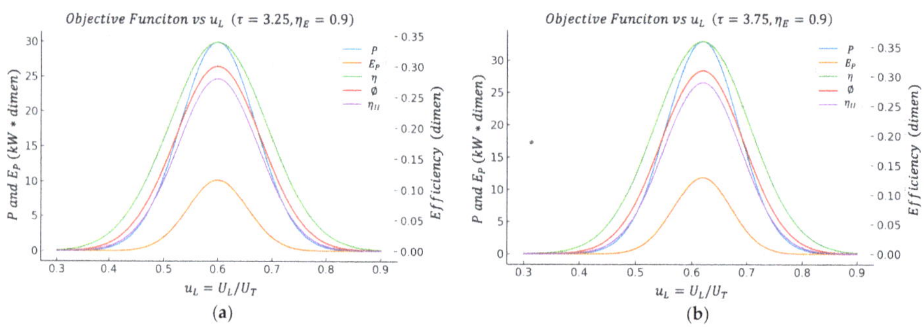

Figure 7. (**a**) Effect of U_T on the objective vs. u_L functions when $\tau = 3.25$. (**b**) Effect of U_T on the objective functions vs. u_L when $\tau = 3.75$.

4.1. Net Power (P) vs. Fraction of Conductance on the Cold Side

The figure shows the net power P as a function of $u_L = U_L/U_T$ for different values of U_T in two cases of the temperature ratio τ: (a) $\tau = 3.25$ and (b) $\tau = 3.75$. From Figure 2a,b, we can observe that $P(u_L)$ has a bell shape; for very low values of u_L, the power is reduced, increases to a clear maximum, and then falls again when u_L gets too close to 1.

This maximum means that it is not advisable to allocate all the thermal conductance to the heater or all the thermal conductance to the cooler; there is an optimal distribution.

When U_T increases (for example, from 2.5 kW/K to 10 kW/K), the height of the curve increases; the cycle can deliver more kW of total net power. Furthermore, the region near the maximum widens, implying that the cycle becomes more tolerant of deviations from u_L.

This reproduces the type of behavior reported in classical analyses of the irreversible Lenoir cycle. There is an optimal allocation of area/conductance between the two heat exchange stages that maximizes the useful power delivered by the engine.

4.2. Thermal Efficiency η Versus u_L

Figure 3a,b show the thermal efficiency $\eta = P/\dot{Q}_{12}$ of the cycle as a function of u_L, again for two cases of the temperature ratio τ, (a) $\tau = 3.25$ and (b) $\tau = 3.75$, and for the same values of U_T.

The results show that $\eta(u_L)$ also has a well-defined maximum. The value of u_L that maximizes η is very close, although not exactly at the same point, to the value of u_L that maximizes the power P. This observation provides a direct and quantitative characterization of the inherent trade-off between power output and thermal efficiency in real thermal engines. Specifically, operating the cycle precisely at the point of maximum power yields a thermal efficiency that, although not maximal, remains relatively high—often within a few percentage points of the maximum attainable value. Conversely, shifting the operation toward the point of maximum thermal efficiency results in only a moderate reduction in power output, not a dramatic loss as might be expected in more sharply peaked systems.

4.3. Efficient Power Regime Versus u_L

Figure 4a,b introduce the efficient power objective function $E_P = P\eta$, which is designed for practical engineering and thermo-economic evaluation, rewarding states where the power is large and, at the same time, the thermal efficiency is not ridiculously low.

In the following figure, it can be observed that the maximum of E_P appears practically in the same region of u_L that maximizes the net power P.

This indicates that the optimal power operating point also has attractive performance when simultaneously requiring "how much power I get out" and "how well I convert the heat from the hot source into useful work."

4.4. Bounded Ecological Function Regime Versus u_L

Figure 5a,b show the so-called bounded ecological efficiency ϕ, defined by Equation (27). This definition is positive compared to other formulations, since it is the definition of the classical ecological function. If one defines the function as $E_F = P - T_L\dot{S}_{gen}$, then this quantity can become negative in very dissipative regimes (because the entropic penalty dominates), which sometimes makes direct physical interpretation difficult.

The figure shows that $\phi(u_L)$ also exhibits a maximum value, which again falls within the same band as $P(u_L)$ and $P_E(u_L)$, where they have maxima. This means that the conductance assignment that maximizes the cycle's power does not generate disproportionate exergy destruction. From an environmental/second law perspective, the power optimum is acceptable. This is particularly significant, as it offers both a physical and environmental interpretation of the optimal operating point. The system not only achieves high power but also maintains thermodynamically efficient behavior, with moderate levels of internal irreversibility.

4.5. Second Law Efficiency Regime Versus u_L

Figure 6a,b show the behavior of the second law efficiency regime η_{II} given by Equation (29). Like the previous behaviors, $\eta_{II}(u_L)$ has a maximum located in the u_L region, which is similar to the other previous regimes.

This result confirms the argument: the same conductance distribution range between the hot and cold heat exchangers that is optimal for power output is also very good from the perspective of overall exergy utilization. This is important when justifying design under second law criteria or energy/industrial audits, because high power output alone is not enough; it must be demonstrated that the exergy from the hot reservoir is not wasted.

5. Conclusions

Finally, we constructed Tables 1 and 2, which show a summary listing; for each combination (τ, U_T), the value of u_L that maximizes each objective function includes net power P, thermal efficiency η, efficient power E_P, bounded ecological efficiency ∅, and second law efficiency η_{II}. The corresponding maximum values are also reported (e.g., P^*

in kW, η^*, and similarly for the other regimes). From the results shown in the table above, we can draw the following conclusions.

1. For each τ and each U_T, the optimal value of u_L that maximizes P is typically around $u_L \approx 0.6$ (for example, in the case of $\tau = 3.25$ and $U_T = 2.5$ kW/K, an optimum was obtained around $u_L \cong 0.59$, with maximum power on the order of tens of kW, efficiency $\eta \sim 0.2$, and so on). These results can be clearly seen in Figure 7.

2. The u_L that maximizes η (pure thermal efficiency), the one that maximizes E_P, the one that maximizes $\varnothing$, and the one that maximizes η_{II} are all extremely close to this same optimal power value.

3. When U_T increases (for example, from 2.5 kW/K to 10.0 kW/K), the maximum power P^* increases appreciably, but the optimum zone of u_L remains approximately in the same band. That is, providing the system with more heat exchange capacity improves the quantitative efficiency (more kW) but does not drastically change the conductance distribution strategy between the hot ($1 \rightarrow 2$) and cold ($3 \rightarrow 1$) heat exchangers.

From an engineering point of view, this has a very useful interpretation since there is a robust operating band around $u_L \sim 0.6$ that simultaneously maximizes net power, maintains high thermal efficiency, maximizes efficient power, minimizes relative entropic impact (high $\varnothing$), and maximizes exergy utilization (high η_{II}).

Table 1. Numerical results for the different operating regimes (P^*, η^*) and effectiveness.

τ	U_T (kW/K)	η_E	u_L^* (P)	P^* (kW)	u_L^* (η)	η^*	u_L^* (E_P)
3.25	2.5	0.9	0.584	19.928	0.627	0.062	0.601
3.25	5	0.9	0.575	46.394	0.614	0.105	0.592
3.25	7.5	0.9	0.579	55.342	0.605	0.117	0.588
3.25	10	0.9	0.579	57.625	0.601	0.120	0.588
3.75	2.5	0.9	0.588	29.057	0.640	0.075	0.610
3.75	5	0.9	0.578	66.340	0.622	0.123	0.597
3.75	7.5	0.9	0.579	78.950	0.610	0.136	0.592
3.75	10	0.9	0.579	82.167	0.610	0.139	0.588

Table 2. Numerical results for the different operating regimes ($E_P^*, \varnothing^*, \eta_{II}^*$) and effectiveness.

τ	U_T	η_E	E_P^*	u_L^*	$\varnothing^*$	u_L^*	η_{II}^*
	(kW/K)			($\varnothing$)		(η_{II})	
3.25	2.5	0.9	1.244	0.631	0.095	0.627	0.091
3.25	5	0.9	4.864	0.618	0.161	0.614	0.152
3.25	7.5	0.9	6.453	0.605	0.180	0.605	0.169
3.25	10	0.9	6.885	0.601	0.184	0.601	0.173
3.75	2.5	0.9	2.168	0.644	0.107	0.640	0.103
3.75	5	0.9	8.141	0.627	0.178	0.622	0.168
3.75	7.5	0.9	10.747	0.614	0.198	0.609	0.186
3.75	10	0.9	11.454	0.605	0.203	0.605	0.190

This convergence between classical regimes (power, efficiency) and more ambient or second law regimes ($\varnothing$, η_{II}) is one of the strongest results of this study because it suggests that the design optimum does not critically depend on choosing a single arbitrary figure of merit. Instead, several reasonable figures of merit point to the same design range, making this recommendation more defensible in real-world applications (for example, in waste heat recovery in industrial plants, where useful power is of interest but overall exergy performance is also monitored). Instead, several reasonable figures of merit point to the same design range, making this recommendation more defensible in real-world applications, such as waste heat recovery units in industrial plants, pulse combustion or pulsed-detonation engines, and small gas turbine or micro-CHP systems, where not only useful power but also exergy performance and environmental impact are monitored.

Author Contributions: Conceptualization, J.C.P.-P.; methodology, formal analysis, and investigation, J.C.C.-E., R.T.P.-H., J.C.P.-P., J.C.-S. and D.L.-L.; visualization, writing—review and editing, and data curation, J.C.P.-P. and R.T.P.-H.; software, validation, writing—original draft preparation, resources, and supervision. All authors have read and agreed to the published version of the manuscript.

Funding: This work was funded in part by the economic support program of the Secretaría de Ciencia, Humanidades, Tecnología e Innovación (Secihti), and the Secretaría de Investigación y Posgrado (SIP) of the Instituto Politécnico Nacional under grant no. SIP-20251347 and Proyecto multidisciplinario no. SIP-3398.

Institutional Review Board Statement: Not applicable.

Data Availability Statement: The data that support the findings of this study are available from the corresponding author upon request.

Acknowledgments: The authors would like to thank the Instituto Politécnico Nacional of México (Secretaría Académica, SIP, COFAA, CIC) and the SECIHTI for their support in the development of this work.

Conflicts of Interest: The authors declare no conflicts of interest.

References

1. Curzon, F.; Ahlborn, B. Efficiency of a Carnot engine at maximum power output. *Am. J. Phys.* **1975**, *43*, 22–24. [CrossRef]
2. Berry, R.S.; Salamon, P.; Andresen, B. How it all began. *Entropy* **2020**, *22*, 908. [CrossRef] [PubMed]
3. Andresen, B.; Salamon, P.; Berry, R.S. Thermodynamics in finite time: Extremals for imperfect engines. *J. Chem. Phys.* **1997**, *66*, 157. [CrossRef]
4. Bejan, A. Models of power plants that generate minimum entropy while operating at maximum power. *Am. J. Phys.* **1996**, *64*, 1054. [CrossRef]
5. Feidt, M. The history and perspectives of efficiency at maximum power of the Carnot engine. *Entropy* **2017**, *19*, 369. [CrossRef]
6. Sieniutycz, S.; Salamon, P. *Finite-Time Thermodynamics and Thermoeconomics*; Taylor & Francis: New York, NY, USA, 1990.
7. De Vos, A. *Endoreversible Thermodynamics of Solar Energy Conversion*; Oxford University Press: Oxford, UK, 1992.
8. Bejan, A. Entropy generation minimization: The new thermodynamics of finite size devices and finite-time processes. *J. Appl. Phys.* **1996**, *79*, 1191–1218. [CrossRef]
9. Hoffmann, K.H.; Burzler, J.M.; Shubert, S. Endoreversible thermodynamics. *J. Non-Equilib. Thermodyn.* **1997**, *22*, 311–355.
10. Wu, C.; Chen, L.; Chen, J. (Eds.) *Recent Advances in Finite Thermodynamics*; Nova Science: New York, NY, USA, 1999.
11. Velasco, S.; Roco, J.M.M.; Medina, A.; Hernández, A.C. A generalization of the ecological optimization criterion. *Phys. Rev. E* **2000**, *61*, 5578–5581.
12. Barranco Jiménez, M.A.; Pacheco Paez, J.C.; Angulo Brown, F. Optimization of heat engines using different heat transfer laws by means of the method of saving functions, X International Congress of Physics Engineering (X CIIF) 2020. *J. Phys. Conf. Ser.* **2021**, *1723*, 012066.
13. Novikov, I. The efficiency of atomic power stations (A review). *J. Nucl. Energy II* **1958**, *7*, 125–128. [CrossRef]
14. Sullivan, C. Newton's law of cooling. A critical assessment. *Am. J. Phys.* **1990**, *58*, 956–960. [CrossRef]
15. Agrawal, D.C. A simplified version of the Curzon–Ahlborn engine. *Eur. J. Phys.* **2009**, *30*, 1173–1179. [CrossRef]
16. Yan, Z.; Chen, J.; Andresen, B. Finite time thermodynamic optimization of the power output and efficiency for a finite-time Carnot cycle. *J. Appl. Phys.* **1990**, *68*, 3740–3744.
17. Angulo-Brown, F. An ecological optimization criterion for finite-time heat engines. *J. Appl. Phys.* **1991**, *69*, 7465–7469. [CrossRef]
18. Yasunaga, T.; Ikegami, Y. Finite-Time Thermodynamic Model for Evaluating Heat. *Entropy* **2020**, *22*, 211. [CrossRef]
19. Chen, L.; Zhou, J.; Sun, F.; Wu, C. Ecological performance optimization of a thermoacoustic heat engine. *Rev. Mex. Física* **2010**, *56*, 386–393.

20. Levario-Medina, S.; Arias-Hernández, L.A. The $P\Phi$-Compromise Function as a criterion of merit to optimize irreversible thermal engines. *arXiv* **2019**, arXiv:1908.11861.
21. Hoffmann, K.H. Recent Developments in Finite Time Thermodynamics. *Tech. Mech.* **2002**, *22*, 14–25.
22. Hoffmann, K.H.; Andresen, B.B.; Salamon, P. Finite-time thermodynamics tools to analyze dissipative processes. In *Proceedings of the 240 Conference: Science's Great Challenges*; Advances in Chemical Physics; Wiley: New York, NY, USA, 2014; Volume 157, pp. 57–67.
23. Hoffmann, K.H. Endoreversible Thermodynamics. In Proceedings of the 18th Joint European Thermodynamics Conference (JETC 2025), Belgrade, Serbia, 26–30 May 2025.
24. Chimal-Eguía, J.C.; Páez-Hernández, R.T.; Pacheco-Páez, J.C.; Saldana-Pérez, M.; Ladino-Luna, D. Linear Irreversible Thermodynamics: A Glance at Thermoelectricity and the Biological Scaling Laws. *Entropy* **2023**, *25*, 1575. [CrossRef]
25. Chen, L.; Tang, C.; Feng, H.; Ge, Y. Power, Efficiency, Power Density and Ecological Function Optimization for an Irreversible Modified Regenerative Brayton Cycle (isothermal heating). *Energies* **2020**, *13*, 5133. [CrossRef]
26. Tang, C.Q.; Chen, L.; Feng, H.; Ge, Y. Four-Objective Optimizations for an Improved Irreversible Closed Modified Simple Brayton Cycle. *Entropy* **2021**, *23*, 282. [CrossRef] [PubMed]
27. Rojas-Gamboa, D.A.; Rodríguez, J.I.; Gonzalez-Ayala, J.; Angulo-Brown, F. Ecological efficiency of finite-time thermodynamics. *Phys. Rev. E* **2018**, *98*, 022130. [CrossRef] [PubMed]
28. Chimal-Eguía, J.C. FTT variational: MPO, EF, PD. *Entropy* **2012**, *14*, 2611–2627. [CrossRef]
29. Pacheco-Paez, J.C.; Angulo-Brown, F.; Barranco-Jiménez, M.A. Thermoeconomic Optimization of an Irreversible Novikov Plant Model under Different Regimes of Performance. *Entropy* **2017**, *19*, 118. [CrossRef]
30. Antar, M.A. Thermoeconomic Considerations in the Optimum Allocation of Heat Exchanger Inventory. *ASME J. Energy Resour. Technol.* **2002**, *124*, 28–35. [CrossRef]
31. Barranco-Jiménez, M.A. Finite-time thermoeconomic optimization of a non-endoreversible heat engine model. *Rev. Mex. Física* **2009**, *55*, 211–220.
32. Valencia-Ortega, G.; Levario-Medina, S.; Barranco-Jiménez, M.A. The role of internal irreversibilities in the performance and stability of Power Plant Models Working at Maximum ϵ-Ecological Function. *J. Non-Equilib. Thermodyn.* **2021**, *46*, 413–429. [CrossRef]
33. Juárez-Huerta, V.H.; Sánchez-Salas, N.; Chimal-Eguia, J.C. Optimization Criteria and Efficiency of a Thermoelectric Generator. *Entropy* **2022**, *24*, 1812. [CrossRef]
34. Lichty, C. *Combustion Engine Processes*; McGraw-Hill: New York, NY, USA, 1967.
35. Georgiou, D.P. Useful work and the thermal efficiency in the ideal Lenoir with regenerative preheating. *J. Appl. Phys.* **2008**, *88*, 5981–5986. [CrossRef]
36. Shen, X.; Chen, L.G.; Ge, Y.L.; Sun, F.R. Finite-time thermodynamic analysis for endoreversible Lenoir cycle coupled to constant-temperature heat reservoirs. *Int. J. Energy Environ.* **2017**, *8*, 272–278.
37. Ahmadi, M.H.; Nazari, M.A.; Feidt, M. Thermodynamic analysis and multi-objective optimisation of endoreversible Lenoir heat engine cycle based on the thermo-economic performance criterion. *Int. J. Ambient Energy* **2019**, *40*, 600–609. [CrossRef]
38. Wang, R.; Ge, Y.; Chen, L.; Feng, H.; Wu, Z. Power and Thermal Efficiency Optimization of an Irreversible Steady-Flow Lenoir Cycle. *Entropy* **2021**, *23*, 425. [CrossRef]

entropy

MDPI

Article

Thermodynamics of Intrinsic Reaction Coordinate (IRC) Chemical Reaction Pathways

Frank Weinhold

Theoretical Chemistry Institute and Department of Chemistry, University of Wisconsin-Madison, Madison, WI 53706, USA; weinhold@chem.wisc.edu

Abstract: We address the scientific "time" concept in the context of more general relaxation processes toward the *Wärmetod* of thermodynamic equilibrium. More specifically, we sketch a construction of a conceptual ladder of chemical reaction steps that can rigorously bridge a description from the microscopic domain of molecular quantum chemistry to the macroscopic materials domain of Gibbsian thermodynamics. This conceptual reformulation follows the pioneering work of Kenichi Fukui (Nobel 1981) in rigorously formulating the *intrinsic reaction coordinate* (IRC) pathway for controlled description of non-equilibrium passages between reactant and product equilibrium states of an overall material transformation. Elementary *chemical reaction steps* are thereby identified as the logical building-blocks of an integrated mathematical framework that seamlessly spans the gulf between classical (pre-1925) and quantal (post-1925) scientific conceptions and encompasses both static and dynamic aspects of material change. All modern chemical reaction rate studies build on the apparent infallibility of quantum-chemical solutions of Schrödinger's wave equation and its Dirac-type relativistic corrections. This infallibility may now be properly accepted as an added"inductive law" of Gibbsian chemical thermodynamics, the only component of 19th-century physics that passed *intact* through the revolutionary quantum upheavals of 1925.

Keywords: thermodynamics; intrinsic reaction coordinate; natural resonance theory; quantum cluster equilibrium

1. Introduction

> *What then is time? If no one asks me, I know what it is. If I wish to explain what it is to him who asks me, I do not know.*
>
> —St. Augustine (Fifth Century).

The kind invitation to contribute to this Special Issue of *Entropy* gives opportunity to interweave related strands of the author's personal scientific interests over the decades since their intersection with the original finite-time thermodynamics (FTT) studies of the 1970s [1]. The discussed topics include the original 1975–1976 papers on thermodynamic geometry [2–4], analytic extensions of frequency-dependent polarizabilities to the complex plane of imaginary-frequency values [5], as well as related methods for obtaining the lifetime of unstable species as the imaginary component of system energy eigenvalue after complex-coordinate rotation of the Schrödinger Hamiltonian operator [6]. In each case, these studies touch upon Einstein's extensions of Euclidean geometry into the large-scale

Riemannian structure of general relativity and $3d$-spatial coordinates into the $4d$ space–time coordinates of special relativity, spanning microscopic and macroscopic domains of physical and chemical description.

The identification of the measurable lifetime of a metastable species with the imaginary component of an analytically continued solution of Schrödinger's equation naturally draws attention to the parallels with other physical descriptors of spontaneous temporal change. These include the *entropy* property, whose monotone *increase* in spontaneous changes of isolated systems underlies the second law of thermodynamics. These parallels also hint at fundamental connections between time-like measures of spontaneous change as perceived in microscopic and macroscopic domains of chemical reaction dynamics, the focus of present study.

In the present work, we suggest how to build formal bridges that can better connect the logic, mathematical formalisms, and conceptual perspectives of the microscopic (Schrödinger/Dirac/Pauling) domain of atoms and molecules with the macroscopic (Gibbs/Boltzmann) domain of bulk material properties in equilibrium or non-equilibrium states. In each case, our viewpoint is that of a *chemical* practitioner, in recognition of chemistry's role as the "central science" [7] that bridges many aspects of marine, geological, biological, atmospheric, and interstellar studies in the natural sciences.

Of particular interest are the optimal mathematical and conceptual building-blocks of a unified theoretical framework that can smoothly bridge the ca. 10^{23}-fold transition from microscopic to macroscopic distance and energy scales. Such a framework necessarily deals with *change* (t-dependence) as well as the time-independent *Wärmetod* of the terminal equilibrium-state limit. It must also grapple with the *logical* gap that separates the perceptions of classical (pre-1925) scientists from those following discovery of the quantal Schrödinger/Dirac equations (and equivalent matrix-algebraic formulations) in 1925–1928 [8]. At a still deeper level, it must strive to reconcile contending preferences for inductive (axiom-free) vs. deductive (axiom-based) modes of scientific reasoning. In each case, bridge-building begins from one side or the other of a conceptual divide. In the following, we deal with these divides in approximate chronological order.

2. Inductive Scientific Tenets and Their Survival in the Face of Conceptual Revolution

We begin (as does every newborn child) with a rigorously inductive (axiom-free) approach to describing the nature that surrounds us. What are gradually perceived as experiential *patterns* or regularities of past natural events become the basis for increasingly confident prediction of future events. Any violation of such inductive expectation is cause for its dismissal from consideration as a possible law of nature, and only those regularities that remain strictly exception-*free* to all competent observers are considered proper starting points for rigorous deductions of other consequences. In this manner, one gains consensus for the universality of conclusions from laws that are strictly self-consistent with direct observations from nature.

Uniquely among the natural sciences, *equilibrium chemical thermodynamics* epitomizes such inductive foundations. As masterfully formulated by J. W. Gibbs [9], the first and second laws of thermodynamics that govern the thermal behavior of all chemical materials can be rigorously expressed in terms of low-order differential relationships among the small set of measurable properties [U (internal energy), S (namesake of this journal), V (volume), and N_i (mole number of the ith pure substance)] that fully specify the *equilibrium state* of the macroscopic chemical system. As Gibbs recognized, the two alternative but fundamentally equivalent representations of thermodynamic relationships (involving 1st-

and 2nd-order derivatives only) can be based solely on the following energy-based (1a) or entropy-based (1b) functional expressions:

$$U = U(S,V,\{N_i\}) \tag{1a}$$

$$S = S(U,V,\{N_i\}) \tag{1b}$$

whereas the (*nearly* equivalent) Legendre transformations of various types [10] require subsidiary assumptions of higher-order differentiability and lower generality.

Also uniquely among the natural sciences, Gibbsian equilibrium thermodynamics passed *seamlessly* through the quantum revolutions of 1925 without need for an iota of revision. If anything, the discovery of bizarre quantal phenomena that defied classical description (including Heisenberg's uncertainty principle and associated non-vanishing vibrational fluctuations that persist even at T = 0 K) only *solidified* Gibbs's conclusions concerning unwarranted third-law and higher-order differentiability assumptions that were entertained by others. These considerations suggest that building on thermodynamic-like inductive notions is a *robust* foundation for a properly unified description of electronic interactions from microscopic to macroscopic limits.

Consistent with all post-1925 measurements of physical and chemical properties, one can say without exception that *all* such critical measurements, carried out to the error limits of current technology, are found to be in *exact* compliance with corresponding solutions of the quantum-mechanical Schrödinger equation and its Dirac-type corrections. This conclusion applies not only to the small number of cases for which exact analytic solutions of Schrödinger's equation are known, but more importantly to the numerical solutions determined by systematically improvable algorithms and currently available computational resources, including rigorous upper- and/or lower-bound error limits [11].

Therefore, until proven otherwise, one can take the *truth of the Schrödinger equation* (and its proper relativistic generalization) as a veritable inductive law of chemical thermodynamics. This allows state-of-the-art quantum chemistry methods to acquire axiomatic-like status in deducing authentic patterns of chemical behavior across the broad span between microscopic (electronic) vs. macroscopic (thermodynamic) levels of description.

Currently, the broad suite of modern density functional theoretic (DFT) methods, based on fundamental contributions of Walter Kohn (Nobel 1998) and others [12], are the practical quantum chemistry tools of choice for realistic approximations to authentic solutions of Schrödinger's equation in broad areas of chemical research. DFT-based approaches are based on rigorous theorems for the adequacy of correct *electron density* [$\rho(\mathbf{r})$] description to provide exact variational solution of Schrödinger's equation.

This in turn calls conceptual attention to the electronic *orbitals* $\{\varphi_i(\mathbf{r})\}$ and non-negative *populations* $\{n_i\}$ (compliant with the Pauli exclusion principle) that quite generally allow the expression of the electron density as a *convex* combination of squared-orbital densities,

$$\rho(\mathbf{r}) = n_1 \,|\, \varphi_i(\mathbf{r}) \,|^2 + n_2 \,|\, \varphi_2(\mathbf{r}) \,|^2 + \ldots \geq 0 \tag{2a}$$

with orbital occupancies summing to the total number of electrons (N),

$$\textstyle\sum_i n_i = N \tag{2b}$$

For conceptual purposes, optimal orthonormal orbitals are frequently chosen to emulate the 1-center (lone-pair) and 2-center (bond) electron-pairs of familiar localized Lewis-structural bonding patterns that long pre-date the discovery of quantum mechanics, but this choice [13] is a matter of convenience. Mathematically optimal population analysis [14] of such DFT or higher-level density distributions in terms of localized atomic orbitals is now a routine by-product of electronic-structure calculations conducted by chemists in every area of specialization. All such efforts are supported by inductive faith in the accuracy of current methods for obtaining systematically improvable solutions of Schrödinger's equation that lead reliably to agreement between theory and measurement (within respective error limits) in all known cases.

3. Quantum Cluster Equilibrium Theory of Fluid Properties

Still another bridge between the Schrödinger-based description of molecular species and the thermodynamic-level description of the associated fluid phase behavior is provided by the *quantum cluster equilibrium* (QCE) model of molecular cluster interactions [15]. QCE methodology was initially developed for simple phase-transition properties of pure fluids but subsequently extended by Kirchner and coworkers to binary mixtures and more complex solution properties [16].

As the name implies, QCE methodology employs the quantum chemistry of molecular *clusters* (of pure or binary-mixed types) in (*T*,*P*)-dependent equilibrium mixtures to predict the properties of the associated macroscopic fluid-phase diagram. In the case of water, the vapor-phase region of the equilibrium QCE distribution is dominated (as expected) by H_2O monomers, while the liquid-phase distribution is dominated by *ring*-like clusters of 5- and 6-member topology [17], as significantly favored by strongly *cooperative* (*non*-pairwise-additive) aspects of hydrogen bonding. Inclusion of sparsely populated *ionic* water clusters in the equilibrium QCE mixture also allows one to also predict (*T*,*P*)-dependent *pH* values for the aqueous fluid phase [18]. With inclusion of clathrate-like bucky-ball clusters, one also finds formation of metastable *solid* phases that connect to the fluid phases in characteristic triple-point fashion [19]. All these (systematically improvable) features of the QCE model partition function are qualitatively or semi-quantitatively consistent with known experimental properties of the macroscopic phase diagram. The QCE model can therefore be seen as an ever-improvable formal "bridge" between supramolecular clusters of the quantal regime and the macroscopic phases of classical Gibbsian thermodynamics.

4. Non-Equilibrium Thermodynamics on IRC Reaction Pathways: An Integrated Resonance-Mechanistic View

Far the greatest challenge to further the extension of Gibbsian-like inductive conceptions is to find a corresponding bridge that spans the full range of equilibrium (static; *t*-independent) states and non-equilibrium (dynamic; *t*-dependent) phenomena. As conceptual building-blocks for such non-equilibrium transformations that intrinsically involve time-like variability, we employ Fukui's *intrinsic reaction coordinate* (IRC) [20] concept for each elementary chemical reaction underlying the material change. The IRC coordinate allows the entire pathway of an elementary chemical reaction (from equilibrium reactant- to final product complex) to be *uniquely* determined from the stationary *transition*-state geometry that lies at the highest energy along the IRC pathway. In this manner, complex material changes can be systematically broken down into IRC-based pathways through the non-equilibrium regions that connect each equilibrium reactant and product species of the overall process.

Note that the practical application of the IRC concept to describe each non-equilibrium transformation (bond-rearranging, configurational, torsional, ...) of a general chemical process relies only on standard quantum-chemical methodology and requires *no* invocation of an empirically constructed *t*-dependent "molecular dynamics" or control operator to describe the chemical reaction process. IRC-based description of chemical relaxation events thereby differs in fundamental respects from dissections into perturbatively small deviations from ideal adiabaticity such as assumed, e.g., in Kosloff's quantum thermodynamics viewpoint [21], which aims to "insert" quantum dynamics into thermodynamics, even though Gibbsian thermodynamics remains *identical* to its origins that long preceded the 1925 quantal revolution.

By dissecting the overall macroscopic process into the underlying IRC-based chemical reactions in this manner, we gain equilibrium footholds at the terminal *states* of reactant and product species in each non-equilibrium reactive step, which thereby form a connected ladder of reactive-type events that comprehensively compose the perceived overall material change. Well-defined properties of the macroscopic process can thereby be based on measurable *changes of state* between the microscopic reactant and product species for each underlying step of the composite manifold of IRC pathways. As remarked above, modern electronic structure programs routinely provide the (T,P)-dependent enthalpic and free-energy values of chosen reactant and product chemical species of each step, allowing direct comparison with thermodynamic measurements on the corresponding macroscopic process under chosen temperature and pressure conditions.

Still closer connections can then be drawn between the Schrödinger-based descriptors of a chemical reaction and the measurable progress variables of the corresponding macroscopic process. In the Born–Oppenheimer approximation (which generally suffices for practical chemical purposes), Pauling's resonance-theoretic concepts for describing the electronic delocalization of individual molecular species can be readily extended to *natural resonance theory* (NRT) analysis [22] of individual points along a chosen pathway on the chosen $E(\{\boldsymbol{R}_\mathbf{A}\})$ potential energy surface (PES). Similar NRT evaluations could also be performed along the extremal-type PES pathway that underlies the original FTT methodology of Andresen and coworkers. As in Pauling's original heuristic formulation of resonance theory, NRT analysis expresses the delocalized electron-density distribution $\rho(\mathbf{r})$ as a convex combination

$$\rho(\mathbf{r}) = \sum\nolimits_\alpha w_\alpha \, \rho_\alpha(\mathbf{r}) \tag{3a}$$

of those for possible Lewis-like bonding patterns $\{\rho_\alpha(\mathbf{r})\}$, with corresponding resonance weightings $\{w_\alpha\}$ satisfying

$$1 \geq w_\alpha \geq 0, \text{ all } \alpha \tag{3b}$$

and summing to unity

$$1 = \sum\nolimits_\alpha w_\alpha \tag{3c}$$

If $b_{AB}{}^{(\alpha)}$ denotes the integer number of bonds between atoms A and B in the Lewis-like bonding pattern for electron-density contribution $\rho_\alpha(\mathbf{r})$, the overall *natural bond order* (b_{AB}) for each interatomic A and B pair is evaluated as

$$b_{AB} = \sum\nolimits_\alpha w_\alpha \, b_{AB}{}^{(\alpha)} \tag{3d}$$

in accordance with Pauling's original resonance precepts. Note, however, that the NRT definition (3d) extends seamlessly to *all* atoms of the chosen system, whether considered to be members of the same molecule or not.

Similar to Equations (2a) and (2b) above, the resonance-theoretic Equations (3a)–(3d) thereby identify each such resonance-hybrid electronic distribution along a reactive pathway as a "resonance hybrid" of those for contributing Lewis-like bonding patterns. This in turn allows the resonance weightings $\{w_\alpha\}$ to be understood as population-type *percentages* of total electron density at the chosen point of the PES, corresponding to the "charge transfer" terminology that is often employed to describe resonance-type electronic delocalization. Further mathematical and computational details of NRT algorithms [22] are described elsewhere.

The resonance-theoretic expressions seem to provide the most general mechanistic representation of chemical transformation from one bonding pattern to another. All such quantal descriptors are in sharp contrast to the classical-type dipole–dipole rationalizations that were often invoked for such intrinsically stabilizing resonance-type phenomena. The convexity properties (3b) and (3c) of NRT weightings may also be contrasted with the more familiar LCAO-MO-type orbital mixings (involving linear combinations of complex coefficients of opposite signs and arbitrary magnitudes) that are often invoked from the unitarity property of electronic wavefunction theory.

5. Illustrative Example: Nitrosamine Degradation to Diazonium Ion

As a rather randomly chosen example, we illustrate the IRC-based "NRT portrait" of an elementary chemical transformation pertaining to carcinogenic properties of nitrosamines. Figure 1 depicts the sequence of reactions $\boldsymbol{R}_1$–$\boldsymbol{R}_3$ by which N-nitrosodimethylamine (NDMA) is converted to a nitrosonium cation that can methylate cytosine and obstruct its proper pairing with guanine in DNA. Reaction $\boldsymbol{R}_1$ represents the schematic replacement of H by OH at the methyl site, and $\boldsymbol{R}_3$ is similar to *E*1-elimination mechanism for acid-catalyzed ethanol dehydration. Here, we focus attention on the central $\boldsymbol{R}_2$ (formaldehyde-elimination) step of diazonium conversion.

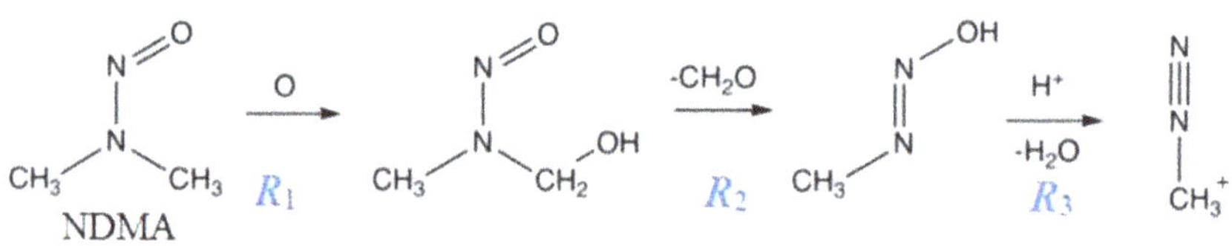

Figure 1. Schematic mechanism of metabolic activation of nitrosamine (NDMA) to diazonium methylating agent in three formal reaction steps $\boldsymbol{R}_1$, $\boldsymbol{R}_2$, and $\boldsymbol{R}_3$ (adapted from *Wikipedia* for "nitrosamine").

Simple DFT-level description of reaction step $\boldsymbol{R}_2$ is summarized in Figures 2–4. Figure 2 displays ground-state-reactant (*r*0), transition-state (*ts*0), and product (*p*0) geometries and relative energies for the $\boldsymbol{R}_2$ reaction step, as calculated at a common DFT level (B3LYP/6-311++G**/D3). Figure 3 displays the full *E*(*IRC*) energy profile, and Figure 4 displays the corresponding variations of NRT bond orders $\{b_{AB}(IRC)\}$ along the IRC reaction pathway. [Sample *Gaussian-16* input (.gau) and output files for calculating the energy and NRT bond orders for individual points on the IRC are available from the author upon request. Many details of the tandem *G16/NBO7* programs that perform these calculations—now commonly hosted in chemistry research centers but available also in laptop configurations—are described in the respective *Gaussian* or *NBO* websites.]

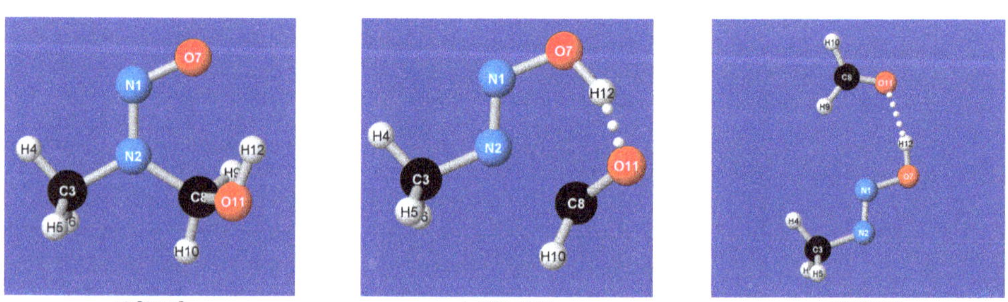

*r*0 **[0.00]** *ts*0 **[26.24]** *p*0 **[13.03]**

Figure 2. Optimized ground-state-reactant (*r*0; **left**), transition-state (*ts*0; **center**), and product (*p*0; **right**) geometries and relative energies (kcal/mol) for $\boldsymbol{R}_2$ reaction step of NDMA conversion to mutagenic methyldiazonium cation with expulsion of formaldehyde.

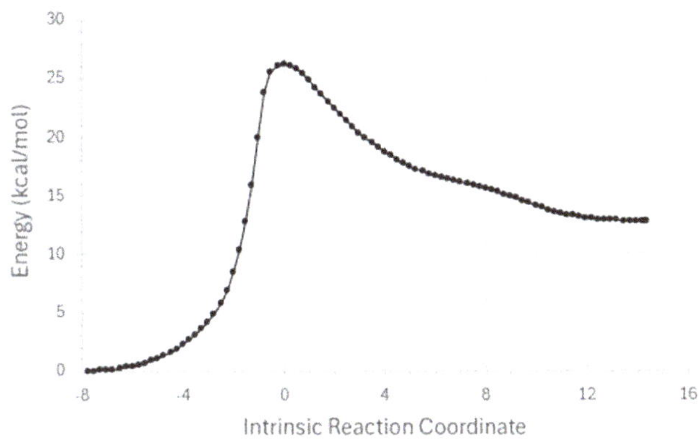

Figure 3. Calculated *E*(*IRC*) energetic variations along the IRC pathway for $\boldsymbol{R_2}$ reaction step of NDMA → diazonium conversion.

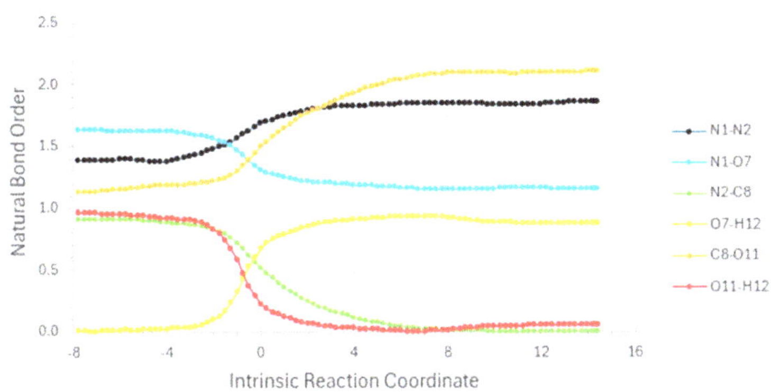

Figure 4. NRT bond-order variations along the IRC pathway for $\boldsymbol{R_2}$ reaction step.

The interatomic curves in Figure 4 represent the formation [$b_{O(7)H(11)}$, yellow] or dissociation [$b_{O(11)H(12)}$, red; $b_{N(2)C(8)}$, green] of bonds to the main CH_3NNO substrate, as well as the principal resonance bond-shifts [$b_{C(8)O(11)}$, gold; $b_{N(1)N(2)}$ black; $b_{N(1)O(7)}$, blue] within the substrate. Note that the *sub*-integer NRT bond orders for "inter"molecular interactions differ in *no* significant respect from conventional Pauling-type *supra*-integer bond orders for "intra" molecular interactions. The former correspond to the extension of resonance conceptions into the "no-bond" domain of H-bonding and related X-ogen interactions, which Pauling stoutly opposed until late in his career [23].

Note that the IRC coordinate of the horizontal axis in Figure 3 is indeed a valid "time-like" measure of progress from one stationary state of reaction to another. This follows from the fact that the vertical *E*(*IRC*) energy curve as plotted in Figure 3 conforms to the idealized reaction-energy profile that underlies rigorous transition-state theory of chemical kinetics [24] as employed in every modern laboratory study of molecular reaction dynamics. The "lifetime" associated with the metastable transition-state feature at *IRC* = 0 might be separately evaluated with the complex-coordinate rotation methodology of Ref. [6]. But more to the present point, every NRT bond-order value as displayed in the vertical axis of Figure 4 is found to satisfy the expected [25] strong bond order-bond length (BOBL) and related correlations that can successfully predict a variety of MW/IR/UV/NMR properties that are measurable over the course of chemical reaction. For example, if spectroscopic monitoring establishes that the point of *equal* O(7)H(12) and O(11)H(12) signal ["*half*-transfer of H(12) from O(7) to O(11)"] occurred Δt UTC clock-ticks prior to the highest-energy point (*IRC* = 0) of the reactive sequence, one can affix Δt as the "standard-*t*" value at the corresponding red/yellow curve-crossing near $IRC \approx -1.0$ in Figure 4.

Note also that the IRC pathway as defined by Fukui [20] has no "near-adiabatic" character nor other resemblance to the pathway envisioned by Kosloff, as can be verified from calculated values of energy, enthalpy, and Gibbs free energy that are included in *Gaussian* output for each step of the IRC search. However, the *intrinsic* property of the IRC insures optimal utility for practical applications throughout the modern chemical and biochemical research domain. As found in many examples—including those for configurational, torsional, or enantiomeric isomerism and long-range supramolecular aggregation—the IRC displays the essential *continuity* of cooperatively-coupled resonance-type interactions throughout the intra- and intermolecular domain in a chemically intuitive manner. Further details of NRT-based mechanistic description follow the pedagogical logic of previously published examples [26].

6. Mathematical and Physical Logic: Seeking Unity

The preceding discussion points to what may be perceived as a fundamental logical dichotomy between mathematical and physical sciences. The mathematical sciences epitomize the realm of *deductive* logic, where a chosen set of axioms (not to be questioned further) lead by the strict rules of deductive logic to an entire subject-area of mathematical relationships. Many such mathematical niches are valued for their near-magical connections to physical phenomena, but others are not, described instead as pure mathematics not to be tinged with associations to practical applications.

The physical sciences have frequently benefited from finding their close associations with one or more established fields of mathematical theory. However, each area of physical research is intrinsically constrained by facts of nature as obtained through the specialized research methodologies of its experimental practitioners. If experimental facts are found to conflict with an adopted mathematical framework, that framework must be abandoned or revised until satisfactory agreement between mathematical assumptions and experimental reality is recovered. As one example, the mathematics of Euclidean geometry (a prototype of axiomatic deductive logic) long appeared sufficient until experimental astronomical observations [27] led to its replacement by more general Riemannian geometry in the early 20th century.

The science of equilibrium thermodynamics offers an exception to how mathematics typically interfaces with physical science. Key underlying regularities ("laws") of macroscopic thermodynamics were first recognized and formalized by Carnot and others [28] in attempts to understand the scientific principles of steam locomotion in the early 19th century. In the elegant formulation of J. W. Gibbs, the science of thermodynamics came to be firmly based on the *inductive* logic of such laws of experience whose violation has never been observed.

Foregoing sections of this work have suggested how successive conceptual bridges can be employed to further extend the chain of rigorous inductive (thermodynamics-like) inferences from their macroscopic 19th-century origins to the quantal perceptions of the present. This extended "inductive" viewpoint encompasses all materials and natural processes of our observable terrestrial surroundings, down to the level of known atomic, molecular, and supramolecular species.

The coherence and unity of an inductively extended viewpoint is implicitly dependent on the coherence and unity of nature itself, and what can be considered observable (measurable) in nature. This unity suggests that entirely unexpected advances of "pure" mathematics may be initiated when the inductively inferred mathematics of one sub-area is found to be *equivalent* to that which describes the inductive facts of another sub-area. One example is the discovery of quantum mechanics itself, which led to subsequent recognition [29] that the mathematical theory of certain low-order differential equations is *equivalent* to the non-commutative algebra of matrix operations, both of which can be viewed in the more general framework of Hermitian operators in Hilbert space. Another example is summarized in the conclusion, "Thermodynamics *is* geometry" [30], that allows thermodynamic inferences to be drawn in the isomorphic mathematical framework of Riemannian geometry.

7. Summary and Conclusions

We have outlined a conceptual framework for extending the inductive logic of macroscopic thermodynamics to the microscopic electronic level of modern computational quantum chemistry, illustrating the underlying computational methodology with a simple application related to chemical mutagenesis. This extension is based on inductive recognition of the infallibility of accurate solutions of the Schrödinger equation in yielding *exact* predictions (within respective computational and experimental error limits) of experimentally measured properties of atomic, molecular, and supramolecular species in all known cases—effectively, an additional *law of nature* since the discovery of the quantum mechanical equations a century ago. The steps of the conceptual ladder connecting macroscopic material changes to molecular-level transformations are the *elementary chemical reactions*, each described by the associated IRC reaction pathway that traverses the non-equilibrium region between equilibrium reactant and product species. Such a stepwise progression through the stationary points of a macroscopic change of state somewhat resembles the finite-time thermodynamics strategy [1] first suggested nearly a half-century ago.

At a more speculative level, we have also suggested that the specific mathematical tools (of thermodynamics, quantum chemistry, statistical mechanics, . . .) that are found to be effective in dealing with spontaneous natural change may themselves reflect a deeper underlying mathematical unity. Prominent examples include the isomorphism between matrix-algebraic and differential representations that led to recognition of the Hilbert-space underpinnings of quantum mechanics, as well as more recent recognition of the metric geometry underlying equilibrium thermodynamics.

Funding: This research received no external funding.

Institutional Review Board Statement: Not applicable.

Data Availability Statement: The original contributions presented in this study are included in the article. Further inquiries can be directed to the author.

Acknowledgments: I thank Eric Glendening for assistance with Figures 3 and 4, Clark Landis and Fleming Crim for discussions, and the many co-workers cited below for the inspiration and pleasures of collaboration over the decades.

Conflicts of Interest: The author declares no conflict of interest.

Abbreviations

IRC	Intrinsic Reaction Coordinate
DFT	Density Functional Theoretic
NRT	Natural Resonance Theory

References

1. Andresen, B.; Berry, R.S.; Nitzan, A.; Salamon, P. Thermodynamics in finite time. I. The step-Carnot cycle. *Phys. Rev. A* **1977**, *15*, 2086–2093, Salamon, P.; Andresen, B.; Berry, R.S. Thermodynamics in finite time. II. Potentials for finite-time processes. *Phys. Rev. A* **1977**, *15*, 2094–2102. Salamon, P.; Andresen, B.; Gait, P.D.; Berry, R.S. The significance of Weinhold's length. *J. Chem. Phys.* **1980**, *73*, 1001–1002. Salamon, P.; Berry, R.S. Thermodynamic length and dissipated availability. *Phys. Rev. Lett.* **1983**, *51*, 1127–1130.
2. Weinhold, F. Metric geometry of equilibrium thermodynamics. *J. Chem. Phys.* **1975**, *63*, 2479–2483, Weinhold, F. Metric geometry of equilibrium thermodynamics. II. Scaling, homogeneity, and generalized Gibbs-Duhem relations. *J. Chem. Phys.* **1975**, *63*, 2484–2487. Weinhold, F. Metric geometry of equilibrium thermodynamics. III. Elementary formal structure of a vector-algebraic representation of equilibrium thermodynamics. *J. Chem. Phys.* **1975**, *63*, 2488–2495. Weinhold, F. Metric geometry of equilibrium thermodynamics. IV. Vector-algebraic evaluation of thermodynamic derivatives, *J. Chem. Phys.* **1975**, *63*, 2496–2501. Weinhold, F. Metric geometry of equilibrium thermodynamics. V. Aspects of heterogeneous equilibrium. *J. Chem. Phys.* **1976**, *65*, 559–564.
3. Weinhold, F. Geometric representation of equilibrium thermodynamics. *Acc. Chem. Res.* **1976**, *9*, 236–240.
4. Weinhold, F. *Classical and Geometrical Theory of Chemical and Phase Thermodynamics*; John Wiley: Hoboken, NJ, USA, 2009.
5. Glover, R.M.; Weinhold, F. Imaginary-frequency polarizability and van der Waals force constants of two-electron atoms, with rigorous bounds. *J. Chem. Phys.* **1977**, *66*, 191–198, Weinhold, F.; Robinson, P.D. Bivariational calculations of bounds on complex-frequency polarizabilities. *J. Chem. Phys.* **1978**, *68*, 2915–2921.
6. Moiseyev, N.; Certain, P.R.; Weinhold, F. Resonance properties of complex-rotated Hamiltonians. *Molec. Phys.* **1978**, *36*, 1613–1630, Moiseyev, N.; Certain, P.R.; Weinhold, F. Complex-coordinate studies of helium autoionizing resonances. *Intern. J. Quantum Chem.* **1978**, *14*, 727–736. Moiseyev, N.; Certain, P.R.; Weinhold, F. Complex-coordinate calculations with complex basis sets. *Phys. Rev. A* **1981**, *24*, 1254–1259. Weinhold, F. Time-conjugation in a unified quantum theory for Hermitian and non-Hermitian electronic systems under time-reversal symmetry. *Symmetry* **2013**, *13*, 808. The complex-coordinate-rotation algorithms of quantum chemistry are based on the fundamental Balslev-Combes theorem: Balslev, E.; Combes, J.M. Spectral properties of many-body Schrödinger operators with dilatation-analytic interactions. *Commun. Math. Phys.* **1971**, *22*, 280–294. Simon, B. Resonances in n-body quantum systems with dilatation analytic potentials and the foundations of time-dependent perturbation theory. *Ann. Math.* **1973**, *97*, 247–274.
7. Brown, T.; LeMay, H.; Bursten, B.; Murphy, C.; Woodward, P.; Stoltzfus, M. *Chemistry: The Central Science*, 14th ed.; Pearson: New York, NY, USA, 2017.
8. Born, M.; Heisenberg, W.; Jordan, P. Zur quantenmechanik II. *Zeits. f. Phys.* **1925**, *34*, 557–615, Schrödinger, E. Quantisierung als Eigenwertproblem. *Ann. Phys.* **1926**, *384*, 361–376. Dirac, P.A.M. The quantum theory of the electron. *Proc. R. Soc. Lond. A* **1928**, *117*, 610–624.
9. Gibbs, J.W. On the equilibrium of heterogenous substances. *Trans. Conn. Acad.* **1876**, *3*, 108–248, **1878**, 343–524.
10. Callen, H.B. *Thermodynamics and an Introduction to Thermostatics*, 2nd ed.; Wiley: New York, NY, USA, 1991.
11. Weinhold, F. Upper and lower bounds to quantum-mechanical properties. In *Advances in Quantum Chemistry*; Löwdin, P.-O., Ed.; Academic Press: New York, NY, USA, 1972; Volume 6, pp. 299–331.
12. Hohenberg, P.; Kohn, W. Inhomogeneous electron gas. *Phys. Rev. B* **1964**, *136*, 864–871, Kohn, W.; Sham, L.J. Self-consistent equations including exchange and correlation effects. *Phys. Rev. A* **1965**, *140*, 1133–1138. Parr, R.G.; Yang, W. *Density-Functional Theory of Atoms and Molecules*; Oxford U. Press: New York, NY, USA, 1989. Becke, A.D. Perspective: Fifty years of density-functional theory in chemical physics. *J. Chem. Phys.* **2014**, *140*, 18A301. Note that the formal vaariational foundations for modern $\rho(\mathbf{r})$-based DFT methods are *superior* to those underlying many popular wavefunction-based approximation methods such as MP(2) or CCSD and their many extensions. Although such methods provide a definite energy value E from a stipulated recipe of computational steps, they cannot generally provide the associated N-representable density matrices that assure *existence* of a corresponding antisymmetric N-electron wavefunction $\Psi(\mathbf{r}_1,\mathbf{r}_2,\ldots,\mathbf{r}_N)$; see Beste, A., Runge, K., Bartlett, R., Ensuring N-representability; Coleman's algorithm. *Chem. Phys. Lett.* **2002**, *355*, 263–269.
13. Weinhold, F.; Landis, C.R. *Discovering Chemistry with Natural Bond Orbitals*; Wiley: New York, NY, USA, 2012.
14. Reed, A.E.; Weinstock, R.B. Weinhold, Natural population analysis. *J. Chem. Phys.* **1985**, *83*, 735–746.
15. Weinhold, F. Quantum cluster equilibrium theory of liquids: General theory and computer implementation. *J. Chem. Phys.* **1998**, *109*, 367–372, Ludwig, R.; Weinhold, F.; Farrar, T.C. Quantum cluster equilibrium theory of liquids: Molecular clusters and thermodynamics of liquid ethanol. *Mol. Phys.* **1999**, *97*, 465–477. Hansen, M.J.; Wendt, M.A.; Weinhold, F. Tests of quantum cluster equilibrium (QCE)-based computational methods for describing formic acid clustering. *Mol. Phys.* **2002**, *101*, 1147–1153.
16. Kirchner, B.; Weinhold, F.; Friedrich, J.; Perlt, E.; Lehmann, S.B.C. Quantum cluster equilibrium. In *Many-Electron Approaches in Physics, Chemistry and Mathematics*; Bach, V., Site, L.D., Eds.; Springer Mathematical Physics Studies: New York, NY, USA, 2014; pp. 77–96, Ingenmey, J.; Domaros, M.V.; Kirchner, B. Predicting miscibility of binary liquids from small cluster QCE calculations. *J. Chem. Phys.* **2017**, *146*, 154502.

17. Weinhold, F. Quantum cluster equilibrium theory of liquids: Illustrative application to water. *J. Chem. Phys.* **1998**, *109*, 373–384. [CrossRef]
18. Perlt, E.; Domaros, M.v.; Kirchner, B.; Ludwig, R.; Weinhold, F. Predicting the ionic product of water. *Sci. Rep.* **2017**, *7*, 10244. [CrossRef] [PubMed]
19. Ludwig, R.; Weinhold, F. Quantum cluster equilibrium theory of liquids: Freezing of QCE/3-21G water to tetrakaidecahedral Bucky-Ice. *J. Chem. Phys.* **1999**, *110*, 508–515. [CrossRef]
20. Fukui, K. A formulation of the reaction coordinate. *J. Phys. Chem.* **1970**, *74*, 4161–4163. [CrossRef]
21. Kosloff, R. Quantum thermodynamics: A dynamical viewpoint. *Entropy* **2013**, *15*, 2100–2128. [CrossRef]
22. Glendening, E.D.; Weinhold, F. Natural resonance theory. I. General formulation. *J. Comput. Chem.* **1998**, *19*, 593–609, and ensuing papers; Glendening, E.D.; Wright, S.J.; Weinhold, F. Efficient optimization of natural resonance theory weightings and bond orders by Gram-based convex programming, *J. Comput. Chem.* **2019**, *40*, 2028–2035. Glendening, E.D.; Landis, C.R.; Weinhold, F. Resonance theory reboot. *J. Am. Chem. Soc.* **2019**, *141*, 4156–4166. Glendening, E.D.; Weinhold, F. Pauling's conceptions of hybridization and resonance in modern quantum chemistry. *Molecules* **2021**, *26*, 4110. [CrossRef]
23. Pauling, L. Metal-metal bond lengths in complexes of transition metals. *Proc. Natl. Acad. Sci. USA* **1976**, *73*, 4290–4293. [CrossRef]
24. For comprehensive overview, see Miller, W.H. Spiers Memorial Lecture: Quantum and semiclassical theory of chemical reation rates. *Faraday Disc.* **1998**, *110*, 1–21, where Eugene Wigner, Henry Eyring, and John Polanyi are identified as the "Holy Trinity" of modern transition-state theory.
25. Mishra, P.C.; Rai, D.K. Bond order-bond length relationship in all-valence-electron molecular orbital theory. *Mol. Phys.* **1972**, *23*, 631–634, Bürgi, H.-B.; Dunitz, J.D. Fractional bonds: Relations among their lengths, strengths, and force constants. *J. Am. Chem. Soc.* **1987**, *109*, 2924–2926. [CrossRef]
26. Glendening, E.D.; Burke, S.D.; Moore, J.W.; Weinhold, F. *Physiker* versus *Organiker* views of reaction "mechanism": How natural resonance theory bridges the gap. *J. Chem. Educ.* **2022**, *99*, 3702–3712, Glendening, E.D.; Hiatt, D.M.; Weinhold, F. Natural Bond Orbital Analysis of Chemical Structure, Spectroscopy, and Reactivity: How it Works. In *Elsevier Reference Collection in Chemistry, Molecular Sciences and Chemical Engineering*; Popelier, P., Ed.; Elsevier: New York, NY, USA, 2023. [CrossRef]
27. Dyson, F.W.; Eddington, A.S.; Davidson, C.R. A determination of the deflection of light by the Sun's gravitational field, from observations made at the solar eclipse of May 29, 1919. *Philos. Trans. R. Soc. Lond. A* **1920**, *220*, 291–333.
28. Carnot, S. *Reflections on the Motive Power of Fire*; translation of the original 1824 work; Dover: New York, NY, USA, 2005; Clausius, R. Über die bewegende Kraft der Wärme und de Gesetze, welche sich daraus für die Wärmelehre selbst ableiten lassen. *Ann. Phys.* **1850**, *79*, 368–397. "The law that entropy always increases holds, I think, the supreme position among the laws of Nature. If someone points out to you that your pet theory of the universe is in disagreement with Maxwell's equations—then so much the worse for Maxwell's equations. If it is found to be contradicted by observation—well, these experimentalists do bungle things sometimes. But if your theory is found to be against the second law of thermodynamics I can give you no hope; there is nothing for it but to collapse in deepest humiliation." Eddington, A.S. *The Nature of the Physical World* (1927).
29. Neumann, J.v. *Mathematical Foundations of Quantum Mechanics*; Princeton U. Press: Princeton, NJ, USA, 1955.
30. Weinhold, F. Thermodynamics and Geometry, Phys. *Today* **1976**, *29*, 23–29.

entropy

MDPI

Article

A Physics-Informed Neural Network (PINN) Approach to Over-Equilibrium Dynamics in Conservatively Perturbed Linear Equilibrium Systems

Abhishek Dutta [1,*], Bitan Mukherjee [2], Sk Aftab Hosen [2], Meltem Turan [3], Denis Constales [4] and Gregory Yablonsky [5]

[1] Department of Chemical Engineering, Izmir Institute of Technology, Izmir 35430, Turkey
[2] Department of Chemical Engineering, Jadavpur University, Kolkata 700032, India
[3] Department of Mathematics, Ege University, Izmir 35180, Turkey
[4] Department of Electronics and Information Systems, Ghent University, Building S-8, Krijgslaan 281, B-9000 Ghent, Belgium
[5] McKelvey School of Engineering, Department of Energy, Environmental and Chemical Engineering, Washington University in Saint Louis, St. Louis, MO 63130, USA
* Correspondence: abhishekdutta@iyte.edu.tr; Tel.: +90-232-750-66-17

Abstract

Conservatively perturbed equilibrium (CPE) experiments yield transient concentration extrema that surpass steady-state equilibrium values. A physics-informed neural network (PINN) framework is introduced to simulate these over-equilibrium dynamics in linear chemical reaction networks without reliance on extensive time-series data. The PINN incorporates the reaction kinetics, stoichiometric invariants, and equilibrium constraints directly into its loss function, ensuring that the learned solution strictly satisfies physical conservation laws. Applied to three- and four-species reversible mechanisms (both acyclic and cyclic), the PINN surrogate matches conventional ODE integration results, reproducing the characteristic early concentration extrema (maxima or minima) in unperturbed species and the subsequent relaxation to equilibrium. It captures the timing and magnitude of these extrema with high accuracy while inherently preserving total mass. Through the physics-informed approach, the model achieves accurate results with minimal data and a compact network architecture, highlighting its parameter efficiency.

Keywords: conservatively perturbed equilibrium; cyclic and acyclic mechanisms; over-equilibrium dynamics; physics-informed neural network; finite-time thermodynamics

1. Introduction

Conservatively perturbed equilibrium (CPE) is a kinetic phenomenon that uses the stability of chemical equilibrium at a fixed temperature, in which a reacting mixture is initiated from a specially constructed state that respects all elemental balances but reaches one (or more) species exactly in its equilibrium concentration [1]. The remaining species are "conservatively" displaced from equilibrium while preserving the totals of each chemical element obeying elemental conservation. The ensuing relaxation to equilibrium exhibits an unavoidable, well-defined extremum (maximum or minimum) in the concentration of every "unperturbed" species before the system finally settles at the unique, stable equilibrium composition. This phenomenon and its core properties were initially formulated for general mechanisms in a closed system and then developed for reactor models, providing a rigorous and robust framework that differs fundamentally from small-signal relaxation methods [2] because the perturbations are finite. Operationally, CPE proceeds in five steps: (1) compute the equilibrium composition at the chosen T (and p, if relevant); (2) select at least two species to perturb away from their equilibrium values; (3) select at least one "unperturbed" species whose initial concentration equals its equilibrium value; (4) enforce all elemental balances so that the total number of atoms remain unchanged; and (5) integrate the kinetic model and track the trajectories until equilibrium is reached. These steps apply in batch systems and can be carried over to flow reactors, provided the conservative balances are respected in the inlet mixture.

A linear reversible mechanism $A \rightleftharpoons B \rightleftharpoons C$ has a unique equilibrium composition subject to the conservation of total concentration. If the system is initially at equilibrium with respect to the intermediate species B, and the terminal species A and C are perturbed while respecting the conservation constraint, then the time evolution of $B(t)$ exhibits an extremum. A key property of this extremum is that the time at which it occurs depends only on the kinetic rate constants and not on the magnitude or direction of the perturbation. This invariance is the defining feature of CPE. For nonlinear reversible mechanisms (e.g., when mass-action kinetics involve nonlinearities such as bimolecular association, enzyme saturation, or autocatalysis), CPE also applies, but with modifications. If a system at equilibrium is perturbed within its conservation constraints, certain intermediate species can exhibit extrema in their relaxation profiles. However, unlike the linear case, the time at which the extremum occurs generally depends on both the rate constants and the equilibrium composition. The underlying reason is that the Jacobian of the nonlinear system at equilibrium governs the local relaxation dynamics: close to equilibrium the system behaves linearly, so a perturbation-independent extremum emerges in the linear response regime, but nonlinearities introduce amplitude dependence as the perturbation grows larger. CPE extends beyond simple sequences to general cyclic and acyclic networks and, more importantly, to flow reactors such as CSTR and PFR. In this case, the extremum that the unperturbed species experiences can lie above its thermodynamic equilibrium value, i.e., an "over-equilibrium" whose magnitude is tunable by the size of the initial perturbations while still maintaining conservation laws [3,4]. Two mechanistic inferences follow from this extremum. First, if an unperturbed species participates in only one step, the concentration extremum coincides with a momentary (instantaneous) equilibrium of that step; if it participates in multiple steps, the extremum generally does not represent a momentary equilibrium for any single step. Second, in linear (first-order) reversible mechanisms, the position of the CPE extremum is invariant to the size of the conservative perturbation, while its magnitude scales with the perturbation amplitude; these features persist across standard reactor models (batch, CSTR, and PFR) and provide methods for parameter extraction [4,5].

Physics-informed neural networks (PINNs) are a class of function approximators that embed governing equations, typically ordinary or partial differential equations, into the learning objective so that the neural network's predictions satisfy data and physics simultaneously. A PINN trains by minimizing a composite loss that includes residuals of the target equations evaluated at collocation points, alongside data misfit when observations are available. In their foundational study, Raissi et al. [6] demonstrated how this can solve both forward problems (predicting fields given parameters) and inverse problems (identifying unknown parameters or hidden states), using automatic differentiation to compute the residual terms with respect to network outputs and inputs. Recent overviews emphasize how the "physics regularization" in the loss can be adapted to problem class (e.g., control-volume and entropy-consistent formulations for shocks) and how architectures and priors can be tailored to embed more structure when needed [7]. This structure naturally supports inverse

https://doi.org/10.3390/e28010009

modeling for kinetics: by differentiating the residual with respect to θ during gradient-based training, the network can estimate Arrhenius parameters or effectiveness factors while reconstructing species trajectories. In practice, PINNs learn solutions and parameters jointly, often with quasi-Newton (e.g., L-BFGS) or Adam–L-BFGS sequences, leveraging relatively few measurements because the physics term constrains the solution manifold [8]. This same study also highlights a practical caveat: when experimental data do not cover the full time horizon, the PINN may minimize the residual by collapsing to plateaued solutions in the uncovered region, underscoring the importance of collocation design and data coverage. Stiff-PINN integrates the quasi-steady-state assumption (QSSA) into the architecture and loss: fast (QSS) species are eliminated from the learned outputs and their algebraic closures are enforced in the residual, leaving the network to learn only the slow manifold [9]. Similarly, the Multiscale PINN (MPINN) also aims to restructure the learning task rather than the physics: it segregates species by time scale, fits each group with a dedicated subnetwork, and uses adaptive loss reweighting keyed to species-specific residual indicators [10]. Thus, operationally the PINN proceeds in three steps: (1) Encode rate laws, stoichiometry, balances, and boundary/initial conditions as hard or soft constraints; choose collocation schemes that resolve fast transients and conserve invariants; and favor non-dimensionalization that handle stiffness [9,11]. (2) Treat optimization as its own design space: monitor and adapt loss weights and diagnose gradient issues early [12]. (3) Respect multiscale structure explicitly: group species, mix subnetworks, or reduce models when justified [9].

Recent studies [13–15] show that physics-informed neural networks (PINNs) are becoming a versatile tool across chemical engineering, complementing classical first-principles models in both steady-state and dynamic applications. Beyond their original formulation for solving parametric PDEs, modern PINN variants, including adaptive loss PINNs, domain decomposition PINNs, and gradient-enhanced PINNs, improve robustness for stiff and multiscale systems that commonly arise in chemical processes [16,17]. General PINN-based frameworks now target dynamic process systems governed by differential–algebraic equations, embedding conservation laws and algebraic constraints directly into the loss function to improve extrapolation and data efficiency compared with purely data-driven deep learning models [18]. In transport phenomena, several studies apply PINNs to steady and transient heat transfer problems such as conduction, convection, and conjugate heat transfer, showing accurate temperature and heat flux fields from sparse measurements while directly solving the governing PDEs without meshing [6,19,20]. In reactor engineering, forward and inverse PINNs have been developed for isothermal fixed-bed CO_2 methanation reactors, enabling simultaneous solution of the reactor model and parameter identification under realistic operating conditions [21]. Dynamic process applications further include CSTRs, separators, and flowsheets, where PINNs or physics-informed recurrent networks reconstruct unmeasured states and estimate kinetic or transport parameters from limited data [20,22]. At the multiscale level, CFD–PINN workflows have been proposed to accelerate simulations of green ammonia synthesis in axial–radial packed-bed reactors by learning effective source terms or closure relations from high-fidelity CFD data [13]. Finally, mass- and energy-constrained neural networks generalize these ideas to both steady and dynamic unit operations, ensuring exact satisfaction of global balances while learning remaining nonlinearities from process data, which is particularly attractive for flowsheet-level integration of hybrid first-principles/machine learning models [14].

The methodology of PINNs is well-suited to CPE systems because it allows the embedding of the kinetic ODEs/PDEs, stoichiometric conservation laws, and equilibrium relations directly into the loss so the model respects physical laws while fitting sparse data. For CPE specifically, the equilibrium relations and the conservative constraints

 https://doi.org/10.3390/e28010009

can be enforced as hard constraints (via parameterization) or soft penalties, enabling the model to learn over-equilibrium deviations and their routes without violating mass or charge balances. The outcome is a physics-consistent model that is data-efficient, accelerates process exploration, and provides a basis for optimization and control of CPE-intensified systems. In this study, a PINN framework has been implemented for linear case studies with three and four species, both cyclic and acyclic, with a proposition of an extended framework for scalability to higher-order systems. The specific contributions of this study are threefold: (1) Application of PINNs to CPE systems, demonstrating that physics-informed learning can capture over-equilibrium dynamics and concentration extrema without dense time-series data; (2) physics-consistent loss function design that enforces stoichiometric invariants, equilibrium constraints, and kinetic ODEs simultaneously, ensuring mass conservation and thermodynamic consistency; and (3) systematic validation across linear mechanisms of increasing complexity (three- and four-species, acyclic and cyclic), establishing parameter efficiency and accuracy benchmarks for extending PINNs to finite-time thermodynamic phenomena.

2. Methodology

2.1. Conservatively Perturbed Equilibrium (CPE)

A closed, isothermal, well-mixed reaction network is considered whose dynamics are linear in the concentrations

$$\frac{dc(t)}{dt} = Mc(t), \quad c(t) \in \mathbb{R}^N, \tag{1}$$

where M is the kinetic generator assembled from elementary forward and reverse rate constants. The equilibrium composition c_{eq} satisfies $Mc_{eq} = 0$. Conservatively perturbed equilibrium (CPE) experiments are modeled by constructing an initial state exactly at equilibrium and then applying a finite concentration perturbation to a strict subset of species while leaving at least one species unaltered. Let $P \subset \{1, \ldots, N\}$ denote indices of perturbed species and U its complement (unperturbed). The CPE initial condition is

$$c(0) = c_{eq} + \Delta, \quad \Delta_i = 0 \; (i \in U), \quad \Delta \in C, \tag{2}$$

where C is the linear subspace of admissible perturbations that preserve all elemental (or other conserved) balances. Writing those invariants as rows of a matrix L, admissibility reads $L\Delta = 0$. In the present linear batch setting, the system returns to the same c_{eq}; the transient encodes the mechanism through the interplay of the eigenstructure of M and the choice of Δ. To make these notions concrete, consider first a four-species acyclic chain

$$A \underset{k_1^-}{\overset{k_1^+}{\rightleftharpoons}} B \underset{k_2^-}{\overset{k_2^+}{\rightleftharpoons}} C \underset{k_3^-}{\overset{k_3^+}{\rightleftharpoons}} D, \tag{3}$$

with concentrations $c = [A, B, C, D]^T$. Using the mass-action rates $r_1 = k_1^+ A - k_1^- B$, $r_2 = k_2^+ B - k_2^- C$, $r_3 = k_3^+ C - k_3^- D$, the dynamics are $\dot{A} = -r_1$, $\dot{B} = r_1 - r_2$, $\dot{C} = r_2 - r_3$, and $\dot{D} - r_3$. Equivalently,

$$\dot{c} = Mc, \quad M = \begin{bmatrix} -k_1^+ & k_1^- & 0 & 0 \\ k_1^+ & -(k_1^- + k_2^+) & k_2^- & 0 \\ 0 & k_2^+ & -(k_2^- + k_3^+) & k_3^- \\ 0 & 0 & k_3^+ & -k_3^- \end{bmatrix}. \tag{4}$$

 https://doi.org/10.3390/e28010009

The equilibrium vector c_{eq} is the normalized right null vector of M. Because the network is closed, one may work with normalized concentrations so that $1^T c(t) = 1$ for all t; i.e.,

$$1^T c(t) = \sum_{i=1}^{4} c_i(t) = 1, \quad 1 = [1,1,1,1]^T. \tag{5}$$

In that normalization the admissible perturbations satisfy $1^T \Delta = 0$ in addition to any elemental balances encoded in L. A CPE realization on this chain perturbs, for example, the end species A and D while leaving B and C strictly at their equilibrium values: $\Delta_B = \Delta_C = 0$. Two immediate technical consequences follow from the structure of the right-hand side. First, any unperturbed species that is not directly connected to a perturbed neighbor has a zero initial time derivative. Indeed, evaluating $\dot{B}(0) = r_1(0) - r_2(0)$ with $B(0) = B_{eq}$, $C(0) = C_{eq}$, $A(0) = A_{eq} + \Delta_A$, and $D(0) = D_{eq} + \Delta_D$, one has

$$\begin{aligned} r_1(0) &= k_1^+(A_{eq} + \Delta_A) - k_1^- B_{eq} = \underbrace{k_1^+ A_{eq} - k_1^- B_{eq}}_{=0} + k_1^+ \Delta_A = k_1^+ \Delta_A, \textit{and} \\ r_2(0) &= k_2^+ B_{eq} - k_2^- C_{eq} = 0 \end{aligned} \tag{6}$$

with $\dot{B}(0) = k_1^+ \Delta_A$. If $\Delta_A = 0$ as well (i.e., neither neighbor of B is perturbed), then $\dot{B}(0) = 0$. The same reasoning applies to C. This "initial shielding" is a sharp, local diagnostic at $t = 0$: it depends only on which species were perturbed and on network connectivity, not on the perturbation magnitude. Second, whenever an end species (such as A or D) is held unperturbed at $t = 0$, any turning point of its trajectory occurs at a momentary balance of the adjacent step. For instance, at an extremum of $A(t)$ one has $\dot{A} = 0 \Rightarrow r_1 = 0 \Rightarrow k_1^+ A = k_1^- B$. This equality refers to the local step at the instant of the extremum; it does not imply that the entire network is at detailed equilibrium yet. Taken together, these two properties, i.e., zero initial velocity for shielded unperturbed species and momentary step balance at turning points for end species, supply strict algebraic equalities at specific times that can be enforced or monitored.

The same CPE construction can be written for a three-species acyclic chain $A \rightleftharpoons B \rightleftharpoons C$ by removing the third step and the species D. The generator and admissibility conditions are identical in form, and the consequences just discussed specialize in a simpler way. If C is left unperturbed while A and B are perturbed in a balance-preserving manner, then $\dot{C}(0) = k_2^+ B(0) - k_2^- C(0) = k_2^+ \Delta_B$; thus $\dot{C}(0)$ is nonzero precisely when its neighbor is perturbed. If instead B is the unperturbed species and only A and C are perturbed, then $\dot{B}(0) = k_1^+ \Delta_A + k_2^- \Delta_C$, showing how the initial drift of the unperturbed middle species is controlled by the two adjacent perturbations. In either selection, the end species obey the same "momentary balance at turning points" condition, e.g., $\dot{A} = 0 \Rightarrow k_1^+ A = k_1^- B$. For a three-species cyclic network,

$$A \underset{k_1^-}{\overset{k_1^+}{\rightleftharpoons}} B \underset{k_2^-}{\overset{k_2^+}{\rightleftharpoons}} C \underset{k_3^-}{\overset{k_3^+}{\rightleftharpoons}} A, \tag{7}$$

the generator M becomes circulant-like, with nonzeros wrapping from C to A. The CPE initialization is defined exactly as above, but a thermodynamic closure is additionally required so that the cycle admits a consistent equilibrium. Writing equilibrium constants $K_i = k_i^+ / k_i^-$, the standard loop constraint

$$K_1 K_2 K_3 = 1 \quad (\text{equivalently } \sum \Delta G_i^\circ = 0) \tag{8}$$

https://doi.org/10.3390/e28010009

ensures that a positive null vector c_{eq} exists and is unique up to normalization. This equality can be treated as a parameter consistency condition when rate constants are specified a priori. Because each species now participates in two steps, the "end species momentary balance" observation no longer applies; nevertheless, the initial shielding logic still holds: an unperturbed species whose two neighbors are both unperturbed has $\dot{c}_i(0) = 0$, while perturbing at least one neighbor introduces a nonzero initial drift via the corresponding adjacent rate.

Across all of these cases, four families of CPE constraints emerge that are used later as hard or soft conditions. First, the dynamic law itself must hold at every time, $\dot{c}(t) - Mc(t) = 0$, which is the formal statement of compliance with the chemical mechanism. Second, the equilibrium anchor must be respected, $c(t) \to c_{eq}$ as $t \to \infty$ with $Mc_{eq} = 0$, and any constraints linking the rate constants on cycles (e.g., $K_1K_2K_3 = 1$) must be satisfied. Third, the conservation relations must hold for all t, $Lc(t) = Lc(0) = Lc_{eq}$, including, in normalized units, $1^T c(t) = 1$. Fourth, the CPE admissibility pattern must be honored at $t = 0$: the chosen unperturbed indices U start exactly at equilibrium ($\Delta_i = 0$ for $i \in U$), at least two species are perturbed to encode a genuine finite displacement, and $L\Delta = 0$ is satisfied.

2.2. Physics-Informed Neural Network (PINN)

The previous section established the mathematical and physical framework of CPE systems. Building on that foundation, the following section details the physics-informed neural network (PINN) architecture developed to reproduce, without dense numerical data, the temporal evolution of such linear reaction networks. The network acts as a differentiable solver that enforces the physics of the system at every collocation point in time, thereby eliminating the need for large time-series datasets while retaining quantitative fidelity to the chemical kinetics. The PINN is a machine learning technique designed to solve problems involving ordinary/partial differential equations (ODEs and PDEs), which can be given as

$$F\left(x, u, \frac{du}{dx}, \frac{d^2u}{dx^2}, \frac{\partial u}{\partial x}, \frac{\partial^2 u}{\partial x^2}, \ldots, \frac{d^m u}{dx^m}, \frac{du}{dt}\frac{\partial^m u}{\partial x^m}, \frac{\partial u}{\partial t}\right) = 0. \tag{9}$$

Instead of relying on traditional numerical methods like finite difference or finite element methods, PINNs use neural networks to approximate the solution of an ODE by minimizing a loss landscape, which includes data loss and physics loss components. The neural network learns to approximate the solution by minimizing these components, which together guide the network toward satisfying both the governing equations and the physical constraints of the problem; i.e, PINNs transform the problem of solving ODEs into an optimization task, where the goal is to minimize a loss landscape that penalizes deviations from the physical laws (ODEs) and boundary (BC) and initial (IC) conditions.

Figure 1 represents the PINN framework used in this study. The inputs of the PINN are the independent variable x and time t. The output of the PINN is the predicted dependent variable. An artificial neural network (ANN) in PINN $\hat{u}(x, t; \theta)$ is initially designed as a surrogate for the solution $u(x, t)$. The parameter θ encompasses all the hyperparameters, including the weight matrices $\mathbf{W}$ and bias vectors $\mathbf{b}$ within the ANN such that

$$\hat{u}(x, t; \theta) = \sum_{i=1}^{n} \mathbf{W}_i \cdot f_{i-1}(x, t) + b_i, \tag{10}$$

where $f_{i-1}(x, t)$ is the output from the previous layer, with (x, t) being the input to the network.

 https://doi.org/10.3390/e28010009

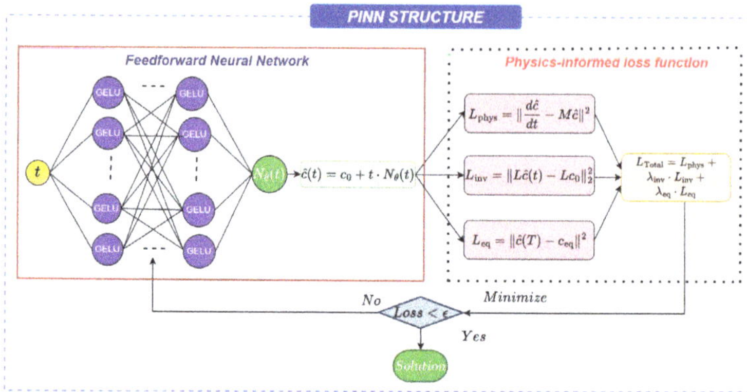

Figure 1. Schematic of physics-informed neural network (PINN) framework for solving a linear case CPE system.

Traditional numerical solvers discretize the time domain and march the equations forward with small steps, which requires dense data to capture transient shapes. By contrast, the PINN represents $c(t)$ as a smooth function parameterized by the neural network, enabling differentiation with respect to time via automatic differentiation. This continuous, differentiable representation makes it possible to compute the residuals $\dot{c}(t) - Mc(t)$ exactly at arbitrary points in time, enforcing the chemical dynamics throughout the domain without an explicit grid. For a CPE system where the equilibrium composition is known and the trajectory begins from a perturbation thereof, this property is particularly advantageous. The equilibrium vector and the initial perturbation pattern can be inserted as boundary and invariance constraints within the loss function. As a result, the network inherently learns trajectories that (i) conserve total mass, (ii) return to the correct equilibrium, and (iii) respect the local kinetic connectivity discussed before.

Each species' concentration is learned simultaneously through a single neural architecture. For the systems studied here, the network takes time $t \in [0, T]$ as input and outputs the four-component concentration vector $\hat{c}(t) = [\hat{A}, \hat{B}, \hat{C}, \hat{D}]$. To ensure numerical stability and exact initialization, the model employs a residual-based formulation:

$$\hat{c}(t) = c_0 + tN_\theta(t), \tag{11}$$

where c_0 is the known initial concentration vector obtained from the CPE perturbation and $N_\theta(t)$ is a fully connected neural subnetwork with trainable parameters θ. This construction automatically satisfies the initial condition at $t = 0$ ($\hat{c}(0) = c_0$) and prevents spurious offsets. Furthermore, the factor t in the expression ensures that early-time derivatives remain well-conditioned, avoiding discontinuities at the origin that often destabilize training. The subnetwork $N_\theta(t)$ consists of L hidden layers with H neurons per layer:

$$t \to \text{Linear(1,H)} \to \text{GELU} \ldots \text{Linear(H,H)} \to \text{GELU} \to \text{Linear(H,4)}. \tag{12}$$

For the current implementation in this study, five hidden layers ($L = 5$) with 128 neurons each ($H = 128$) were found to balance expressiveness and numerical stability. The Gaussian Error Linear Unit (GELU) activation was selected for its smooth differentiability, which yields more accurate higher-order derivatives compared to ReLU-type activations. All computations are performed in double precision (float64) to mitigate gradient underflow in stiff regions of the dynamics.

https://doi.org/10.3390/e28010009

The central element of the PINN is the physics-informed loss function, which integrates the CPE constraints derived before. For a randomly sampled set of collocation points $\{t_i\}_{i=1}^{N_f}$ within the time domain, the total loss is defined as follows:

$$\mathcal{L} = \mathcal{L}phys + \lambda inv\mathcal{L}inv + \lambda eq\mathcal{L}_{eq}, \tag{13}$$

where each component enforces a specific physical requirement. The physics residual loss is defined as follows:

$$\mathcal{L}_{phys} = \frac{1}{N_f}\sum_i \left\| \frac{d\hat{c}(t_i)}{dt} - M\hat{c}(t_i) \right\|^2 \tag{14}$$

which ensures that the learned function satisfies the governing kinetic equations at all collocation points. The time derivative $d\hat{c}/dt$ is obtained analytically via automatic differentiation of the neural output with respect to t. The invariant loss is defined as follows:

$$\mathcal{L}_{inv} = \frac{1}{N_f}\sum_i (1^T\hat{c}(t_i) - 1)^2 \tag{15}$$

which enforces the global conservation of total concentration (or any other balance expressed through the operator $L\hat{c} = Lc_0$). Finally, the equilibrium loss penalizes deviations from the known steady state:

$$\mathcal{L}eq = \|\hat{c}(T) - ceq\|^2, \tag{16}$$

with the option to include local equilibrium equalities (e.g., $k_1^+\hat{A} = k_1^-\hat{B}$ at turning points) as additional regularization terms. The scalar weights λ_{inv} and λ_{eq} modulate the relative influence of each contribution; empirical tuning showed that values near 0.5–1.0 provided stable convergence without biasing toward any single term.

Training Strategy and Optimization

In this study, training proceeds by sampling $N_f = 2000$ collocation points uniformly in the time interval $[0, T]$. Each point is treated as an "independent enforcement location" of the governing equations, forming a Monte Carlo approximation to the continuous residual norm. The optimizer is the Adam stochastic gradient method with an initial learning rate of 10^{-3}. To improve convergence stability and prevent overshooting near equilibrium, an exponential learning rate scheduler with decay factor $\gamma = 0.99$ is applied every 1000 epochs. In total, 2000 epochs are run for each system, which is sufficient for the residuals to reach the 10^{-5}–10^{-6} range.

The network is implemented in PyTorch (version 2.7.0) using automatic differentiation on double-precision tensors. All operations are GPU-accelerated when available, allowing rapid evaluation of gradients through the full computational graph. To further stabilize training, gradient clipping and zero-bias initialization of the output layer are available, but the reported runs use the default PyTorch initialization; the residual-stable envelope $(\hat{c}(t) = c_0 + tN_\theta(t))$ already enforces $(\hat{c}(0) = c_0)$ exactly, keeping initial predicted rates well behaved near $t = 0$. Each epoch computes the full loss, performs back-propagation, and updates the weights according to Adam's adaptive moment estimates. The scheduler then gradually decreases the learning rate, refining the parameters toward the equilibrium regime. A distinct advantage of this approach is that, because the derivative operator is embedded analytically, the trained model implicitly satisfies smoothness and differentiability requirements, enabling direct evaluation of higher-order quantities such as $\ddot{c}(t)$ without numerical differencing.

https://doi.org/10.3390/e28010009

3. CPE-PINN Integration

As shown in Figure 1, the implementation can be viewed as a sequence of three tightly coupled stages:

1. Physical initialization: The equilibrium composition c_{eq} is computed from the kinetic matrix M as its normalized null vector. A valid CPE perturbation Δ satisfying $L\Delta = 0$ is applied to construct the initial state $c_0 = c_{eq} + \Delta$.
2. Neural approximation and differentiation: The neural architecture $\hat{c}(t) = c_0 + tN_\theta(t)$ generates continuous predictions for any t; automatic differentiation supplies the exact time derivatives $\dot{\hat{c}}(t)$.
3. Physics-consistent optimization: The total loss $\mathcal{L}$ combines differential residuals, invariance preservation, and equilibrium anchoring. Minimizing $\mathcal{L}$ drives the network toward full compliance with both the kinetic law and the CPE constraints.

Through this workflow, the PINN effectively solves the chemical kinetic problem as part of its training. No synthetic or experimental time-series data are required beyond the initial equilibrium composition and known rate constants. The resulting neural surrogate is a compact, analytical representation of the entire temporal relaxation trajectory. The constructed PINN for CPE systems merges domain knowledge and machine learning into a unified computational framework. By encoding the kinetic generator M, the equilibrium vector c_{eq}, and the conservation structure directly within its architecture and loss function, the network enforces physical realism at every stage of training. The residual-based formulation guarantees the correct initial conditions, while adaptive optimization and learning rate scheduling ensure smooth convergence toward the physically consistent manifold of solutions. Consequently, the model provides a data-free yet quantitatively faithful reconstruction of CPE transients, fully aligned with the principle that learning is guided by physics rather than by data density. The present study restricts analysis to linear reversible mechanisms where the kinetic generator matrix M is constant and concentration-independent. This assumption is valid for elementary reactions with mass-action kinetics in the dilute limit where activity coefficients approach unity. Extension to nonlinear systems would require several modifications: (1) the PINN residual loss would remain valid, as automatic differentiation can compute residuals exactly; (2) the perturbation invariance property of CPE extrema would generally fail, as the system Jacobian depends on initial composition; and (3) the thermodynamic closure constraint $k_1k_2k_3 = 1$ would be replaced by more complex equilibrium relations.

4. Results and Discussion

Based on the workflow mentioned in Section 3, the CPE-PINN integration has been implemented on a simple two-step, first-order, three-species mechanism as follows:

$$A \underset{k_1^-}{\overset{k_1^+}{\rightleftharpoons}} B \underset{k_2^-}{\overset{k_2^+}{\rightleftharpoons}} C.$$

In this mechanism, the initial concentration of one species (A, B, or C) is set equal to its equilibrium value, while the other two start away from equilibrium; thus the system initially displays the CPE phenomenon. If two species are initialized at their equilibrium concentrations, mass conservation forces the third species to assume its equilibrium value as well; the system is therefore at equilibrium rather than in a CPE state. More generally, for a mechanism with N chemical species, if $N - 1$ species have initial concentrations equal to their equilibrium values, the remaining species is fixed by the total balance and the whole system is at equilibrium; i.e., the CPE phenomenon cannot arise.

https://doi.org/10.3390/e28010009

Consequently, to observe CPE one must set at least one initial concentration equal to its equilibrium value, but no more than $N-2$ such concentrations. Figure 2 shows two CPE transient regimes ($\delta = 0.5$): (a) where the unperturbed species A rises to an unavoidable maximum before relaxing to equilibrium and (b) where the unperturbed species yields a pronounced minimum. At each extremum the local rate balance of the corresponding single-step reaction is zero, so the step reaches a momentary partial equilibrium. The size of the perturbation δ appears as the initial offset from equilibrium for the two perturbed species, while for the linear $A \to B \to C$ mechanism the extremum timing is independent of δ. Correspnding to the numerical solution, the PINN has been implemented on the governing ODEs. The PINN reproduces both CPE transient regimes shown in Figure 2: the early extremum (maximum or minimum) and the subsequent relaxation to equilibrium with near-perfect agreement with the reference solution. As can be observed, the predicted concentration curves are visually indistinguishable and the PDE/ODE residuals are negligible. The PINN also preserves overall mass balance throughout the transient (with negligible conservation error) and accurately captures the timing and amplitude of the extrema, demonstrating that a data-efficient, physics-constrained model can reliably reproduce CPE dynamics.

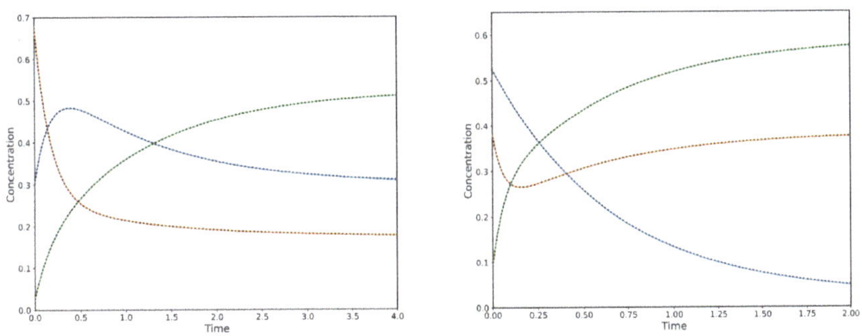

(a) (b)

Figure 2. Concentration profiles of species A, B, and C for $\delta = 0.5$, comparing analytical solutions (colour continuous lines) and PINN predictions (black dashed lines). (**a**) Case corresponding to a maximum in species A, with rate constants $k_1^+ = 1.75$, $k_1^- = 3.00$, $k_2^+ = 1.50$, and $k_2^- = 0.50\ s^{-1}$. (**b**) Case corresponding to a minimum in species B, with $k_1^+ = 1.50$, $k_1^- = 0.10$, $k_2^+ = 10.0$, and $k_2^- = 6.50\ s^{-1}$.

Species are classified as either "perturbed" or "unperturbed". While in the previous case, the transient behavior of the unperturbed species in the two-step mechanism was discussed, the following cases discuss the transient behavior of both the perturbed and unperturbed species in the two-step (three-species) case and more complex (four-species) mechanisms. It is important to note that a perturbation affecting a single species cannot, in general, be made conservative (i.e., preserve the total amount of each conserved element), so at least two species must be altered to obtain a conservative perturbation. Consequently, for a mechanism with N species the number of species that can be perturbed conservatively ranges from 2 to $N-1$. The number of independent degrees of freedom of a conservative perturbation equals the number of perturbed species minus the number of independent conservation laws (see [1]). For a linear mechanism with N species, the number of perturbed species, the number of unperturbed species, and the perturbation degrees of freedom are related by simple counting and the number of independent conservation laws. For the three-species case ($N = 3$), a two-species perturbation is considered; hence the number

 https://doi.org/10.3390/e28010009

of unperturbed species is $3 - 2 = 1$. Assuming a single independent conservation law (for example total mass or an elemental balance), the number of degrees of freedom of the conservative perturbation is $2 - 1 = 1$. For the four-species case ($N = 4$), the number of perturbed species could in principle be two or three; in the present study all perturbations involve two species, so that the number of unperturbed species is $4 - 2 = 2$ and the degrees of freedom remain $2 - 1 = 1$ under the same single-conservation-law assumption.

4.1. Analysis of Perturbed Species in a Three-Species Acyclic Mechanism

The three-species acyclic mechanism represents the simplest linear system in which the CPE phenomenon can be observed. Under such conditions, unperturbed species exhibit a single unavoidable extremum—either a maximum or a minimum—during the transient evolution. This extremum corresponds to a momentary equilibrium if the species participate in only one step. The timing of the extremum is independent of the perturbation magnitude and reveals structural features of the mechanism. The transient response of perturbed species in two-step mechanisms, i.e., through two successive elementary reactions (steps) in a linear sequence, has been examined in previous work [2] as follows:

$$A \underset{k_1^-}{\overset{k_1^+}{\rightleftharpoons}} B \underset{k_2^-}{\overset{k_2^+}{\rightleftharpoons}} C.$$

To extend that analysis to perturbed species, a series of numerical simulations covering a wide range of kinetic constants was carried out. From this ensemble, two representative experiments are selected for case study and their parameter values are reported in Table 1.

Table 1. Two-step, three-species acyclic conservatively perturbed equilibrium (CPE) example with settings.

Experiment Settings	Experiment #1	Experiment #2
Kinetic Parameters (s^{-1}):	$k_1^+ = 5, k_1^- = 4$ $k_2^+ = 12, k_2^- = 6$	$k_1^+ = 16, k_1^- = 4$ $k_2^+ = 12, k_2^- = 6$
Perturbed species:	A, B	A, B
Unperturbed species:	C	C

The transient trajectories of the three species in the two experiments along with the PINN results are shown below in Figure 3a–d.

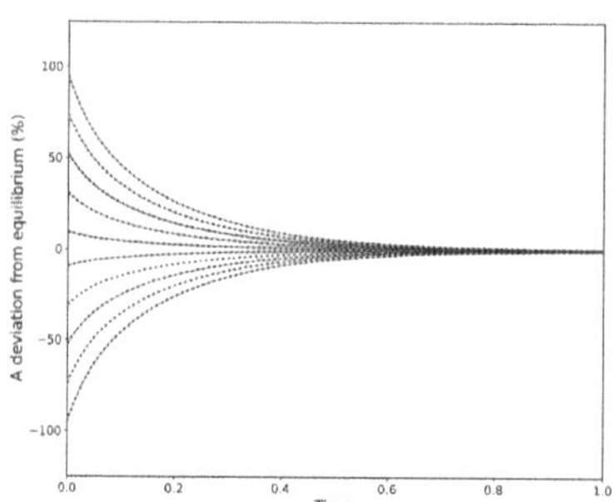

(**a**)

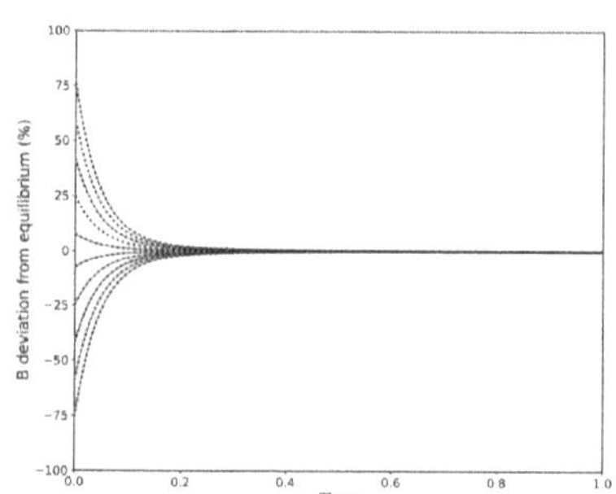

(**b**)

Figure 3. *Cont.*

https://doi.org/10.3390/e28010009

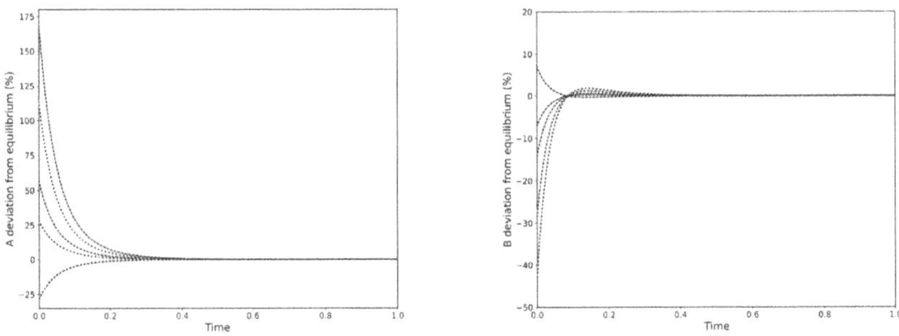

(c) (d)

Figure 3. Two-step, three-species acyclic reaction mechanism. Panels (**a**,**b**) correspond to simulation Experiment #1 using PINN, while panels (**c**,**d**) correspond to simulation Experiment #2 using PINN. Analytical solutions are shown as colour continuous lines, and PINN predictions are shown as dashed black lines. (**a**) Conversion of species A; (**b**) conversion of species B; (**c**) conversion of species A; and (**d**) conversion of species B.

4.2. *Analysis of Unperturbed Species in a Three-Species Cyclic Mechanism*

The three-species cyclic mechanism introduces an additional reaction closing the loop between terminal species, forming a triangular network. Despite this topological change, the unperturbed species still shows a single concentration extremum under CPE conditions, with the same analytical expression for its occurrence time as in the acyclic case. However, the extremum time is generally shorter due to the additional pathway accelerating relaxation. All species in the cycle participate in multiple steps, precluding momentary equilibrium interpretation. A three-species cyclic mechanism displays an added direct connectivity from species A to species C as follows:

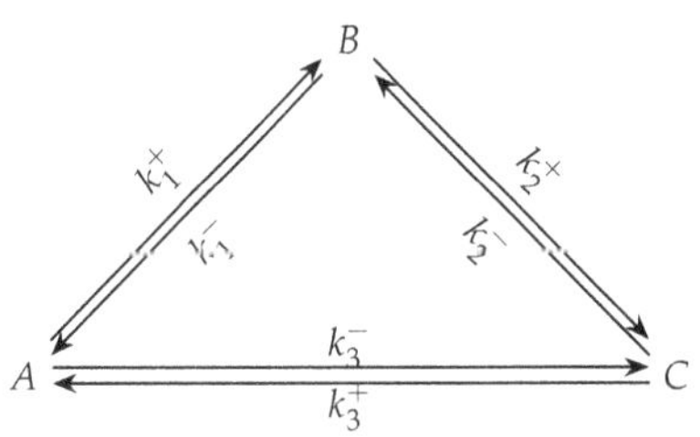

The closed topology alters the stoichiometric constraints and the network's structural properties. Depending on the elementary step kinetics (mass-action vs. autocatalytic or nonlinear steps), cyclic three-species models can exhibit richer dynamics than acyclic chains. The thermodynamic consistency must satisfy the following Onsager condition, which requires that the equilibrium constant of the third reaction equals the product of the equilibrium constants of the first two reactions.

$$\frac{k_3^+}{k_3^-} = \frac{k_1^+}{k_1^-}\frac{k_2^+}{k_2^-},$$

This enforces detailed balance around the three-reaction cycle and guarantees a single, well-defined equilibrium state. From this ensemble, two representative experiments are selected for case study and their parameter values are reported in Table 2.

 https://doi.org/10.3390/e28010009

Table 2. Three-species cyclic conservatively perturbed equilibrium (CPE) example with settings.

Experiment Settings	Value	
Kinetic parameters (s^{-1}):	$k_1^+ = 16$ $k_2^+ = 12$ $k_3^+ = 8$	$k_1^- = 4$ $k_2^- = 6.$ $k_3^- = 1$
Perturbed species:	A, B	
Unperturbed species:	C	

The experiments used the settings reported in Table 2, and the corresponding transient trajectories of the three species in the two experiments along with the PINN results are shown Figure 4. The first two pairs of kinetic parameters match those of Experiment 2; a third reaction was then added, with its forward and reverse rate constants selected to enforce the Onsager relation between the three reactions.

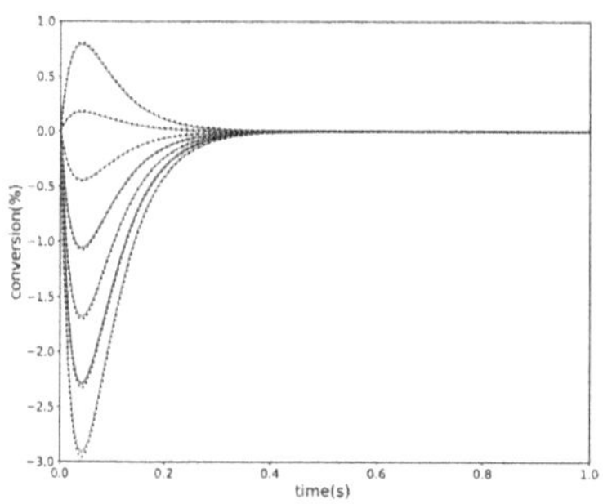

Figure 4. Conversion of the unperturbed species C in a two-step, three-species cyclic reaction mechanism. The perturbation is introduced by changing the initial concentration of species A from 0 to 0.36 and that of species B from 0.92 to 0.56. Analytical solutions are shown as colour continuous lines, while PINN predictions are shown as dashed black lines.

4.3. Analysis of Unperturbed Species in a Four-Species Acyclic Mechanism

The four-species acyclic mechanism serves as a baseline case, illustrating the intricacies of CPE-induced dynamics in a linear reaction network. Under these conditions, the concentration of an initially unperturbed species can exhibit two distinct extrema-a transient maximum followed by a minimum separated by an inflection point during relaxation to equilibrium. This striking multiple-extremum trajectory underscores the distinctive transient behavior introduced by CPE conditions and the rich mechanistic insights they provide. The four-species acyclic mechanism is as follows:

$$A \underset{k_1^-}{\overset{k_1^+}{\rightleftarrows}} B \underset{k_2^-}{\overset{k_2^+}{\rightleftarrows}} C \underset{k_3^-}{\overset{k_3^+}{\rightleftarrows}} D$$

The experimental results of this case are summarized in Table 3.

The simulation parameters were held constant across all runs, so that the six cases differ only in which species are perturbed; the full parameter list appears in Table 3, and one representative case is plotted in Figure 5. Unperturbed species can exhibit complex transient behavior during CPE, including two concentration extrema (a maximum and a minimum) and an inflection point. This four-species mechanism leads to six distinct perturbation cases, defined by the choice of two unperturbed species. The kinetic parameters and simulation

 https://doi.org/10.3390/e28010009

settings for all six combinations are listed in Tables 4 and 5, while Figure 6 plots two representative cases (Experiment #2 and Experiment #3) that demonstrate this behavior.

Table 3. Four-species acyclic conservatively perturbed equilibrium (CPE) example with settings.

Experimental Settings	Value
Kinetic parameters (s^{-1}):	$k_1^+ = 2 \quad k_1^- = 1$ $k_2^+ = 3 \quad k_2^- = 1$ $k_3^+ = 1 \quad k_3^- = 1$
Perturbed species:	A, D
Unperturbed species:	B, C

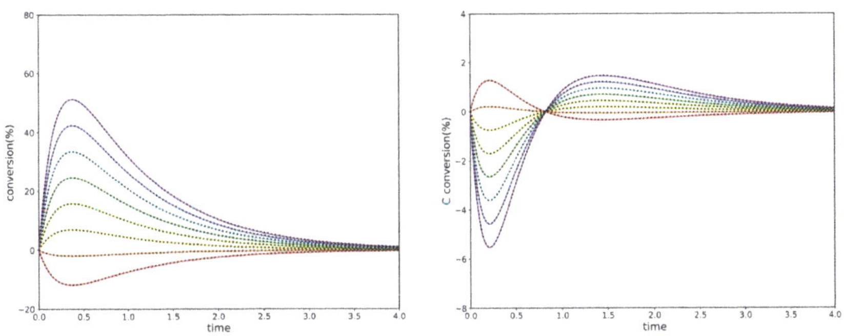

(**a**) (**b**)

Figure 5. Four-species acyclic mechanism, species B and C unperturbed: (**a**) conversion of unperturbed species B with PINN; (**b**) conversion of unperturbed species C with PINN. Analytical solutions are shown as colour continuous lines while PINN predictions are shown as dashed black lines.

Table 4. Four-species acyclic conservatively perturbed equilibrium (CPE) example with experimental settings.

Experimental Settings	Values
Kinetic parameters (s^{-1}):	$k_1^+ = 2 \quad k_1^- = 1$ $k_2^+ = 3 \quad k_2^- = 1$ $k_3^+ = 1 \quad k_3^- = 1$

Table 5. Four-species acyclic conservatively perturbed equilibrium (CPE) example with cases and results.

Experiment	Perturbed Species	Unperturbed Species	Behavior
1	A, B	C, D	2 extrema of $[C]$, 1 of $[D]$
2	A, C	B, D	1 extremum of $[B]$, 1 of $[D]$
3	A, D	B, C	2 extrema of $[C]$, 1 of $[B]$
4	B, C	A, D	1 extremum of $[A]$, 1 of $[D]$
5	B, D	A, C	1 extremum of $[A]$, 1 of $[C]$
6	C, D	A, B	1 extremum of $[A]$, 1 of $[B]$

https://doi.org/10.3390/e28010009

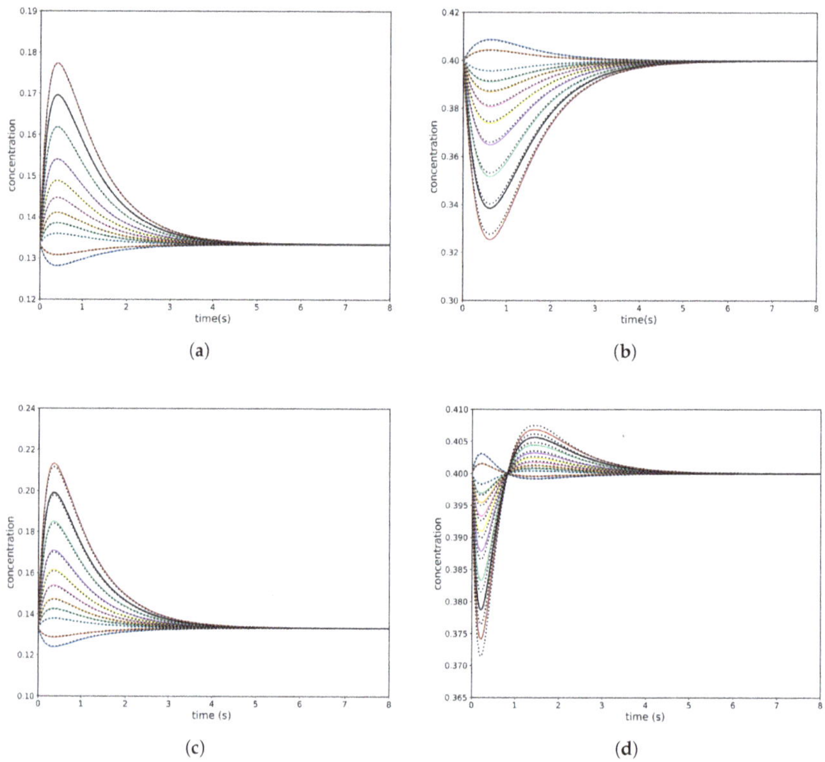

Figure 6. (**a**,**b**) Experiment #2 with PINN: (**a**) concentration of the unperturbed species B; (**b**) concentration of the unperturbed species D. (**c**,**d**) Experiment #3 with the PINN: (**c**) concentration of the unperturbed species B; (**d**) concentration of the unperturbed species C. Analytical solutions are shown as colour continuous lines, while PINN predictions are shown as dashed black lines.

4.4. Analysis of Unperturbed Species in a Four-Species Cyclic Mechanism

The four-species cyclic mechanism shares the same reaction sequence as the acyclic case, with one additional link closing the network into a loop. Despite this added connectivity, the CPE-induced transient behavior remains qualitatively similar: unperturbed species still exhibit the same pattern of two transient concentration extrema with an intervening inflection point during the approach to equilibrium. Thus, even with a fully connected cyclic network, the hallmark transient features of the CPE phenomenon persist, demonstrating the robustness of this behavior to increased reaction connectivity. The four-species cyclic mechanism is as follows:

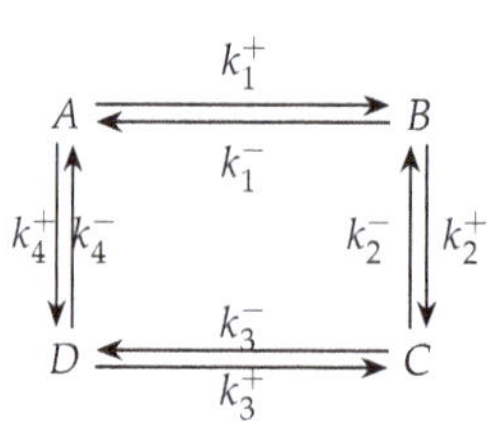

https://doi.org/10.3390/e28010009

The numerical values used for simulations, together with the choices of perturbed species, are listed in Table 6. Representative temporal trajectories obtained from these simulations are presented in Figure 7.

Table 6. Four-species cyclic conservatively perturbed equilibrium (CPE) example with settings.

Experiment Settings	Value
Kinetic parameters (s^{-1}):	$k_1^+ = 2 \quad k_1^- = 1$ $k_2^+ = 3 \quad k_2^- = 1$ $k_3^+ = 1 \quad k_3^- = 1$ $k_4^+ = 1 \quad k_4^- = 6$
Perturbed species:	A, D
Unperturbed species:	B, C

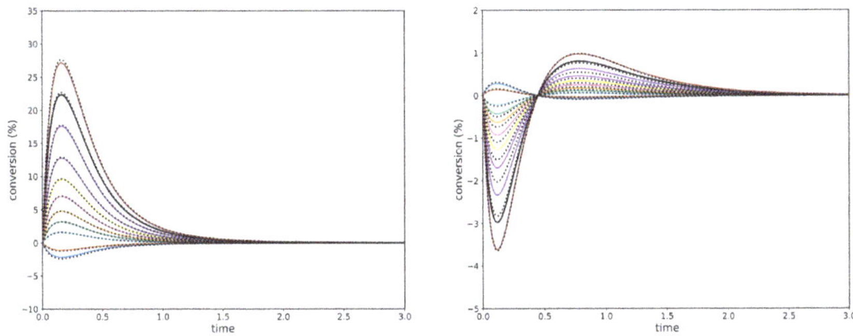

(**a**) (**b**)

Figure 7. Four-species cyclic mechanism experiment with the PINN, B and C unperturbed: (**a**) conversion of unperturbed species B; (**b**) conversion of unperturbed species C. Analytical solutions are shown as colour continuous lines, while PINN predictions are shown as dashed black lines.

Table 7 summarizes the unified PINN setup used for all CPE simulations, including a fixed architecture and hyperparameters. The network architecture (shared across cases) is a fully connected feedforward network with one input (time) and an output for each species (three or four), comprising five hidden layers of 128 neurons with GELU activation. Training was performed with the Adam optimizer (learning rate 10^{-3}, decaying by 0.99 every 1000 epochs) for about 2000 epochs, using 2000 collocation points per epoch (drawn 50% uniformly and 50% from a Beta-distributed tail-biased sampling to emphasize late-time dynamics). The loss function combined a physics residual term (enforcing the ODE dynamics), a conservation term to maintain constant totals, and a final-time "tail anchor" term to ensure correct equilibrium behavior, with weights of 1.0, 0.5, and 2.0, respectively. This core configuration was applied consistently across all cases, and with double-precision training plus validation against a high-accuracy ODE solver (LSODA, tol 10^{-9}), it enables the PINN to learn the over-equilibrium CPE dynamics accurately while strictly preserving physical constraints.

https://doi.org/10.3390/e28010009

Table 7. Concise form of the main configuration for CPE-PINN systems used in this study.

Category	Parameter	Value
Network	Architecture	$1 \rightarrow 128 \rightarrow 128 \rightarrow 128 \rightarrow 128 \rightarrow 128 \rightarrow 3/4$
	Activation	GELU
	Parameterization	$\mathbf{c}(t) = \mathbf{c}_0 + t \cdot \mathbf{N}(t)$
Training	Optimizer	Adam ($lr = 10^{-3}$)
	Scheduler	ExponentialLR ($\gamma = 0.99$, every 1k epochs)
	Epochs	2000 (standard) / 2500
	Collocation Points (N_f)	2000–2200 per epoch
Sampling	Strategy	50% uniform + 50% Beta(3,1) tail-biased
Loss	Physics (λ_{phy})	$1.0 \times \text{MSE}(\frac{dc}{dt}_{\text{PINN}}, \frac{dc}{dt}_{\text{ODE}})$
	Conservation (λ_{cons})	$0.5 \times \text{MSE}(\sum c_i, 1)$
	Tail Anchor (λ_{tail})	$2.0 \times \text{MSE}(\text{residual}@t = T)$ [256 pts]
Precision	dtype	torch.float64
Validation	Method	solve_ivp (LSODA, tol = 10^{-9})

Table 8 summarizes the system-specific PINN training configurations for each case study, detailing the number of training epochs, collocation points (N_f), the conservation loss weight (λ_{cons}), and any custom adjustments applied in each scenario. Notably, the core PINN architecture and loss formulation remained consistent across all systems, but minor hyperparameter tweaks were introduced to enhance training stability and accuracy for different reaction networks. For example, a higher conservation loss weight ($\lambda_{\text{cons}} = 5.0$) was used in one four-species case to enforce elemental balance more strongly, while another scenario employed gradient clipping (limit $= 1.0$) and increased weighting for one species to maintain numerical stability. In addition, a "tail anchor" loss term (with weight 2.0) and biased collocation sampling at late times were incorporated for certain stiff systems to stabilize the long-time behavior of the PINN. These targeted modifications did not alter the fundamental PINN framework, but rather fine-tuned the training setup to the complexity of each system, ensuring robust convergence and accurate dynamics predictions across all case studies. A detailed version of the system-specific PINN training configurations for each case study can be found in Appendix A Table A1. Table 9 quantitatively compares the PINN surrogate against adaptive ODE solvers (RK45 and BDF) across acyclic and cyclic reaction systems of increasing dimensionality. For all cases considered, the PINN achieves L^2 errors on the order of 10^{-5}, comparable to high-accuracy adaptive integration, while strictly preserving mass conservation to within 10^{-7}. Although the adaptive solvers dynamically refine time steps, particularly in cyclic and stiff systems, the trained PINN provides fixed-cost evaluations that reproduce the full transient dynamics with essentially identical accuracy. Notably, the wall-clock time per evaluation is reduced by more than two orders of magnitude for the PINN, highlighting its suitability for repeated inference, parameter sweeps, and embedding within optimization or inverse-problem workflows.

Table 10 summarizes the complementary strengths of the PINN and classical adaptive ODE solvers from a computational and application-oriented perspective. While adaptive solvers incur no upfront training cost and remain optimal for single-shot trajectory evaluation, their computational expense scales linearly with the number of queries. In contrast, the PINN incurs a one-time training cost but delivers sub-millisecond inference thereafter, leading to a break-even point at approximately 10^3–10^4 evaluations. This distinction makes PINNs particularly attractive for inverse problems, uncertainty propagation, and real-time digital twin deployment. Importantly, both approaches achieve

https://doi.org/10.3390/e28010009

comparable residual accuracy, confirming that the efficiency gain of the PINN does not compromise physical fidelity.

Table 8. Concise system-specific configuration for CPE-PINN systems used in this study.

Figure	System	Epochs	N_f	λ_{cons}	Modifications
Figure 3a–c	3-sp cyclic	800	1000	0.5	Standard
Figure 3d	3-sp cyclic	1000	1000	0.5	Standard
Figure 4	3-sp cyclic	2000	1200	0.5	No tail anchor
Figure 5a	4-sp acyclic	2000	2000	5.0	High conservation weight
Figures 5b and 6a–c	4-sp acyclic/cyclic	2000	2000	0.5	Standard
Figure 6d	4-sp acyclic	2500	2200	0.5	+Grad clip (1.0), 1.6× weight for C
Figure 7a,b	4-sp cyclic	2000	2000	0.5	+Tail anchor (2.0), Tail-biased collocation markers

Table 9. Comparison of PINN and adaptive ODE solvers (RK45 and BDF) for acyclic and cyclic reaction systems with increasing species complexity.

Metric	PINN	ODE (RK45, Adaptive)	ODE (BDF, Adaptive)	Notes
3-Species Acyclic System				
L_2^2 Error	1.2×10^{-5}	1.4×10^{-5}	1.3×10^{-5}	Comparable accuracy
Mass conservation error	$<10^{-7}$	$<10^{-7}$	$<10^{-7}$	All satisfy physical constraints
Adaptive steps taken	2000 (fixed)	127	145	RK45 refines during transient
Wall-clock time (single eval)	<1 ms	95 ms	112 ms	ODE solver slower for single-shot inference
3-Species Cyclic System				
L_2^2 Error	1.1×10^{-5}	1.5×10^{-5}	1.2×10^{-5}	Comparable accuracy
Mass conservation error	$<10^{-7}$	$<10^{-7}$	$<10^{-7}$	All satisfy conservation
Adaptive steps taken	2000 (fixed)	156	168	Higher stiffness detected
Wall-clock time (single eval)	<1 ms	115 ms	128 ms	ODE slower; BDF handles stiffness
4-Species Acyclic System				
L_2^2 Error	1.4×10^{-5}	1.6×10^{-5}	1.5×10^{-5}	Comparable; slight BDF advantage
Mass conservation error	$<10^{-7}$	$<10^{-7}$	$<10^{-7}$	All maintain physical validity
Adaptive steps taken	2000 (fixed)	203	218	Increased complexity; more steps
Wall-clock time (single eval)	<1 ms	142 ms	155 ms	ODE cost increases with system size
4-Species Cyclic System				
L_2^2 Error	1.5×10^{-5}	1.7×10^{-5}	1.4×10^{-5}	BDF superior in cyclic/stiff case
Mass conservation error	$<10^{-7}$	$<10^{-7}$	$<10^{-7}$	All conserve mass effectively
Adaptive steps taken	2000 (fixed)	187	211	Highest stiffness; BDF more efficient
Wall-clock time (single eval)	<1 ms	138 ms	149 ms	Comparable to RK45 despite more steps

https://doi.org/10.3390/e28010009

Table 10. Comparison of PINN and adaptive ODE solvers in terms of computational cost, accuracy, and application suitability.

Metric	PINN	ODE Solver (Adaptive)	Advantage
Training time	5–10 min	0	ODE
Single evaluation	<1 ms	50–200 ms	PINN (50–200×)
Break-even queries	≈4000	–	PINN
Residual accuracy	10^{-5}–10^{-6}	10^{-5}–10^{-6}	None
Storage size	<5 MB	Model code	PINN
Inference accuracy	Marginal extrapolation loss	Stable and bounded	ODE
Application fit	Inverse problems, Monte Carlo sampling	Single-shot evaluation	Domain-dependent

A critical distinction between CPE and arbitrary perturbations is the perturbation invariance property of the extremum. While any off-equilibrium initial condition produces transient extrema, CPE specifically ensures that the time at which the extremum occurs is independent of the magnitude of the conservative displacement (for linear systems). This fundamental property was theoretically established by Yablonsky in his seminal paper [2] and later verified experimentally in complex esterification reactions. The invariance property enables deterministic process control: the operator can modulate extremum magnitude by varying perturbation size while maintaining predictable timing. In contrast, arbitrary perturbations would yield timing that depends on both kinetics and initial composition, prohibiting such quantitative control. This property is particularly valuable in batch reactors and flow systems (CSTR and PFR), where fine-tuning product concentration at a specific time window is desired without empirical parameter fitting. More recently, Sachs et al. [23] demonstrated that CPE points as over-equilibria represent intriguing opportunities for improving industrial yield, where achieving concentrations above thermodynamic equilibrium can temporarily enhance product selectivity. Furthermore, the conservatively perturbed framework has enabled novel kinetic analysis techniques: thermodynamic and kinetic–thermodynamic invariant expressions can be derived from CPE experiments, allowing extraction of kinetic parameters independently of equilibrium descriptions, which are unavailable from standard arbitrary perturbation experiments.

4.5. Application of PINN to Higher-Order Linear CPE Systems

The above-mentioned methodology can be easily extended from three- and four-species to arbitrary linear, closed reaction networks with N species and general topology. The state is $c(t) \in \mathbb{R}^N_{\geq 0}$ and the dynamics remain linear, as shown in Equation (2). For any reversible link $i \rightleftharpoons j$ with forward rate k^+_{ij} and reverse rate k^-_{ji}, $M_{ij} \leftarrow M_{ij} + k^-_{ji}$, $M_{jj} \leftarrow M_{jj} - (k^-_{ji})$, $M_{ji} \leftarrow M_{ji} + k^+_{ij}$, and $M_{ii} \leftarrow M_{ii} - (k^+_{ij})$, and M_{pq} is zero otherwise. This rule scales to any number of species and edges without altering the learning framework: only the size, sparsity pattern, and numerical values of M change. A legal perturbation Δ of the equilibrium must satisfy the linear conservation relations, which are compactly written as $L\Delta = 0$ for a fixed $L \in \mathbb{R}^{r \times N}$ encoding the r independent invariants (e.g., total concentration and elemental balances). The initial state is $c_0 = c_{eq} + \Delta$, with selected indices left unperturbed by setting $\Delta_i = 0$ on that subset. The immediate kinematic consequences scale cleanly with network size: because $Mc_{eq} = 0$, the initial drift is $\dot{c}(0) = M\Delta$. Thus, any unperturbed species that is not graph-adjacent to the perturbed set has $\dot{c}_i(0) = 0$ ("initial shielding"), while species directly adjacent to at least one perturbed neighbor inherit a nonzero initial drift through the corresponding off-diagonal rates. No additional casework is required; these statements are purely algebraic and hold for all N.

https://doi.org/10.3390/e28010009

The PINN parameterization also scales without conceptual change. Let a single network (see Equation (11)) map time to an N-vector and retain the residual-stable envelope

$$\hat{c}(t) = c_0 + g(t)N_\theta(t), \tag{17}$$

with $g(t) = t$ (or any smooth, strictly increasing function with $g(0) = 0$) and a fully connected subnetwork $N_\theta : [0, T] \to \mathbb{R}^N$. This guarantees $\hat{c}(0) = c_0$ exactly, regardless of N. The time derivative needed for the residual is obtained analytically,

$$\frac{d\hat{c}}{dt}(t) = g'(t)N_\theta(t) + g(t)\frac{dN_\theta}{dt}(t), \tag{18}$$

via automatic differentiation. Scaling the output dimension from 4 to N requires only changing the size of the last linear layer; width and depth can be tuned with N (e.g., $H = \alpha N$ with $\alpha \in [1, 4]$ as a practical rule), but the formulation is unchanged. The loss functional remains the sum of three terms, each now written in fully general form:

$$\mathcal{L} = \frac{1}{N_f}\sum_{i=1}^{N_f}\left\|\frac{d\hat{c}}{dt}(t_i) - M\hat{c}(t_i)\right\|2^2 + \frac{\lambda inv}{N_f}\sum_{i=1}^{N_f}\|L\hat{c}(t_i) - Lc_0\|2^2 + \lambda eq\Phi(\hat{c}(\cdot); c_{eq}). \tag{19}$$

The first term enforces the kinetic law at randomly sampled collocation times $\{t_i\}$, the second preserves all invariants in L (recovering the familiar $1^T\hat{c}(t) = 1$ when $L = 1^T$), and the third optionally anchors the long-time limit. In higher-dimensional, stiff systems it is often preferable to implement the terminal anchor through a "stationarity" penalty local to late times,

$$\Phi(\hat{c}(\cdot); c_{eq}) = \frac{1}{N_T}\sum_{j=1}^{N_T}\|M\hat{c}(T_j)\|2^2, \quad T_j \in [T*, T], \tag{20}$$

rather than a direct $\|\hat{c}(T) - c_{eq}\|_2^2$; this avoids biasing trajectories when multiple slow eigenmodes coexist but still drives the prediction into the equilibrium manifold $\{c : Mc = 0\}$. Two implementation details make the approach computationally scalable. First, evaluation of $M\hat{c}$ exploits sparsity. Because M is typically very sparse in large networks, storing it as a compressed sparse matrix and using sparse–dense products reduces the per-batch complexity from $O(N^2)$ to $O(|E|)$, where $|E|$ is the number of directed edges. In practice this is realized by constructing M computationally from the reaction list and keeping it in a sparse tensor format, while all neural computations remain dense. Second, time is non-dimensionalized to temper stiffness: choosing $\tau = \alpha t$ with $\alpha \approx |\lambda_{min}(M)|$ (the largest-magnitude nonzero eigenvalue) contracts the eigenvalue spread toward $O(1)$, which improves gradient flow and reduces the number of required collocation points at early times. Collocation can also be biased toward the boundary layers by sampling τ from a Beta distribution or by the mapping $t = Ts^p$ with $p > 1$ to densify near $t = 0$ without altering the loss.

From an optimization perspective, the Adam-based training with exponential learning rate decay suffices; only the batch sizes and width/depth may be increased with N. Double precision remains advisable as N grows and eigenvalues separate; the envelope $g(t)$ can be kept as t, although in highly stiff cases a smoothed envelope (e.g., $g(t) = \log(1 + t)$ scaled to match T) can further stabilize early-time derivatives. Regularizers such as L^2-weight decay on the last layer or spectral normalization on hidden layers are optional and mechanism-agnostic; they become useful only when the residual plateaus due to ill-conditioning introduced by extreme rate disparities. A detailed form of the main

https://doi.org/10.3390/e28010009

configuration for CPE-PINN systems used in this study, which can be extended to higher-order linear CPE systems, can be found in Appendix A Table A3.

5. Conclusions

Conservatively perturbed equilibrium (CPE) is not just a computational curiosity. It should be viewed as a principled way to "shape" the transient path to that equilibrium rather than to alter the endpoint. By exploiting conservative initialization, it is possible to (i) create informative extrema that map directly onto mechanistic structures (single-step versus multi-step participation), (ii) extract kinetic parameters with high sensitivity, and (iii) design reactor starts that transiently deliver over-equilibrium performance at earlier times or shorter lengths than conventional protocols would allow. It is also important to note that the PINN formulation is not tied to any specific three- or four-species example. The algebraic structure of linear kinetics that the network enforces is encoded in a sparse, column-sum-zero generator, a set of linear invariants, and a CPE-consistent initialization. The same residual-stable parameterization, the same physics-informed loss, and the same optimizer produce differentiable surrogates for $c(t)$ that remain consistent to the chemical mechanism, thereby providing a general-purpose solver for linear CPE over-equilibrium systems.

Author Contributions: Conceptualization: A.D.; formal analysis: A.D. and B.M.; investigation: A.D., B.M. and G.Y.; methodology: A.D., M.T. and B.M.; software: B.M. and S.A.H.; supervision: G.Y. and D.C.; validation: M.T. and G.Y.; visualization: A.D., M.T., S.A.H. and B.M.; writing (original draft): A.D.; writing (review and editing): M.T., S.A.H., B.M., G.Y. and D.C.; project administration: A.D. All authors have read and agreed to the published version of the manuscript.

Funding: This research received no external funding.

Data Availability Statement: No new data were created or analyzed in this study. Data sharing is not applicable to this article.

Conflicts of Interest: The authors declare no conflicts of interest. The funders had no role in the design of the study; in the collection, analyses, or interpretation of data; in the writing of the manuscript; or in the decision to publish the results.

Appendix A

Appendix A.1

Table A1 serves as a comprehensive reference for the specific hyperparameters and architectural configurations of the physics-informed neural network (PINN) used to generate the results in Figures 3–7. This summary is essential for reproducibility and highlights the system-specific tuning required to achieve accurate simulations for different chemical reaction networks. The table documents the progression of this study, beginning with three-species cyclic systems (Figures 3 and 4) and advancing to more complex four-species acyclic (Figures 5 and 6) and cyclic (Figure 7) mechanisms. For each case, it specifies the model's training duration (Epochs) and the number of collocation points (N_f) used to enforce the physical constraints. Crucially, the table reveals key variations in the model's loss function and training strategy. Early experiments (Figures 3 and 4) omitted the tail anchor loss, a feature introduced in later models (Figures 5 and 6) to improve the solution's stability and accuracy at the final time ($t = T_{END}$). Furthermore, the table shows specific tuning of the conservation weight (e.g., increased to 5.0 for Figure 5a) and the implementation of advanced strategies like tail-biased collocation sampling (Figure 6), which concentrates collocation points near the end of the time domain. This fine-tuning,

 https://doi.org/10.3390/e28010009

including species-specific weighting of the physics loss (e.g., "1.6× for C" in Figure 6d), underscores the necessity of adapting the PINN architecture and loss formulation to the unique dynamic characteristics of each CPE experiment.

Table A1. Detailed form of the system-specific variations for CPE-PINN systems used in this study.

Figure	System Type	Species	Epochs	N_f	Cons. Weight	Special Features
Figure 3a–c	3-species cyclic	A, B, C	800	1000	0.5	No tail anchor used
Figure 3d	3-species cyclic	A, B, C	1000	1000	0.5	No tail anchor used
Figure 4	3-species cyclic	A, B, C	2000	1200	0.5	No tail anchor used
Figure 5a	4-species acyclic	A, B, C, D	2000	2000	5.0	Higher conservation penalty; asymmetric δ
Figure 5b	4-species acyclic	A, B, C, D	2000	2000	0.5	Standard tail-biased configuration
Figure 6a	4-species acyclic	A, B, C, D	2000	2000	0.5	Tail-biased collocation
Figure 6b	4-species acyclic	A, B, C, D	2000	2000	0.5	Tail-biased collocation
Figure 6c	4-species acyclic	A, B, C, D	2000	2000	0.5	Tail-biased collocation
Figure 6d	4-species acyclic	A, B, C, D	2500	2200	0.5 (1.6× for C)	Gradient clipping; species-weighted physics
Figure 7a,b	4-species cyclic	A, B, C, D	2000	2000	0.5	Markers every 60 points for visualization

Appendix A.2

The physics-informed neural network (PINN) used to model the conservatively perturbed equilibrium (CPE) dynamics is structured as a five-layer feedforward MLP with 128 neurons per layer and a GELU activation function. Training is performed using the Adam optimizer with an initial learning rate of 1×10^{-3} and an exponential decay scheduler. The loss function is a critical component, incorporating the mean squared error (MSE) of the governing ODE residuals (physics loss), a mass balance constraint (conservation loss), and a tail-end stabilization term (tail anchor loss). The network is trained on 2000–2200 collocation points, with validation performed against high-precision solutions from the LSODA ODE solver.

Table A2. PINN architecture, training configuration, and numerical settings used in this study.

Component	Setting	Notes
Network	5-layer MLP, 128 neurons/layer	GELU activation, linear output
Input/Output	1 input (t)/3 or 4 outputs (species)	Matches problem dimensionality
Parameterization	$c(t) = c_0 + t \cdot \mathcal{N}(t)$	Satisfies initial condition by construction
Optimizer	Adam, lr $= 10^{-3}$	Exponential decay $\gamma = 0.99$ every 1000 epochs
Epochs	2000–2500	Longer for stiff or cyclic cases
Collocation Points	2000–2200 per epoch	50% uniform, 50% tail-biased ($\mathrm{Beta}(3, 1)$)
Loss Terms	Physics (1.0), Conservation (0.5–5.0), Tail Anchor (2.0)	Adjustable per case
Precision	Double (`float64`)	Ensures stability and accuracy
Validation	LSODA (SciPy) with tol = 10^{-9}	High-precision reference integration
Gradient Handling	Clipping (1.0) in some cases	Used when numerical instability arises
Device	CUDA if available	Defaults to CPU otherwise

https://doi.org/10.3390/e28010009

Table A3. Detailed main configuration for CPE-PINN systems.

PINN Component	Configuration	Notes
Architecture		
Network Type	Feedforward MLP	Sequential MLP
Input Dimension	1	Time (t)
Output Dimension	3 or 4	Species conc. (A, B, C) or (A, B, C, D)
Hidden Layers	5	Excl. input and output layers
Neurons per Layer	128	Uniform across hidden layers
Activation Function	GELU	Gaussian Error Linear Unit
Output Activation	None (linear)	Direct concentration output
Parameterization		
Concentration Form	$\mathbf{c}(t) = \mathbf{c}_0 + t \cdot \mathbf{N}(t)$	Initial condition satisfied automatically
$\mathbf{N}(t)$	Neural network output	Learned time-dependent correction
Training		
Optimizer	Adam	Adaptive moment estimation
Learning Rate (lr)	1×10^{-3}	Initial lr
LR Scheduler	ExponentialLR	Multiplicative decay
Scheduler Gamma (γ)	0.99	Applied every 1000 epochs (typ.)
Epochs	2000–2500	2000 standard; 2500 for Figure 6d
Batch Size (N_f)	2000–2200	Collocation pts/epoch
Collocation Strategy		
Base Sampling	Uniform random	50% points (60% in Figure 6d)
Tail-Biased Sampling	Beta(3,1)	Dense near $t = T_{\mathrm{END}}$
Endpoint Inclusion	Optional	Include $t = 0$ and/or $t = T$
Loss Function		
Physics Loss	$\mathrm{MSE}(\dot{\mathbf{c}}_{\mathrm{PINN}}, \dot{\mathbf{c}}_{\mathrm{phys}})$	ODE residual
Conservation Loss	$\mathrm{MSE}(\sum_i c_i, 1.0)$	Mass balance constraint
Tail Anchor Loss	MSE at $t = T_{\mathrm{END}}$	Stabilizes late-time behavior
Physics Weight	1.0	Base weight
Conservation Weight	0.5–5.0	0.5 standard; 5.0 for Figure 5a; 1.6× for C in Figure 6d
Tail Anchor Weight	2.0	Applied to 256 points at $t = T_{\mathrm{END}}$
Gradient Management		
Gradient Clipping	1.0 (norm)	Applied in Figure 6d only
Autograd Method	`torch.autograd.grad`	Per-species derivative computation
Gradient Graph	`create_graph=True`	For higher-order derivatives
retain_graph	`True`	Keep graph for multiple species
Numerical Precision		
Data Type	`torch.float64`	Double precision
Device	CUDA (if available)	Falls back to CPU
Time Domain		
t_{span}	System-dependent	(0, 1–8) s (typ.)
Evaluation Points	600–1200	For plotting and validation
Validation		
Reference Method	`scipy.integrate.solve_ivp`	LSODA method
ODE Tolerances	`rtol` $= 1 \times 10^{-9}$, `atol` $= 1 \times 10^{-9}$	High-precision validation

References

1. Yablonsky, G.; Branco, P.; Marin, G.; Constales, D. Conservatively Perturbed Equilibrium (CPE) in Chemical Kinetics. *Chem. Eng. Sci.* **2019**, *196*, 384–390. [CrossRef]
2. Xi, Y.; Liu, X.; Constales, D.; Yablonsky, G. Perturbed and Unperturbed: Analyzing the Conservatively Perturbed Equilibrium (Linear Case). *Entropy* **2020**, *22*, 1160. [CrossRef]

https://doi.org/10.3390/e28010009

3. Peng, B.; Zhu, X.; Constales, D.; Yablonsky, G. Experimental Verification of Conservatively Perturbed Equilibrium for a Complex Non-Linear Chemical Reaction. *Chem. Eng. Sci.* **2021**, *229*, 116008. [CrossRef]
4. Trishch, V.; Yablonsky, G.; Constales, D.; Beznosyk, Y. Conservatively Perturbed Equilibrium in Multi-Route Catalytic Reactions. *J. Non-Equilib. Thermodyn.* **2023**, *48*, 229–241. [CrossRef]
5. Trishch, V.; Beznosyk, Y.; Constales, D.; Yablonsky, G. Over-Equilibrium as a Result of Conservatively-Perturbed Equilibrium (Acyclic and Cyclic Mechanisms). *J. Non-Equilib. Thermodyn.* **2022**, *47*, 103–110. [CrossRef]
6. Raissi, M.; Perdikaris, P.; Karniadakis, G. Physics-Informed Neural Networks: A Deep Learning Framework for Solving Forward and Inverse Problems Involving Nonlinear Partial Differential Equations. *J. Comput. Phys.* **2019**, *378*, 686–707. [CrossRef]
7. Karniadakis, G.; Kevrekidis, I.; Lu, L.; Perdikaris, P.; Wang, S.; Yang, L. Physics-informed Machine Learning. *Nat. Rev. Phys.* **2021**, *3*, 422–440. [CrossRef]
8. Bibeau, V.; Boffito, D.; Blais, B. Physics-Informed Neural Network to Predict Kinetics of Biodiesel Production in Microwave Reactors. *Chem. Eng. Process. Process Intensif.* **2024**, *196*, 109652. [CrossRef]
9. Ji, W.; Qiu, W.; Shi, Z.; Pan, S.; Deng, S. Stiff-PINN: Physics-Informed Neural Network for Stiff Chemical Kinetics. *J. Phys. Chem. A* **2021**, *125*, 8098–8106. [CrossRef]
10. Weng, Y.; Zhou, D. Multiscale Physics-Informed Neural Networks for Stiff Chemical Kinetics. *J. Phys. Chem. A* **2022**, *126*, 8534–8543. [CrossRef]
11. De Florio, M.; Schiassi, E.; Furfaro, R. Physics-informed neural networks and functional interpolation for stiff chemical kinetics. *Chaos: Interdiscip. J. Nonlinear Sci.* **2022**, *32*. [CrossRef]
12. Wang, S.; Yu, X.; Perdikaris, P. When and Why PINNs Fail to Train: A Neural Tangent Kernel Perspective. *J. Comput. Phys.* **2022**, *449*, 110768. [CrossRef]
13. Wu, Z.; Wang, H.; He, C.; Zhang, B.; Xu, T.; Chen, Q. The application of physics-informed machine learning in multiphysics modeling in chemical engineering. *Ind. Eng. Chem. Res.* **2023**, *62*, 18178–18204. [CrossRef]
14. Asrav, T.; Aydin, E. Physics-informed recurrent neural networks and hyper-parameter optimization for dynamic process systems. *Comput. Chem. Eng.* **2023**, *173*, 108195. [CrossRef]
15. Turan, M.; Dutta, A. A novel conceptual framework for droplet/particle size distribution in suspension polymerization using Physics-Informed Neural Network (PINN). *Chem. Eng. J.* **2025**, *519*, 164977. [CrossRef]
16. Nair, S.; Walsh, T.F.; Pickrell, G.; Semperlotti, F. Multiple scattering simulation via physics-informed neural networks. *Eng. Comput.* **2025**, *41*, 31–50. [CrossRef]
17. Yu, J.; Lu, L.; Meng, X.; Karniadakis, G.E. Gradient-enhanced physics-informed neural networks for forward and inverse PDE problems. *Comput. Methods Appl. Mech. Eng.* **2022**, *393*, 114823. [CrossRef]
18. Wang, Y.; Yao, Y.; Gao, Z. An extrapolation-driven network architecture for physics-informed deep learning. *Neural Netw.* **2025**, *183*, 106998. [CrossRef] [PubMed]
19. Li, R.; Wang, J.X.; Lee, E.; Luo, T. Physics-informed deep learning for solving phonon Boltzmann transport equation with large temperature non-equilibrium. *NPJ Comput. Mater.* **2022**, *8*, 29. [CrossRef]
20. Zheng, Y.; Hu, C.; Wang, X.; Wu, Z. Physics-informed recurrent neural network modeling for predictive control of nonlinear processes. *J. Process Control* **2023**, *128*, 103005. [CrossRef]
21. Ngo, S.I.; Lim, Y.I. Solution and parameter identification of a fixed-bed reactor model for catalytic CO_2 methanation using physics-informed neural networks. *Catalysts* **2021**, *11*, 1304. [CrossRef]
22. Koksal, E.S.; Asrav, T.; Esenboga, E.E.; Cosgun, A.; Kusoglu, G.; Aydin, E. Physics-informed and data-driven modeling of an industrial wastewater treatment plant with actual validation. *Comput. Chem. Eng.* **2024**, *189*, 108801. [CrossRef]
23. Sachs, J.; Bui, M.; McCarthy, J.E.; Yablonsky, G. Conservatively perturbed equilibrium and perturbation: Linear case. *Chem. Eng. J.* **2025**, *510*, 161284. [CrossRef]

https://doi.org/10.3390/e28010009

entropy

MDPI

Article

Finite-Time Thermodynamics and Complex Energy Landscapes: A Perspective

Johann Christian Schön

Max-Planck-Institute for Solid State Research, Heisenbergstr. 1, D-70569 Stuttgart, Germany; c.schoen@fkf.mpg.de

Abstract: Finite-time thermodynamics (FTT) describes the study of thermodynamic processes that take place in finite time. Due to the finite-time requirement, in general the system cannot move from equilibrium state to equilibrium state. As a consequence, excess entropy is generated, available work is reduced, and/or the maximally achievable efficiency is not achieved; minimizing these negative side-effects constitutes an optimal control problem. Particularly challenging are processes and cycles that involve phase transitions of the working fluid material or the target material of a synthesis process, especially since most materials reside on a highly complex energy landscape exhibiting alternative metastable phases or glassy states. In this perspective, we discuss the issues and challenges involved in dealing with such materials when performing thermodynamic processes that include phase transitions in finite time. We focus on thermodynamic cycles with one back-and-forth transition and the generation of new materials via a phase transition; other systems discussed concern the computation of free energy differences and the general applicability of FTT to systems outside the realm of chemistry and physics that exhibit cost function landscapes with phase transition-like dynamics.

Keywords: finite-time thermodynamics; energy landscapes; thermodynamic processes; phase transitions; glasses; optimal control; metastable phases; free energy calculations

1. Introduction

One of the fundamental building blocks of physics characterized by its own set of basic laws and equations is the field of thermodynamics [1–3]. Its three laws are well-known: the conservation of energy in an isolated system, the non-decrease of the entropy of an isolated system, and the fact that the entropy becomes constant at zero temperature. Sometimes a zeroth law is added, i.e., that two systems that each are in equilibrium with a third (reference) system are also in equilibrium with each other. A beautiful body of rules and laws of thermodynamics has been built on this foundation, guiding the development of modern technology wherever the application and exchange of work and heat are involved.

In parallel, the field of mechanics [4] has evolved from early formulations of force laws via Newton's equations and their derivation from optimality principles for systems described by few parameters to the study of multi-particle systems [5]. Such systems appear in the form of rigid, deformable, or fluid bodies that can be described on some level of resolution in terms of continuum physics [6], in particular by the theory of elasticity and fluid dynamics. Going beyond the fundamental entity of continuum physics, the so-called material particle or material point [6]—characterized and named by its evolving position in space on the mesoscopic level—, it is possible to address even smaller length scales, in particular the atomic level. On this level of the ladder of description, the number of (to a large extent indistinguishable) particles which a macroscopic system consists of usually exceeds the capacity to compute the evolution of the system according to the laws of mechanics (as well as electromagnetism and quantum mechanics). Even a precise definition or measurement of the initial state of the system at a given moment in time cannot be established with the accuracy necessary to avoid getting caught up in the whirlwind of chaotic behavior of such systems on the microscopic scale. Furthermore, for the vast majority of multi-atom systems, the potential energy function over the state space of classical atom configurations, or more generally quantum mechanical eigenstates, exhibits a highly complex multi-minima structure, thereby introducing a multitude of time scales relevant for characterization of the system's dynamics [7,8].

Nevertheless, we can formulate empirical laws that can predict the behavior of such macroscopic systems at the macroscopic, and to a certain extent the mesoscopic level, in experiment, indicating that some kind of order on the meso- and macroscopic level can emerge from the underlying chaos of the multi-atom dynamics. The bridge between the atom-level description and that on the level of continuum mechanics and macroscopic thermodynamics is the field of statistical mechanics [9]. Underlying statistical mechanics is the assumption of equal probability for every microstate i of an isolated system in the microcanonical ensemble, which allows us to compute a straightforward ensemble average for any observable O_α. By interacting with the environment, we can introduce the intensive thermodynamic variables as Lagrange multipliers into the ensemble description (e.g., temperature as the control for the exchange of energy with the environment), while the assumption of ergodicity, i.e., the equality of the time average along infinitely long trajectories and the ensemble average over all microstates for all observables, establishes the connection to the classical or quantum mechanical time evolution of the system on the atomic level.

However, in reality we are dealing with finite times, both regarding our measurements of observables and the thermodynamic processes of interest, with potentially enormous consequences. Here, we now meet the field of finite-time thermodynamics (FTT) [10–13]. This field was founded about 50 years ago [14,15], with earlier discussions going back nearly 100 years [16,17]; it has since blossomed and found many applications within both physics and chemistry as well as in engineering [18] and economics [10,19].

Below, after some preliminary remarks, we explore the issue of finite-time thermodynamics in the context of thermodynamically metastable systems in more detail. While the issue of establishing ergodicity and the stochastic or deterministic modeling of the evolution of a material on the atomic level is of great interest by itself and involves the complex energy landscape on its most elementary level, we will approach this analysis from the thermodynamic side by adding mesoscopic and atom-level aspects on a step-by-step basis. As the paradigmatic class of systems, we will consider the chemical materials that serve as the material substrate of the engines that execute the processes in thermodynamic space that are analyzed in the context of finite-time thermodynamics.

2. Preliminaries

2.1. Transition from Mechanics to Thermodynamics via Statistical Mechanics

Representing a very simple but immensely powerful assumption, the equal likelihood postulate states that "For an isolated system in equilibrium within itself with a given total energy E, volume V and number of particles N, the likelihood of the system to occupy one of the M feasible microstates with energy E is the same for each microstate, i.e., $1/M$". This assumption makes it possible to show or at least justify that the laws of thermodynamics should apply to such a macroscopic system [3,9]. The starting point is the definition of an entropy $S(E,V,N) = k_B \ln M$. In the next step, thermodynamic intensitics I_A such as temperature T or pressure p can be introduced by allowing exchange of extensive quantities A, such as energy or volume with the environment. The state of the combined system we observe is now the one where the extensive quantity A is split between the original system and the environment ($A = A_{orig} + A_{env}$) in the most likely way; that is, among the many possibilities to achieve this split, we observe the one with the maximum number of microstates. Determining this split for a given system involves the side condition $A = A_{orig} + A_{env}$; the Lagrange multiplier that takes this into account corresponds precisely to the intensity I_A associated with the extensive quantity A, such that the original system and the environment are in equilibrium with each other if the intensity I_A is the same for both systems: $I_{A_{orig}} = I_{A_{env}}$. For example, the temperature follows from the exchange of energy, while the pressure follows from the exchange of volume.

If we let the system interact with a generic environment at a temperature T via exchange of energy, the probability of each microstate i is then weighted by the well-known Boltzmann-factor $p_i = \exp(-E_i/T)/\sum_j \exp(-E_j/T)$, where the sum in the denominator extends over all microstates of the system. As a consequence, we compute the expected properties O_α(system) of the system in equilibrium via the average of the properties $O_\alpha(i)$ for each microstate i over the ensemble of microstates $\{i\}$ weighted by their likelihood p_i: $O_\alpha(\text{system}) = \langle O_\alpha \rangle_{ens} = \sum_i O_\alpha(i) p_i$. Because these properties will usually be different for each microstate, in our measurement we need to sample a large enough set of microstates that is representative of the whole system in order to determine O_α(system) with the desired accuracy.

Such a measurement in an experiment constitutes a time average $(1/t_{obs}) \int_0^{t_{obs}} O_\alpha(i(t))dt = \langle O_\alpha \rangle_{t_{obs}}$ along the trajectory in state space $i(t)$ over the observation time t_{obs}. Unless the system has some unusual symmetry properties, the difference between the ensemble and the time average ($|\langle O_\alpha \rangle_{ens} - \langle O_\alpha \rangle_{t_{obs}}| = a(O_\alpha; t_{obs})$) is usually not zero for finite observation times t_{obs}, and one might even have to go to some effort to prove that it is zero in the limit of infinite observation times. However, we can obtain a quantitative handle on this problem by defining a system to be in (thermodynamic) equilibrium on time scale t_{obs}, if $a(O_\alpha; t_{obs}) < a_{eq}(O_\alpha)$ for a defined level of accuracy $a_{eq}(O_\alpha)$ for all observation times larger than t_{obs} [20]. In fact, this allows us to define an equilibration time τ_{eq} as the smallest such observation time. Of course, this can only be a probabilistic definition; we must consider averages of the observation times t_{obs} at which the value of $a(O_\alpha; t_{obs})$ falls below the critical value $a_{eq}(O_\alpha)$ for a given trajectory, where the average is taken over all trajectories of the evolving system starting from all possible microstates of the system when defining τ_{eq} (and similarly regarding the definition of the escape time below) [21]. Due to the importance of the agreement between time and ensemble averages in the practical definition of thermodynamic equilibrium, we say that the system is ergodic for the observable O_α on time scales longer than $\tau_{eq}(O_\alpha)$ with accuracy $a_{eq}(O_\alpha)$ if the difference between time and ensemble averages is smaller than this accuracy. In practice, one would prefer $a(O_\alpha; t_{obs})$ to be dimensionless normalized quantities; we can achieve this by dividing $|\langle O_\alpha \rangle_{ens} - \langle O_\alpha \rangle_{t_{obs}}|$ by $|\langle O_\alpha \rangle_{ens}|$, i.e., the ergodicity criterion would be that $a^{norm}(O_\alpha; t_{obs}) = |\langle O_\alpha \rangle_{ens} - \langle O_\alpha \rangle_{t_{obs}}| / |\langle O_\alpha \rangle_{ens}|$ be smaller than $a_{eq}(O_\alpha)$. However, for the purpose of this presentation, we will ignore this more technical aspect.

We note that this definition of equilibrium and ergodicity only applies to the observable O_α, while our common-sense definition of a system being in thermodynamic equilibrium assumes that for all observables of interest the difference between ensemble and time averages is below the required level of accuracy. Furthermore, in principle all of these variables $O_\alpha, O_\beta, \ldots$ could have different prescribed levels of accuracy, i.e., $a_{eq}(O_\alpha) \neq a_{eq}(O_\beta)$.

In particular, this also implies that τ_{eq} is only defined for a given observable or set of observables, i.e., in general $\tau_{eq}(O_\alpha) \neq \tau_{eq}(O_\beta)$. In practice, we would focus on a finite set of observables which are most relevant for our description of the system and define the equilibration time τ_{eq} as the largest among the set of observable dependent equilibration times $\tau_{eq}(O_\alpha)$. In principle, we can achieve such a "complete equilibrium" if we employ the M statistical likelihoods L_i of all M microstates i as the observables which must fulfill the ergodicity requirement; since in the ensemble picture $\langle L_i \rangle_{ens} = p_i$, the requirement would be $|\langle L_i \rangle_{t_{obs}} - p_i| < a_{eq}(L_i)$ for all microstates i. Clearly, if all microstates of a system are visited according to their ensemble equilibrium probability, then the time averages of all their property observables must also fulfill the ergodicity requirement.

The condition of ergodicity is usually non-trivial to prove for a given physical system, at least if one wants to prove that we can make $a_{eq}(O_\alpha)$ arbitrarily small in the limit $t_{obs} \to \infty$. There has been considerable discussion of this issue in the literature [3,22–24], since it is often unclear whether it is possible to reach (or be arbitrarily close to) every other microstate from every initial microstate of the system. For our purposes, we will always assume that $a_{eq}(O_\alpha)$ is small but finite. In particular, this allows us to introduce the concept of local ergodicity [20,25], which in many realistic physical or chemical systems is much more appropriate than the global ergodicity we have considered up to now.

Here, we note that many macroscopic properties are actually only defined on the mesoscopic or continuum level. Typical examples are densities, density related quantities, and transport coefficients such as fluid densities, electric and magnetic polarizations, elastic constants, viscosity, electric conductivity, thermal conductivity, and so on [6,26]. As a consequence, values for such properties are only valid on time and length scales larger than those that define the material point in the continuum approximation. The continuum approximation implicitly assumes that we can treat the atoms contributing to the material particle as if they were locally equilibrated within the region associated with the particle; otherwise, the material point or material particle cannot be properly defined. As a consequence, the material particle exists on a finite length and time scale; in order to achieve smoothness of the system on the macroscopic level, continuum physics-based quantities are only properly defined on length scales larger than at least ten nanometers to ten micrometers (depending on the material, its phase, and the properties of interest). In order for the system to be treated as a continuum, similar estimates need to be obtained for the smallest time scales, depending on the equilibration times for the material point and the speeds with which disturbances propagate in the system; for more details, we refer to the literature [6,26]. Thus, we can usually ignore such physical properties as far as the definition of ergodicity and the conceptual developments below are concerned. However, because these properties include transport coefficients such as thermal and electrical conductivity, they play a role if one considers finite-size effects in thermodynamic models.

2.2. Concepts of Finite-Time Thermodynamics

The field of finite-time thermodynamics deals with the analysis of thermodynamic processes and their applications in finite time, extending to generalizations to other cost function-based processes that can be formulated somewhat analogously to thermodynamic systems [10,27]. We typically prescribe a certain total time t_{total} which is available for a given process, e.g., a thermodynamic cycle (typical for extraction of work via a thermodynamic engine), or a line-type process in which the final location in thermodynamic space differs from the starting point (typical for production of a compound via a chemical reaction). When executing such a process for infinitely long process times, a total well-defined amount of, e.g., entropy is produced together with a certain amount of net work being performed or a certain amount of product produced during the chemical reactions that take place [2,3]. However, if the process is to run in finite time, usually some additional excess entropy is generated, while the amount of work extracted or the desired product generated is smaller than for the infinite-time case [28]. In the latter case of infinite time being available, one can proceed in a quasi-static fashion that is infinitely close to equilibrium, taking as much time as needed to stay close to thermodynamic equilibrium along the trajectory in thermodynamic space.

This contrasts with the FTT case, where the system never reaches true thermodynamic equilibrium for real systems along the path in thermodynamic space. However, it is usually assumed that we can stay close enough to equilibrium such that at each step it is possible to imagine (or justify) that the system relaxes close to equilibrium on some (local) relaxation time scale τ_{relax}, meaning that the next step can be assumed to again start from an equilibrium state [29,30]. This already points to the issue of characteristic time scales of the real system that are relevant for a description, or at least a suitable model of the actual processes taking place during such a cycle. We note that there are exceptions to the rule that the process would ideally run infinitely slowly for an infinitely long time; for example, we might want to capture the intermediary reaction product B in the reaction sequence $A \to B \to C$, where an infinitely long reaction would result in total conversion of A to C.

Given some phenomenological laws or assumptions about how the finite-time condition affects the amount of (excess) entropy produced at each step, we can now perform a global optimization over the possible ways of assigning the available time to the legs of the cycle and how to allocate the time t_{total} to the sub-steps within each leg. As the target of this optimization, one frequently takes the maximum work extracted within the given time t_{total}—or the maximum power where one employs heat and work flow (rates) instead of heat and work [31]—; alternatively, one can aim for maximum efficiency in the spirit of the original Carnot engine [14]. Furthermore, the optimization can also involve the choice of the path itself in thermodynamic space. In this case, one would perform the time allocation optimization along the path as an implicit sub-optimization step for each choice of path, although a combined one-step path-plus-allocation optimization might also be feasible in principle.

We note that these optimization or optimal control problems for a thermodynamic cycle process usually assume that the system must follow the work cycle and that it will be possible to return to the original state of the system. However, this condition is not necessarily fulfilled if we are dealing with realistic systems involving complex energy landscapes that possess metastable macroscopic states.

2.3. Abstract vs. Real Thermodynamic Systems

In many work cycles discussed in classical thermodynamics, it is implicitly assumed that the stability of the material or device that one employs is perfect, and that one can spend as much time as is needed (up to infinity) on each leg of the cycle, ensuring that no excess work, entropy, etc., is generated beyond the amount implied by the laws of equilibrium thermodynamics. In particular, it is assumed that global ergodicity holds, and that metastable phases or marginally ergodic material states such as glassy states or aging phenomena can be ignored. As pointed out in the previous subsection, this assumption is not always true if we are dealing with processes that run in finite time, where the system can be in non-equilibrium states for long times throughout the process. In particular, if the material employed in building the device or used as the working fluid exhibits more than one stable or metastable phase, then bifurcations during phase transitions and nucleation-and-growth processes can occur. Note that for simplicity we will use the term 'working fluid' throughout, even though the practical realization of the working fluids we discuss will often be a solid material and not a liquid or gas.

A second challenge posed by the real materials employed in devices, working fluids, or targets for thermodynamic engines and processes is the spatial dimension. It is typically assumed that we are dealing with a homogeneous material and that the changes in the environment, e.g., temperature, pressure, or volume changes, are instantaneously felt throughout the material even though they are only applied at the surface of the material. This treats the material (which in reality is macroscopic) like a small "point", meaning that heat transport due to a change in the external temperature, propagation of pressure disturbances, etc. can be neglected as far as the analysis of the processes is concerned. Again, while such neglect can be justified for the case of the infinitely slow processes of classical thermodynamics, it cannot be justified for processes in finite time. Thus, we encounter length scales of relevance in thermodynamic engines, where the speed with which an external disturbance spreads throughout the material converts the relevant length scale into the corresponding time scale.

While heat leaks and similar phenomena have been included in FTT models of processes and cycles [32], delay times due to the spatial extent of a working fluid or target material are usually not accounted for. We note that to a certain extent such phenomena are included in the definition of the overall heat or work flow in the model systems employed in FTT as a source of irreversibility and lost efficiency, where we postulate, e.g., a law for heat conduction from the reservoir to the working fluid. In these typical endoreversible models [12] of finite-time thermodynamic processes, we might (for instance) restrict the speed with which heat is transferred to the working fluid; however, we usually do not consider the processes that are taking place inside the working fluid or their limitations and irreversibilities due to its finite size and possible internal structure. As a consequence, the assumption of endoreversibility may be too optimistic, and it is necessary to consider the details of the working fluid's behavior during the process. Sometimes, this can be achieved by splitting the working fluid into subsystems which each can be considered endoreversible or even reversible [33,34], but this adds a lot of complexity and may not always be possible.

Clearly, the challenge is even greater if the microstructure of the material on the mesoscopic scale—single-crystal vs. granular composites vs. powder materials—becomes an issue in the performance of the thermodynamic process. One way to address this issue might be via the introduction of some "effective" or "average" properties of the material, although this might not always be feasible. Here, we need to distinguish between the use of the material as part of the device, as the working fluid, or as the target.

In the first instance, e.g., the development of a rust layer on an engine due to its exposure to the working fluid or environment, would result in limitations on the speed with which we can drive the engine when performing work without risking catastrophic failure. As a consequence, we are dealing with limitations on the time we can spend in various stages of the thermodynamic cycle, i.e., such deviations from the ideal material impose "minimum/maximum time requirements" regarding the time spent in a part of the cycle. If the material is employed as a working fluid, then the cycling through different parts of thermodynamic space will be affected by changes in the microstructure of the material as far as the speed with which the working fluid can react to, e.g., exposure to different temperatures is concerned. In this case, the various phenomenological coefficients characterizing, e.g., the heat conduction between the reservoir and the working fluid in common generic model descriptions would change from cycle to cycle, or even within a cycle. On the other hand, if the material itself is the target of the process of a chemical reaction or transformation—regardless of whether it is actually part of the working fluid or is only acted upon by the working fluid during the process—then the microstructure can be a crucial feature of the target material besides its more general atom level properties such as the composition and selected crystalline modification.

2.4. Concepts of Energy Landscapes

Understanding the behavior and dynamics of materials, including their stability, response to external forces, and various properties on the macroscopic, mesoscopic, and down to the atomic level, leads to the study of their potential energy landscapes as well as various generalizations such as their enthalpy or free energy landscapes. [7,8] The energy landscape of a physical or chemical system is a complex hypersurface or metagraph over a high-dimensional continuous or discrete state space, respectively. This landscape consists of microstates which underlie the microscopic dynamics of the system [7]. Each microstate i is assigned an energy $E(i)$, and the space of microstates exhibits a certain topology, i.e., a neighborhood structure. This topology can represent the natural neighborhood reflecting the microscopic dynamics of the system based on the laws of physics on the microstate level. For example, for a chemical multi-atom system, a typical microstate corresponds to one specific arrangement $\vec{X} = (\vec{x}_1, ...\vec{x}_N)$ of N atoms in 3D space, where $\vec{x}_i \in R^3$ are the coordinates of atom i and a neighbor microstate differs only minimally in the atom positions. Thus, the state space for a system with N atoms would be R^{3N}, and the movement of the system on the landscape can be described according to Newton's laws via the gradient of the potential energy landscape. The topology can also explicitly define the dynamics of the system, e.g., by an exploration algorithm in a computer simulation of via probabilistic transitions among microstates reflecting the quantum mechanical nature of the system (at least partially); in such a case , we would speak of the moveclass of the system when talking about its neighborhood structure.

Regardless of whether we are dealing with a discrete or continuous energy landscape, we can now use the laws for classical or quantum mechanics to compute the time evolution of the system. However, for many systems we will notice that the time averages of observables along a trajectory of the system vary as function of the (randomly picked) starting point of the simulations, even when employing long simulation times, and can exhibit sudden jumps even after the average has seemed to converge to a fixed value.

The reason for this is that many such systems possess a multitude of metastable macroscopic states $\mathcal{R}_i$, which (at least at low temperatures) can persist for very long times that often far exceed the equilibration time $\tau_{eq}(\mathcal{R}_i; a_{eq})$ among all microstates that contribute to this metastable macroscopic state. As a consequence, the system appears to be in thermodynamic equilibrium for observation times t_{obs} between the equilibration time $\tau_{eq}(\mathcal{R}_i; a_{eq})$ and the escape time from the metastable state $\tau_{esc}(\mathcal{R}_i; b_{esc})$, beyond which the likelihood of leaving the region of microstates that belong to the macrostate $\mathcal{R}_i$ exceeds b_{esc}: $\tau_{esc}(\mathcal{R}_i; b_{esc}) \gg t_{obs} \gg \tau_{eq}(\mathcal{R}_i; a_{eq})$ [20,25]. For these observation times, the system in the macrostate $\mathcal{R}_i$ appears to be in perfect equilibrium and the laws of equilibrium thermodynamics apply to the system. For a given set of thermodynamic parameters and important physical observables, we can now determine (usually via numerical simulations of the dynamics and landscape explorations) which subregions $\mathcal{R}_i$ of the (state space of the) landscape can equilibrate on a certain observational time scale while not leaving the region, i.e., we can identify the so-called locally ergodic regions which fulfill the criteria of local ergodicity by computing their equilibration and escape times [20,25].

In the case of chemical systems, each such locally ergodic region $\mathcal{R}_i$ corresponds to a thermodynamically (meta)stable compound on a given observational time scale of interest.

Thus, on short time scales the phase diagram of the system needs to include metastable phases as well [35]; only in the infinite observation time limit do we reach the classical phase diagram [7]. In particular, one can perform classical thermodynamic processes with the system restricted to this metastable compound (adding and extracting heat, applying pressures, etc.) without ever noticing that the system is not in the global thermodynamic equilibrium state, so long as the total time available for the process t_{total} is much smaller than the escape time from this region $t_{total} \ll \tau_{esc}(\mathcal{R}_i; T, p)$ for all the values of pressure and temperature along the process path. Figure 1 shows a plot of the Gibbs free energy vs. the observation time, depicting the time ranges over which two crystalline phases, the melt, and a glassy state, are locally ergodic.

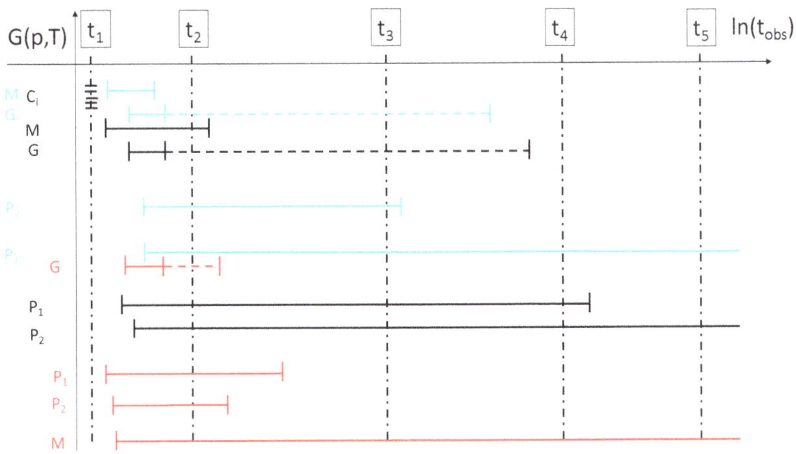

Figure 1. Qualitative sketch of the Gibbs free energies of some typical locally ergodic regions found in a solid for three given temperatures $T_1 < T_2 < T_3$ and standard pressure p, with data for T_1, T_2, and T_3 marked in blue, black, and red, respectively. Here, C corresponds to individual defect configurations, M represents the super-cooled melt for T_1 and T_2 and the actual melt for T_3, respectively, G is a glassy state, and P_1 and P_2 are (real, i.e., including all the local minima that correspond to equilibrium defect configurations) crystalline modifications. At temperature T_1, P_1 is the thermodynamically stable phase and P_2 is a metastable phase, just as the super-cooled melt is. The glassy state is locally ergodic only on very short time scales; for longer observation times it is only marginally ergodic and ages slowly (indicated by the dashed horizontal line). At temperature T_2, the (high-temperature) phase P_2 is now the globally ergodic equilibrium phase, while P_1 is only metastable. For both temperatures, the metastable phase and the glassy state are stable up to quite large observation times. Finally, the high temperature T_3 is above the melting temperature, and the melt is in the thermodynamic equilibrium phase. Note that the escape times from the glassy state and the crystalline phases are now quite short, even though their Gibbs free energies are much lower than at low temperatures. The vertical lines denote various observation times; e.g., at time t_4 we have only two locally ergodic regions left for temperature T_2. Remember that after preparing the system in one of the phases that are metastable for a given observation time, we will not know of the existence of competing phases with possibly lower local free energies.

However, it should be noted that the escape and equilibration times for a metastable state will (possibly strongly) depend on the external influences that the system is exposed to. In particular, high temperatures can affect the stability of a metastable state, i.e., $\tau_{esc}(\mathcal{R}_i)$ can rapidly become small when increasing the temperature. Similarly, the equilibration time will vary with temperature, but usually not as strongly as the escape time. As a consequence, a given work cycle that includes temperatures that greatly reduce the escape time must be executed quite quickly if one wants to return to the original state of the system in thermodynamic space while avoiding a phase transition into a different (meta)stable phase along the way. After all, a metastable state is by definition not in global thermodynamic equilibrium; thus, the system will not return to this state after it has left the metastable state.

In this context, we remark on the concept of marginal ergodicity [7,36–38]. This applies to, e.g., glassy systems and systems that exhibit aging [39–43]. In practice, we find for such systems that they appear to be in equilibrium for some time interval $t_{quasi}(i)$ correlated to the time since their creation $t_{age}(i)$, e.g., by quenching from a melt; this falsely suggests that the landscape region $\mathcal{R}_i$ that they explore is locally ergodic. For larger times, these systems begin to show deviations from equilibrium. However, in contrast to a transition to another metastable state $\mathcal{R}_j$, where the system explores and equilibrates inside a competing region $\mathcal{R}_j$, here the system just explores a larger landscape region $\mathcal{R}_{i+1}$ that encompasses the previous one ($\mathcal{R}_i \subset \mathcal{R}_{i+1}$). If we pretend that we would have started the experiment anew at the point of time after losing the equilibrium in region $\mathcal{R}_i$, except now only starting our check of the equilibration after a time $t_{age}(i+1) \approx t_{age}(i) + t_{quasi}(i)$ had passed, the system would seem to be again in quasi-equilibrium, now in the larger region $\mathcal{R}_{i+1}$. Furthermore, it would remain so for a longer time interval $t_{quasi}(i+1)$, since the time from creation $t_{age}(i+1)$ is longer than in the earlier part of the experiment ($t_{age}(i+1) > t_{age}(i)$). However, instead of stabilizing and truly equilibrating in the region $\mathcal{R}_{i+1}$, the system again falls out of equilibrium when t_{obs} exceeds $t_{age}(i+1)$, and the process repeats itself on a longer time scale. From the point of view of ergodicity, we are dealing with a sequence of ever-larger regions $\mathcal{R}_i$ where $\mathcal{R}_i \subset \mathcal{R}_{i+1}$ and where the equilibration and escape time for region $\mathcal{R}_i$ are approximately equal. This does not allow the system to truly enter a metastable state, while at the same time the escape time from region $\mathcal{R}_i$ is approximately equal to the equilibration time of the subsequent region $\mathcal{R}_{i+1}$: $\ldots \approx \tau_{esc}(\mathcal{R}_{i+1}) \geq \tau_{eq}(\mathcal{R}_{i+1}) \approx \tau_{esc}(\mathcal{R}_i) \geq \tau_{eq}(\mathcal{R}_i) \approx \ldots$. This process is frequently called aging, since the time in quasi-equilibrium is often more or less proportional to the time one has waited, giving the system time to "age" since its creation ($t_{quasi}(i) \sim t_{age}(i)$).

Next, we briefly mention the concept of self-averaging of macroscopic systems [7,44]. It is clearly impossible to completely sample the landscape of an infinite macroscopic system on the microstate level in a finite time such that we can verify its global ergodicity with $a_{eq} = 0$, and even for small finite values of a_{eq} this is essentially impossible for real systems. Thus, we must ask why we can still talk about ergodicity and thermodynamic equilibrium for, e.g., a macroscopic solid, and why such systems can be employed in thermodynamic processes. Recall that classical thermodynamics usually assumes the so-called thermodynamic limit [9], where all extensive quantities of interest go to infinity while their ratios approach finite values. To address this challenge, a chimera-like approach is employed in which one represents the macroscopic system as an ensemble of essentially identical small mesoscopic systems, each of which can reach equilibrium within a finite observation time. At the same time, this set of mesoscopic systems constitutes an ensemble representing the original macroscopic system in an approximate fashion, such that the average over this ensemble's time-averages on the time scale of observation yields a good approximation of the true ensemble average of the (infinite) macroscopic system. Clearly, such a self-average over the macroscopic solid only makes sense if the thermodynamic ground state is homogeneous (exhibiting the same atom distribution and crystalline or amorphous "structure" throughout the material); if the ground state consists of two disjoint phases with different compositions, then self-averaging is not possible (or at least only possible in the metastable phase sense discussed above).

Here, we note that by representing the time evolution of a macroscopic material in the self-averaging approximation, we are effectively constructing an approximate probabilistic evolution of the system. This type of evolution is commonly employed in (abstract) stochastic models such as Markov processes when describing the time evolution of a single (!) system in a probabilistic fashion, although the single system—defined on the microstate level of the atomistic system—should follow its own trajectory in state space that would be different from the stochastic trajectory of the probability distribution over all possible (micro)states of the system. This issue is similar to the question of time scales beyond which the results of molecular dynamics simulations agree with those of Monte Carlo random walk-based simulations [45].

Finally, we must comment on the effect of placing a system described via such a complex energy landscape in contact with an environment [7,46]. Here, we only consider the assignment of thermodynamic intensities, in particular T and p; for more examples of interactions with the environment and their inclusion in the energy landscape description, we refer to the literature [46]. Prescribing a certain temperature does not change the potential energy landscape of the otherwise isolated system, but does influence the existence, time scales, and size of the locally ergodic regions on the landscape [7]. However, the pressure and other intensities such as electromagnetic fields do affect the landscape itself by adding terms such as $+pV$ to the energy function [7,46,47]. In particular, for a non-zero pressure we are now dealing with a potential enthalpy landscape $H_{pot}(\vec{X};p)(\text{``} = \text{''}E(\vec{X};p)) = E_{pot}(\vec{X}) + pV(\vec{X})$ instead of a potential energy landscape $E_{pot}(\vec{X})$. We can also visualize this as a family of such (potential energy/enthalpy) landscapes parameterized by the pressure p, where we switch to neighbor landscapes when changing the pressure. Each of these enthalpy landscapes exhibits its own locally ergodic regions, which correspond to, e.g., high-pressure phases or phases that are (meta)stable at effective negative pressures. In particular, a thermodynamically stable modification can become metastable or unstable in response to a change in pressure; thus, in searching for metastable modifications of a material, one explores not only the potential energy landscape but also a wide spectrum of potential enthalpy landscapes for both negative and positive pressures [48].

In the chemical and physical literature, the complex energy landscapes of material systems have been frequently employed to study (thermodynamic) behavior, particularly of liquids [24,49–52] and glasses [53–58] (e.g., polymers [39,59–65], solids [66,67], spin systems [68–71], etc.) as well as proteins and biopolymers [61,72–79], where the complexity of the barrier structure of the multi-minima landscape is expected to prevent the system from reaching global thermodynamic equilibrium. In the case of glassy materials, this inability to equilibrate and the multitude of competing more-or-less disordered local minima on the energy landscape constitutes nearly the defining quality of a glass. A major goal of such studies has been to understand the way in which such materials can be fitted into the classical phase diagram picture of matter, together with explaining their unusual long-time dynamics exhibiting phenomena such as aging [39,40,80,81] which do not appear in the usual ordered crystalline phases.

Furthermore, the ability of glasses to very rapidly become less viscous over a rather small temperature range that remains below the melting point of the (crystalline) material (the converse of the nearly exponential rise in viscosity upon quenching the glass-forming melt [82]) remains a puzzling aspect of glassy systems. A possible answer has been proposed based on the observation that many glass-forming systems exhibit an exponentially growing local density of states [55,71,83,84]. It can be noted that for such systems we observe a bifurcation behavior called "exponential trapping" when one computes the probability of the system residing in a given range of energies as a function of the temperature [84–86]. The system is either deep inside the local minima basins, meaning that it faces large barriers against transformation into energetically more favorable states (the highly viscous quasi-equilibrium glassy state), or is found at the top of the energy range for which the local densities of states are growing exponentially. In this case, the system only needs to cross much lower energy barriers to neighbor basins with a concurrent much higher mobility of the atoms, i.e., we are now dealing with a low-viscosity melted material. Here, the fascinating aspect is that the range of temperatures over which the switch occurs between these two different (marginally and locally ergodic, respectively) regions of the energy landscape with very distinct rheological behavior is very small regardless of the difference in energy between the top and the bottom of the part of the landscape exhibiting the exponentially growing densities of state.

This is in contrast to the case of proteins, where a main challenge has been understanding the ability of certain proteins to reach their native folded state on comparatively short time scales in spite of their complex landscape, in something like the converse of the glass transition problem. One popular answer to this so-called Levinthal's paradox [72] is the assumption that the energy landscape exhibits a funnel-like structure for these specific proteins [75,87], with rather low energy barriers to neighboring local minima but a clear trend for the pathways on the landscape to guide the probability flow towards the low-energy region. This allows the protein system to reach this region of the landscape corresponding to the native state in spite of the multitude of competing minima which might otherwise trap the system along paths where the large majority of accessible paths lead to the native state.

These issues and other related ones have been extensively discussed in the literature over the past sixty years. For general multi-minima systems in chemistry such as crystals [48,88–101], clusters [102–112], and complex molecules [104,113–118], their complex landscapes have been explored mainly with the goal of identifying promising (meta)stable modifications and isomers. Furthermore, the free energies have been computed for individual locally ergodic regions, usually in the harmonic approximation [119], and (tree) models for so-called free energy landscapes have been constructed [120–122] ("so-called" because the issues of the relevant timescales for the existence of the locally ergodic regions and the free energy barriers between them are typically ignored in their construction [123]). Finally, equilibration trees have been computed that indicate the way in which different sub-regions of the energy landscape equilibrate into larger locally ergodic regions on the level of both microstates [71,83] and minima [124].

These are fascinating research questions in themselves, and have formed the motivation behind the development of many algorithms for studying complex energy landscapes. They have also inspired many concepts and model descriptions of such landscapes; for an overview, see [7,8]. However, although issues such as the glass transition and the folding pathways on biologically relevant timescales highlight the finite-time aspect as part of their formulation, we are not going to discuss these phenomena in this perspective, as the focus here is on classical thermodynamic processes and the optimization of their output in finite time, be it work or heat production and transfer without wasting work or generating excess entropy, the efficiency of such processes, or the amount of material that can be produced with minimal effort.

3. Prototypical Examples

This section briefly discusses prototypical examples of classes of processes for systems where the complex energy landscape adds features that need to be taken into account when setting up (and subsequently solving) the optimization problems of finite-time thermodynamics. For every class of systems, we will point out the time scales of relevance that need to be considered, the quantities one might want to optimize, the type of cycles that are of interest, and the special complications that can arise when modeling and optimizing the relevant processes. Given this focus on the additional features and challenges due to the complex energy landscape involved, we will not discuss standard aspects of the FTT description of thermodynamic processes and their optimization, referring to the literature for such details [10,11,13,27,125]. Similarly, we do not include finite-time studies of thermodynamic engines where a phase transition occurs as part of the cycle but where the actual transformation process is of no major concern, and is dealt with in a generic way as part of the overall heat transfer along the leg where the (isothermal) transition occurs [126].

These descriptions are only rough schematics, and are only meant to provide guidance for dealing with actual problems of interest; realizing and solving such an example in practice is a whole research project in itself [127,128], and goes beyond the purview of this concept-focused presentation.

3.1. Thermodynamic Processes Involving Phase Transitions

The most relevant class of examples for FTT processes involving systems with complex energy landscapes concerns the use of materials as working fluids which can exhibit more than one phase during operation of the engine along its path in thermodynamic space. In a more general sense, this also includes the use of such materials in building the device, as already discussed in Section 2.3, where possible complications due to wear and tear of the real device were noted together with the limitations on the speed with which the thermodynamic intensities T and p can spread throughout the working fluid. Such issues can affect the efficiency and capability of the device, e.g., as limitations on the speed at which the thermodynamic cycles can be performed.

We note that the pure "mathematical thermodynamicist" who wants to focus on analytically solvable problems might prefer to consider issues concerning metastability, instability, irreversibility, etc., as an external mathematical side-condition on the optimization problem, such as adding heat conduction between hot and cold reservoirs leading to leakage of heat from the system proportional to the time spent in the overall cycle. Still, even such a researcher needs to first formulate a mathematical model of this feature that represents real-life engines and working fluids, before such real-life aspects can afterwards be reduced to a formal overall factor encapsulating a high entropy production rate or loss of work at certain parts of the cycle.

3.1.1. No Competing Crystalline/Amorphous Phases

The simplest case is a system with a working fluid that consists of a homogeneous (infinitely large) material that can exist in several equilibrium phases but where neither metastable modifications nor amorphous/glassy states are present. For each point in thermodynamic space $(T, p, \ldots)$, only one phase is present, i.e., on all observational time scales of interest, the energy/enthalpy landscape exhibits only one locally ergodic region, which is identical with the globally ergodic equilibrium phase for this temperature and pressure. For simplicity, we assume that only two equilibrium phases A and B exist in the region of thermodynamic space where the cyclic or line-type path resides; here, we denote a path to be a line-type one if the starting and end points are different and if the path does not intersect itself. Figure 2 shows a simple cyclic path in (T, p) space, where for low and high temperatures the thermodynamic equilibrium phases are A and B, respectively.

The goal of the thermodynamic process is to perform some work in a finite time t_{total}, where the excess entropy/heat production or loss of work due to the finite-time requirement is minimized or the efficiency of the process is maximized. For simplicity, we assume a cyclic process with a prescribed path in (T, p) space, where the time spent in each leg of the path and the way in which the time is allocated along the legs are subject to the optimization. Along this cyclic path, the material switches between two (or more) equilibrium phases, such that at least two phase transitions will occur. We note that of course the case of a path with only one transition that can occur when cycling around a second order phase transition point in (T, p) space—e.g., in some cycles involving a gas–liquid transition—also works, and might be of interest in certain kinds of refrigeration cycles employing the gas–liquid phase transition. In this context, we recall that the phase transitions occur along lines or at points in thermodynamic (T, p) space; depending on which value is varied, one can speak of transitions driven mainly by pressure changes or mainly by temperature changes. In the kinds of materials we are discussing here, most of the transitions are first-order, although second-order transitions can also occur. In solids, the latter are mostly associated with small displacements of the atoms changing the symmetry group of the crystal to a subgroup, such that no nucleation of the new phase followed by growth of the new phase from this nucleus is needed. As a consequence, for second-order phase transitions there is no free energy barrier for the atom rearrangement, and the amount of thermodynamic work needed is small. Of course, if the transformation takes place at points in (T, p) space that are some distance away from the second order transition point or line, then we are dealing with a first-order transition.

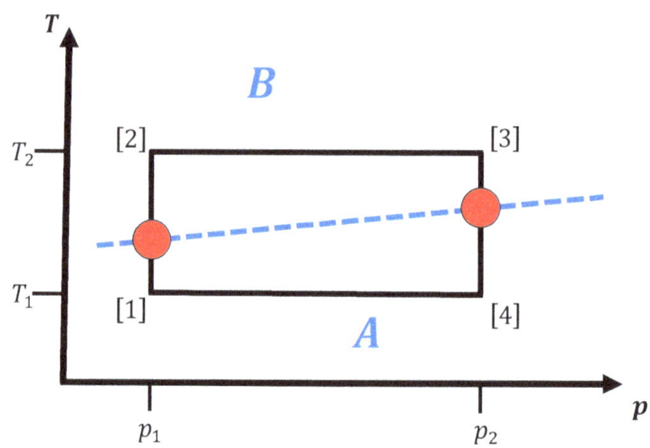

Figure 2. Sketch of a simple cyclic path ([1] → [2] → [3] → [4] → [1]) in (T, p) space, with two phase transitions: from equilibrium phase A to equilibrium phase B in the first leg, and from phase B to phase A in the third leg, respectively. The dashed blue line is the equilibrium phase separation line in the (T, p) phase diagram. The points where the transition would occur for an infinitely slow cycle are indicated by red spots.

Upon a change in temperature and/or pressure at the transition temperature/pressure along the path, the material has to re-organize on the atomic level in order to transform into the new equilibrium phase. To achieve this process in finite time, it is usually necessary to move the system to temperatures/pressures deviating from the equilibrium transition values such that the material enters a (potentially massively) non-equilibrium state. In this state, the first phase is no longer stable, but the new equilibrium phase is not yet established;

in order for the system to settle into this new phase within a finite time, excess entropy/heat generation or extra work is required. Figure 3 shows another schematic cycle, where we now indicate the ranges of metastability of phases A and B in the respective regions where phases B and A are the equilibrium phases.

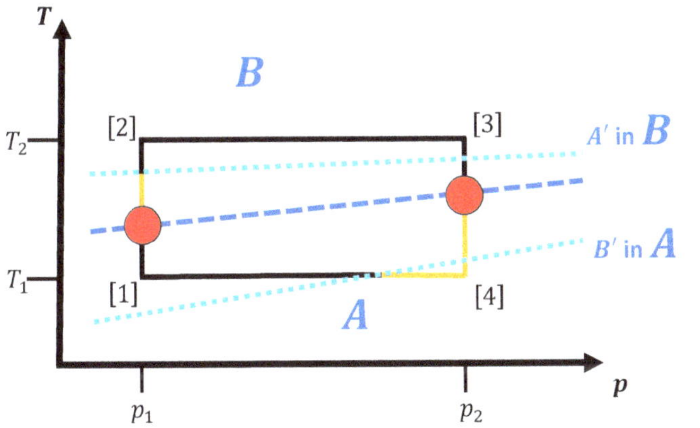

Figure 3. Sketch of a simple schematic path in (T, p) space with two phase transitions: from equilibrium phase A to equilibrium phase B in the first leg, and from phase B to phase A in the third leg, respectively. The dashed dark blue line is the equilibrium phase separation line in the (T, p) phase diagram, while the dotted light blue lines indicate the range in (T, p) space where the metastable phase (marked as A′ and B′) is still locally ergodic on moderately large time scales. The points where the transition would occur for an infinitely slow cycle are indicated by red spots, and the orange colored part of the path shows the range of the path where "non-trivial" times would be needed to perform the transition from the metastable phase to the equilibrium phase in the event that only a finite time t_{total} is available for running the whole cycle. Note that the difficulty of nucleating phase A from the metastable phase B' extends to the fourth leg of the cycle, even though the stability of phase B' is greatly reduced around point [4], which marks the high-pressure and low-temperature region along the path. On the other hand, after the transition has taken place, the metastable phase becomes essentially inaccessible. Typically, the farther away we are from the actual equilibrium phase transition line, the faster the nucleation-and-growth transition can take place. However, we note that the actual length of the orange sections of the path would depend on the speed at which we are moving along the path, relative to which transition times could be considered "non-trivial"; the more slowly we can move along the cycle overall, i.e., the larger t_{total}, the shorter the orange sections will be. This can lead to confusion if we were to interpret the orange color as indicating the section of the path where we still observe the system in the respective metastable phase; such an interpretation would only be appropriate for a given specific choice of time allocation along the path.

In order to design models of this process for the purpose of time-allocation optimization, we note the relevant time scales involved. Obviously, there is the available time for the whole cycle t_{total}, which distinguishes finite-time thermodynamics from classical thermodynamics. Next, we discuss various equilibration time scales; note that in the following we do not worry about a_{eq} or about the observables O_α used to classify ergodicity and equilibrium. The first time scale of interest is the equilibration time $\tau_{eq}(i; T, p)$ for phase i when starting from a generic point inside phase i for general values of T and p for which phase i is locally ergodic. The second time scale is the relaxation time $\tau_{relax}(i; T_{new} - T, p_{new} - p)$ for phase i after a (small) perturbation of the system away from equilibrium. Here, we are starting from a point on the landscape belonging to the locally ergodic region $\mathcal{R}_i$ of the landscape associated with the system in equilibrium at (T, p), except that now the thermodynamic parameters have been shifted to (nearby) values (T_{new}, p_{new}), placing this point at the rim or even slightly outside the new locally ergodic region associated with the thermodynamic point (T_{new}, p_{new}). Although strictly speaking being an equilibration time, τ_{relax} corresponds to the usual relaxation time to equilibrium that is employed when modeling and optimizing the time step allocation for a standard working fluid that needs to be kept in or at least very close to equilibrium. Typically, one assumes that some simplified (phenomenological) model can be used for the excess entropy production as long as the time step associated with moving from (T, p) to (T_{new}, p_{new}) is larger than $\tau_{eq}(i; T_{new}, p_{new})$.

The third group of time scales concerns those associated with the phase transition $A \to B$. The first is the equilibration time inside the locally ergodic region associated with phase B, $\tau_{eq}(B; T, p)$. Next, there is the escape time from inside the unstable/metastable phase A, $\tau_{esc}(A; T, p)$. However, for complex energy landscapes it is not obvious that leaving region A will automatically transfer the system at once into the neighborhood of region B; instead, the system enters a general transition region that connects regions A and B. This region can be quite large, and in general may border on many other (locally ergodic or marginally ergodic) regions of the landscape. In particular, reaching the right "exit" from such a complex transition region crossing all of the generalized barriers involved can require quite a long time [38]. Thus, the escape time from region A together with the movement of the system inside the transition region into the neighborhood of the (stable) target phase B, $\tau_{esc}(A \to B; T, p)$ at the point (T, p) in thermodynamic space where phase B is the equilibrium phase can be considerably larger than $\tau_{esc}(A; T, p)$. Here, we note that during the subsequent equilibration "into" phase B when starting from the region of the landscape formerly associated with phase A or the transition region connecting A and B, the system does not start from a generic point inside phase B, as is usually assumed in the definition of $\tau_{eq}(B; T, p)$; instead, we start from a very specific point on the landscape inside the region associated with the former locally ergodic phase A or the transition region surrounding it. Thus, we denote this time scale by $\tau_{eq}(A \to B; T, p)$. Therefore, $\tau_{eq}(B; T, p)$ will usually be smaller than the time required to move from the exit of the transition region into the locally ergodic region B and equilibrate there, i.e., $\tau_{eq}(A \to B; T, p) > \tau_{eq}(B; T, p)$. These two time scales $\tau_{eq}(B; T, p) < \tau_{eq}(A \to B; T, p)$ and $\tau_{esc}(A; T, p) < \tau_{esc}(A \to B; T, p)$ constitute lower bounds on the time scale within which a successful transition from the newly unstable phase A into the new equilibrium phase B takes place on the energy landscape.

If the system in phase A still has a certain degree of stability, as is often the case in first order phase transitions, it can be considered locally ergodic on some small time scale; thus, the quantity of interest would usually be this escape time $\tau_{esc}(A \to B; T, p) > \tau_{eq}(A \to B; T, p)$. On the other hand, if the system can no longer equilibrate on essentially any time scale of interest in phase A at point (T, p) in thermodynamic space, then the equilibration time of the new phase would often be the relevant quantity, i.e., $\tau_{esc}(A \to B; T, p) < \tau_{eq}(A \to B; T, p)$.

A special very important case in this context consists of nucleation-and-growth processes, which proceed in two stages: the formation of a stable nucleus of phase B inside the metastable phase A, with its time scale $t_n(B \text{ in } A; T, p)$ (a special case of $\tau_{esc}(A \to B; T, p)$), and the growth process of this nucleus into the macroscopic phase B on the time scale $t_g(B \text{ in } A; T, p)$ (a special case of $\tau_{eq}(A \to B; T, p)$). We note that the system might well have left the locally ergodic region A before the nucleus of phase B has come into existence, i.e., $t_n(B \text{ in } A; T, p) \gg \tau_{esc}(A; T, p)$. Furthermore, during first-order phase transitions, $t_g(B \text{ in } A; T, p)$ can far exceed $\tau_{eq}(B; T, p)$; thus, the total time $t_{n+g}(A \to B; T, p)$ needed for such a process can greatly exceed the sum of $\tau_{esc}(A; T, p)$ and $\tau_{eq}(B; T, p)$,

$t_{n+g}(A \to B; T, p) = t_n(B \text{ in } A; T, p) + t_g(B \text{ in } A; T, p)$ ($\approx \tau_{esc}(A \to B; T, p) + \tau_{eq}(A \to B; T, p)) > \tau_{esc}(A; T, p) + \tau_{eq}(B; T, p)$.

Of course, we assume that a phase transition occurs in the first place: if $\tau_{esc}(A; b_{esc}) \gg t_{total}$ for the starting phase A for all values of (T,p) along the path, then no transition happens during the cycle and we stay in the same locally ergodic region A as far as the material is concerned. Analogously, if $\tau_{esc}(B; b_{esc}) \gg t_{total}$, then the transition back to the initial phase will not take place, and we are stuck in the second phase at the end of the cycle. In this case, it is necessary to add a long waiting time before restarting the cycle in order to allow the system to re-initialize and reach equilibrium in phase A. For a cycle that is performed periodically, this waiting time clearly poses a problem. An alternative to deal with either of these events within the available time t_{total} can be the choice of a different cycle, e.g., extending to higher or lower temperatures and/or different pressures, where the escape times from phases A and B are much shorter and the cycle can be performed as desired, though at the expense of a much higher excess entropy production or loss of work.

A related problem can occur if $\tau_{eq}(A \to B; T, p) \gg t_{total}$ for the whole region of the path in (T, p) space where phase B is the equilibrium phase. In this instance, the working fluid remains in a non-equilibrium state for the whole duration of this stage of the path, never quite leaving the transition region connecting regions A and B, and we need to explicitly model the thermodynamic processes during this stage on a non-equilibrium basis. After entering the stage of the path where A is again the equilibrium phase, the problem might be alleviated, since it is to be expected that the relaxation back to the phase A might be rather fast and within the time limits of the process. Again, we can try to address this issue by choosing a different thermodynamic path, along which the two phase transitions occurring quickly enough.

In this context, we recall the probabilistic aspect of the escape time definition, incorporated in the characteristic constant b_{esc}: if we attempt the transition $1/b_{esc}$ times, each time running the simulation/experiment for a length of $\tau_{esc}(b_{esc})$, then we expect that we will make a transition into the other phase at least once. In order to take this feature into account, a probabilistic approach can be employed to model the behavior of the system along the cycle, as outlined below in Section 3.1.5.

Assuming that neither of these serious problem cases occur, it is nevertheless clear that we need to spend enough time in the phase transition region to reach the new phase j from i, i.e., along these legs of the path, the times $t_{n+g}(i \to j; T, p)$ need to be included and subtracted from the total available time t_{total}, and the excess heat or entropy produced during the nucleation-and-growth stage must also be accounted for. As mentioned above, in order to speed up the phase transformation process, we might have to employ values of (T, p) that are (possibly considerably) larger/smaller than the values where the two phases are in equilibrium. Depending on how much time we can afford to spend on the phase transition, we can reduce the excess entropy production by staying closer to the transition values of pressure and temperature. On the other hand, spending too much time at the phase transition forces us to be "too fast" on the rest of the path, again producing excess entropy. One critical issue in the modeling of such rather fast movements along the cycle is that we might move the system so far out of equilibrium that the linear approximation-based models for the entropy production during relaxation to equilibrium (frequently used when computing or estimating optimal schedules [29]) are no longer applicable, considerably complicating the modeling of the thermodynamic cycle. This usual (excess) entropy production along the path in thermodynamic space and its finite-time optimization is not discussed here; we assume that we know or can model the excess entropy production due to the usual finite-time movement along the path for a given material once it is essentially equilibrated inside one phase, since this constitutes the standard FTT problem.

A detailed prototypical example of modeling a first-order phase transition using a finite-time optimal control analysis can be found in the study of the liquid–gas transition by Santoro et al. [127]. There, the authors minimized the excess work needed to perform this phase transition in a finite time slightly away from the phase transition point in thermodynamic space. Their study included explicit models for the nucleation and growth processes. Such models would also be necessary for the study of any thermodynamic cycle that includes phase transitions if one desires to obtain quantitative results for the optimization of a specific phase transition in a given chemical or physical system.

3.1.2. Presence of Amorphous Precursor or Glassy States

In the preceding Section 3.1.1, we have considered the case of a system with only two phases, which transform along a certain line in thermodynamic (T, p) space into each other. For first-order phase transitions $A \to B$, the transition is not instantaneous but commonly takes place through a nucleation-and-growth process on a time scale $t_{n+g}(A \to B; T, p)$ that depends on the thermodynamic conditions (T, p). For such processes, we deal with free energy barriers that require a certain time to cross in order to create a nucleus of critical size large enough to grow into the new phase [129–131]. In order to achieve this in finite time, we must move beyond the transition line in (T, p) space deeper into the region where phase B is the equilibrium phase, which requires excess work or excess entropy. For simplicity, we can visualize the nucleation process being characterized by the above-mentioned escape time $\tau_{esc}(A; T, p)$ from phase A, while the time spent in the growth stage of the transition is related to the equilibration time scale $\tau_{eq}(B; T, p)$ in phase B. We have noted already that $\tau_{eq}(B; T, p)$ is only a lower limit on the time that this growth process can take, as we do not start from a generic point inside the locally ergodic region B on the energy landscape. As a consequence, as already mentioned earlier, the overall nucleation-plus-growth time scale usually exceeds the sum of the above two time scales, $t_{n+g}(A \to B; T, p) > \tau_{esc}(A; T, p) + \tau_{eq}(B; T, p)$.

In practice the situation is more complex, since the nucleation process takes place via the generation of a multitude of nuclei throughout the macroscopic material. The precise number depends on the specific values of the temperature and pressure applied, quite frequently leading to the formation of a polycrystalline material with less well defined macroscopic properties instead of a single crystal. Furthermore, during a general nucleation process the system resides in a transition region on the complex energy landscape, from which in principle several alternative (metastable) phases can be reached, assuming that such additional phases exist on the landscape in the first place.

In particular, in many materials such a transition region—while never being locally ergodic in itself, and as such not corresponding to a thermodynamic (meta)stable phase—is structurally realized in solid form as an amorphous precursor state AP or a glassy state G, which can persist in some fashion for quite a long time at low enough temperatures. We note that there are of course many amorphous solids with surprisingly high stability that can be created employing a variety of alternative routes; the generation of amorphous $Si_3B_3N_7$ via the sol–gel process [132,133] constitutes such a case. However, here we only focus on those amorphous materials that are generated along a thermodynamic path, i.e., via quenching from a melt or the gas phase, or via amorphization through high-pressure treatments.

In the general amorphous state, the material is far from global thermodynamic equilibrium and serves as the matrix within which thermodynamically (meta)stable crystalline phases i can nucleate. This synthesis approach of crystallization from an amorphous precursor is frequently employed to access unusual crystalline modifications from, e.g., an amorphous film deposited from the gas phase, or to reach a high-pressure phase from a material that is being amorphized as an intermediary stage via application of high pressures and temperatures. The corresponding time scale for the specific nucleation-and growth process $t_{n+g}(AP \rightarrow i; T, p)$ now takes the place of the earlier mentioned time scales—$\tau_{esc}(i)$, $\tau_{eq}(j)$, and $t_{n+g}(i \rightarrow j)$ during first-order phase transitions—and must be compared with the total time available for the cyclic or line-type thermodynamic path. We note that the ability to access multiple metastable phases from such a precursor can be an advantage if one wants to reach valuable or just interesting phases that might not be accessible otherwise [134–136]. However, this adds uncertainty to the process with respect to its outcome, in addition to the excess work/entropy associated with the phase transition in finite time. To deal with this uncertainty, we need to introduce a probabilistic scheme into the cycle description, which is described in more detail in Section 3.1.5.

In many regards, the glassy state G can be treated as another instance of the amorphous state, although it exhibits some features that are of particular relevance for the thermodynamic cycles that include phase transitions which we are discussing here. Similar to the amorphous state, the glassy material will also eventually transform into the thermodynamically more stable equilibrium phase via some kind of nucleation and growth process. However, the glassy state tends to be quite stable on long time scales as far as its macroscopic properties are concerned. This also often applies to its general structural features; in particular, both the time scale of nucleation and the time scale for growth of a crystalline phase i inside the glassy state G, $t_n(i \text{ in } G; T, p)$ and $t_g(i \text{ in } G; T, p)$, respectively, can be very large, and may often exceed the total time available for the cycle, $t_{total} < t_n(i \text{ in } G; T, p)$, $t_g(i \text{ in } G; T, p)$.

This is related to the fact that in many materials of interest, glassy states appear when quenching a melt (i.e., cooling the material too fast to a temperature below the freezing temperature for nucleation of the crystalline equilibrium phase to occur). As a consequence, the glassy material resembles a liquid with extremely high viscosity in many aspects. Here, we note that the glassy state should not be confused with a metastable super-cooled liquid; the moment the super-cooled liquid experiences a (local) disturbance above a certain (small) strength, a spontaneous nucleation of many nuclei of the solid equilibrium phase takes place and these nuclei very rapidly grow to produce a polycrystalline material. In contrast, the glassy material is very stable, and even a strong disturbance like being hit with a hammer only results in mechanical damage, not in the initiation of a nucleation and growth process. As a consequence of being a quasi-equilibrium continuation of the melt in the solid state, such glassy systems typically exhibit marginal ergodicity and show aging, as discussed in Section 2.4.

For the thermodynamic cycles in finite time that we are interested in, such glassy states can, e.g., occur if the two phases that are visited along the trajectory in (T, p) space correspond to the solid crystalline modification (A) and the melt (B). Melting a solid crystalline phase is usually quite straightforward as long as only one solid phase exists in the system close to the melting temperature, and this melting transition usually takes place relatively quickly even for temperatures close to the melting point. We note that some complications can occur for unusual systems such as gallium [137]; however, here we focus on the standard case. Furthermore, the formation of the thermodynamically stable crystalline solid phase from the melt when cooling the system usually requires a non-negligible amount of time even if the material is not a glass former. As mentioned above, if glassy states G in the system are possible or likely, then the time to reach the crystalline modification A, $t_{n+g}(G \rightarrow A; T, p) = t_n(A \text{ in } G; T, p) + t_g(A \text{ in } G; T, p)$ can be very large, far exceeding the equilibration time of the crystalline modification $t_{n+g}(G \to A; T, p) \gg \tau_{eq}(A; T, p)$, and might require careful tuning of the thermodynamic conditions. Figure 4 shows the simple cycle already discussed in Figures 2 and 3, except that now phase B is the melt (denoted by M) and we encounter the added complication that the system can enter a glassy state from the melt instead of nucleating into the equilibrium phase A.

For the purposes of our discussion, this implies that if $t_{n+g}(G \to A; T, p)$ exceeds t_{total}, then we will remain in the glassy state of the material for the rest of the trajectory. In particular, if the cycle is run with a starting point in the crystalline phase A, then we do not reach the original starting point of the cycle, and not only in a small fraction of instances; i.e., being stranded in the glassy state is not a low-probability occurrence. If the working fluid at the start of the cycle had been the melt M, i.e., if the cycle were to have started at point [2], and if the cycle included a transformation into the solid phase A before returning back to the melt, then the situation is problematic but not completely lost. In this latter case, the solid phase that we access and perform some of the work and heat transfer with is just the glassy state of the material instead of the crystalline equilibrium phase. As long as the properties of the glassy material are such that all relevant tasks can be performed—although perhaps not with the same efficiency compared to using the crystalline state as the working fluid—then the thermodynamic cycle can still be completed.

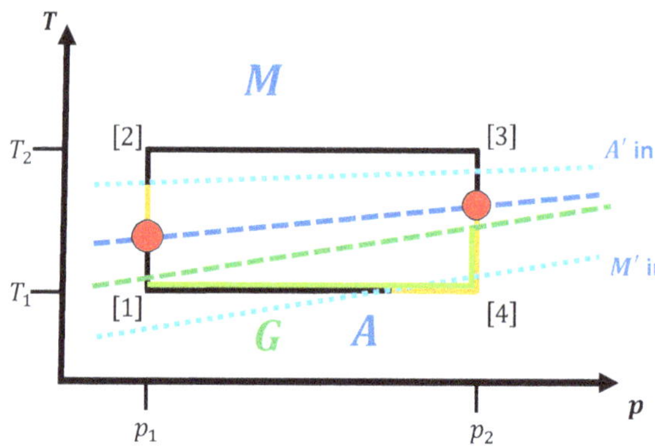

Figure 4. Sketch of a simple schematic path in (T, p) space with two phase transitions: from the crystalline equilibrium phase A to the melt equilibrium phase M in the first leg, and from phase M to phase A in the third leg, respectively. The dashed dark blue line is the equilibrium phase separation line in the (T, p) phase diagram between the melt and the solid, while the dotted light blue lines indicate the range in (T, p) space where the metastable phase (marked as A′ and M′) is still locally ergodic on moderately large time scales. The points where the transition would occur for an infinitely slow cycle are indicated by red spots, while the orange colored part of the path shows the range of the path where "non-trivial" times would be needed to perform the transition from the metastable super-cooled melt or super-heated crystalline phase to the crystalline or melt equilibrium phase, respectively, when only a finite time t_{total} is available for running the whole cycle. As an added complication, the system is assumed to be a glass-forming one, meaning that the glassy state can exist on long time scales (below the dashed dark green line) in competition with the crystalline equilibrium phase. If the nucleation and growth process of phase A from the melt does not take place along the thermodynamic path, e.g., because we are moving too quickly through the phase transition region close to the phase transition point on leg 3 of the cycle, then the system will exhibit a glass transition and remain in the glassy state for the rest of the cycle. This is indicated by the light green path in (T, p) space, which begins slightly below the equilibrium phase transition point on the third leg of the cycle when the glassy state becomes long-time stable, and persists until the ending/starting point [1] of the cycle has been reached.

In principle, we could choose the glassy state and the melt as the two "phases" of interest, since then we can complete the cycle, seemingly returning the working fluid to its original state at point [1]. Nevertheless, even in this case we have to deal with the aging of the glass that will occur while the material of the working fluid is in the glassy state. Depending on the length of time for which the material remains a glass, it will slowly evolve; along each point of the trajectory in (T, p) space, the system "quickly" leaves the marginal quasi-equilibrium state reached for these thermodynamic conditions and continues to respectively "emit" or "collect" (configurational) entropy into or from the universe as time goes on, while approaching (though not reaching) the crystalline equilibrium phase. Here, we note that this change in configurational entropy occurs in addition to the usual excess entropy associated with the usual "equilibration" that we observe in the equilibrium solid (crystalline) phase when perturbed (slightly) out of equilibrium. As a consequence, if we want to use the glassy state as the starting state of the cycle together with the melt as the second phase—a cycle which can be achieved, since melting a glass is no more problematic than melting a crystal—then we have to realize that the final glassy state is unlikely to be in exactly the same thermodynamic state as the starting one, which presumably had been relaxed for a long time before the engine was started. Nevertheless, if we immediately restart the engine after finishing one cycle, then after a few cycles the material will be in the "same" (evolving) glassy state of the same age at point [1] from one cycle to the next.

Here, we note that such a very slow approach to the equilibrium solid state can also appear in systems where the high-temperature phase corresponds to a solid solution instead of the melt and where the solid equilibrium phase at lower temperatures corresponds to a separation into two solids with different concentrations (possibly realized in a polycrystalline fashion). In particular, if the solid solution state is quenched to a temperature much below the critical point of the miscibility gap in the phase diagram of the material, where the thermodynamically stable phase would consist of two separated solid solutions with different compositions [2,138], then it can occur that the system very slowly evolves into the final two-solid phase. As a consequence, the material exhibits a state between a simple homogeneous metastable solid solution and the equilibrium state consisting of two segregated solids with different compositions, each of which is in thermodynamic equilibrium by itself. Conversely, the same can happen if we raise the temperature on the two-solid phase and have to wait a long time before the thermodynamically stable homogeneous solid solution equilibrium phase has been able to form at the higher temperature. Here, one should keep in mind that the solid solution phase is also a solid; thus, the internal atom diffusion needed to establish the equilibrated solid solution phase is likely to proceed quite slowly. Hence, we observe that returning to the original state of the material can become extremely difficult. We need to spend a great deal of time at the right transformation conditions (temperature, pressure, etc.) in order to nucleate the thermodynamically stable phase out of the glassy, amorphous, or solid solution state. This might force us to add a whole thermodynamic cycle or cycles after our original work cycle in order to re-establish the original starting phase of the working fluid material.

In this context, we note that even if we do not transform the crystalline starting material into the liquid state (from where the subsequent transformation into the glassy state would occur) because we stay below the melting temperature along the whole path, softening (or hardening) can still take place when approaching the melting temperatures/transformation pressures, with cyclic fatigue phenomena appearing [139]. In this case, high amounts of disorder can arise in the single-crystal material, which can result in both microscopic and mesoscopic structural changes. Similarly, high pressures, even when quite far below the actual transition pressures needed for the formation of a high-pressure phase, can change the amount and distribution of equilibrium and non-equilibrium defects, and create domain changes which take considerable effort and time to reverse. This is particularly critical if the properties of the material (electronic, mechanical, etc.) actually depend on the number and distribution of the (equilibrium and non-equilibrium) "defects" in the solid.

Quite generally, in many materials we find slow weakening or other changes that can occur during thermodynamic cycles and need to be accounted for during computation of the efficiency of the cycle and the resulting excess entropy or loss of available work. These are especially important if they concern the mesoscopic structure (crystallite size, domain size, grain boundary and dislocation distribution, interfaces of composite materials), as these tend to be more difficult to reverse. This applies even if no competing metastable phases are present. In particular, we acquire large numbers of defects or other changes which are not eliminated upon return to the starting material, even if it is still a single-crystal or polycrystalline material of the only solid phase that is locally ergodic in the system. One cannot magically make it so without spending enormous amounts of effort and work on essentially "re-forming"/"transforming" the material at the end of the cycle back into the original starting equilibrium material. We might treat these large re-initializations of the working fluid as somewhat external features one would prefer to ignore when designing and optimizing a work cycle. Nevertheless, such additional work and entropy production needs to be accounted for when analyzing the finite-time thermodynamics of systems that employ materials exhibiting long-lived amorphous or glassy states as working fluids.

3.1.3. Existence of Multiple Metastable Phases in Parallel

Clearly, the situation is even more critical if the material possesses several competing moderately long-lived metastable phases i at (T, p) values along the path of the thermodynamic cycle. In contrast to the case of the glassy state, which is expected to vary slowly while inexorably transforming more and more into the equilibrium phase, we now deal with well-defined phases that are in local equilibrium on time scales of interest along the cycle, i.e., once we are in such a phase i, equilibrium thermodynamics holds on time scales $\tau_{eq}(i; T, p) \ll t_{obs} \ll \tau_{esc}(i; T, p)$. In the following, we distinguish two main cases: where we start the cycle with the working fluid in the equilibrium phase (case 1), and where we start with it in a metastable phase (case 2).

First, we discuss the case where the cycle starts with the material in the globally ergodic equilibrium phase A. At some point (T, p) along the trajectory in thermodynamic space, we find that the system can/does switch into a different phase B—which can be either the new equilibrium phase or a metastable phase—since at and beyond the point (T, p), A has become metastable/unstable on the time scale of observation, $t_{obs} \geq \tau_{esc}(A; T, p; b_{esc})$. Here, t_{obs} can refer to the total time the system spends in the leg of the path in (T, p) space where A is only metastable at best as well as to the time allocated to a given point in (T, p) space, i.e., the point where the phase transition takes place or should take place. As far as the possible transition from phase A to phase B is concerned, we are again dealing with the balance in the time scales for the nucleation and growth process discussed in the two previous subsections, that is, the time available at any given point along the trajectory and the total time spent in the leg where phase A is metastable. Let us now assume that $t_{obs} \gg \tau_{esc}(A; T, p; b_{esc})$ for this leg; then, the system will be in phase B starting at some point along this leg of the cycle. Of course, we again have to optimize the time allocation to this leg as a whole, and in particular to the transition stage from A to B.

At the end of this portion of the cycle in thermodynamic space, phase B becomes metastable and phase A is again the equilibrium phase. However, we note that B does not necessarily transform nicely back to A; the system might evolve into competing phases analogous to the crystallization of alternative modifications from the glassy state or an amorphous precursor mentioned in Section 3.1.2. Instead, several alternatives can occur. First, B may be a long-lived metastable phase $\tau_{esc}(B;T,p;b_{esc}) \gg t_{total}$ for the (T,p) values of the rest of the cycle; in that case, we will only reach A with a low probability on the order of $(b_{esc}\frac{t_{total}}{\tau_{esc}(B;T,p;b_{esc})})$. Alternatively, B could transform into another metastable phase C (with a certain probability ranging from very small to 100 %), which is a long-lived metastable competitor to phase A. Finally, some kind of glassy or amorphous state might appear, as discussed above in Section 3.1.2. Fine-tuning the path and the time spent in various regions of (T,p)-space along the cycle trajectory might be required in order to avoid either of these three outcomes if we want to return the working fluid to the starting phase A at the end of the cycle.

Second, we start in a metastable state C (perhaps because its properties are just perfect for our purposes). Now, we want to return to this state after finishing the cycle. As long as we spend less time at each point (T,p) on the path than the escape time $\tau_{esc}(C;p,T)$, we might be able to perform the cycle avoiding any phase transition and return to C at the end of the cycle. However, running through the dangerous parts of the cycle—where C is highly unstable—at high speed will presumably extract a considerable price in excess entropy production and lost work compared to the ideal values if we did not have to follow this fast schedule. However, if we accept the transition to a different phase B at some point along the cycle, we are again faced with the four possible outcomes upon return to the starting point of the cycle in thermodynamic space: we remain in the new metastable phase B, end up in the equilibrium phase A, reach the desired metastable phase C, or are stuck in a glassy state. Considering that at least some of the undesired options can occur with a non-vanishing probability, it appears that again a probabilistic approach to the optimization problem of (a set of) cycles will be needed (see Section 3.1.5).

Figure 5 shows how the possibility of several outcomes when dealing with phase transitions in finite times may result in different (meta)stable phases along the path in thermodynamic space. Starting with the equilibrium phase A at point [1], two possible transitions can occur at the phase transition point $TP1$. First, phase A can continue while being metastable until transforming into the new equilibrium phase B where the system now remains until the phase transition point $TP2$ is reached; at that point, phase B becomes metastable and subsequently transforms either into the equilibrium phase A (again after some delay) or the metastable phase C, and it is assumed that no glassy state is present in the system. Alternatively, at $TP1$ we can switch into the metastable phase C and remain in this phase until $TP2$ is reached; there, we either stay in the metastable phase C until we reach the starting point [1] again, or we transform back into the equilibrium phase A. Of course, there are other possibilities, such as the system staying in phase B (now metastable) until the end. Keeping track of these possible bifurcations results in a decision tree, as discussed in Section 3.1.5.

We recall that the material can enter in several ways when employed as a working fluid, and can influence the analysis of the thermodynamic cycles. From a modeling point of view, this can pose an external source of problems that, for one, impose restrictions on the speed along the path or the choice of path as such. For example, one might like to avoid the unstable region C at the end of the cycle (case 1 above), or perhaps not want to transform the working fluid into phase B at all if possible. To achieve this, one could, e.g., take a worse (e.g., more inefficient) path in (T,p)-space, e.g., stay below the phase transition line that separates phases A and B as far as the process is concerned, while still performing the desired work, or alternatively by using a different range of temperatures that has the disadvantage of producing less work than one would like or generating more excess entropy. If this avoids the large amount of extra thermodynamic work needed to recreate the original material, it might be worth the thermodynamic price. Here, we note that if the transformation to phase B is an integral part of the functioning of the thermodynamic engine, then it of course cannot be excluded from the thermodynamic cycle.

As we have seen already, a second issue is that the choice of material can impose an additional fixed cost, e.g., of transforming the material back to A from C. This might occur in case 1 discussed above, and the extra cost needs to be added on top of the otherwise optimized route. We note that the phase transition $A \to B$ is assumed to be already included in the optimization procedure, since it belongs to the "ideal" cycle we are planning to follow.

A third concern might be that we can, e.g., achieve certain goals more easily when the material has the properties of modification B while running the engine, but phase B is environmentally unstable (susceptible to rusting) or unsafe (poisonous), and as such should be transformed back to A after the process is finished and the engine is at rest. An example might be a switch from a stable insulating material A to an unstable but conducting material B, where the electrical conductivity is important for the performance of the cycle, e.g., in a battery.

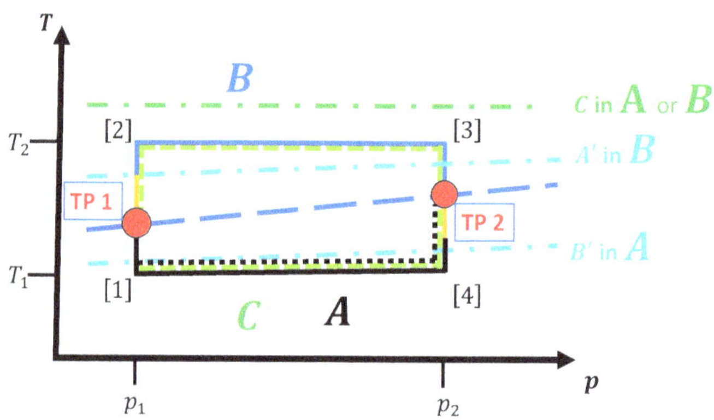

Figure 5. Sketch of a simple schematic path in (T, p) space involving three phases. A is the global equilibration phase below the dark blue line (large dashes) and B is the equilibrium phase above the dark blue line (large dashes), while the blue (dash-dotted) lines indicate the range of metastability of phases A and B in the part of the thermodynamic space where the equilibrium phases are B and A, respectively. The dark green (dash-dotted) line indicates the range up to which the phase C is metastable; note that C is never the equilibrium phase. $TP1$ and $TP2$ indicate the points along the thermodynamic path the system is following, at which the equilibrium (infinite-time) phase transitions between A and B would occur. The colors along the path in (T, p)-space indicate which of the phases are present: black = equilibrium phase A is present at the start and after nucleation from the phase B (after B has become metastable); orange = metastable phase A or metastable phase B is present before the transformation to the equilibrium phase takes place (as in Figure 3); dark blue = equilibrium phase B is present; dashed green = metastable phase C is present; and dotted black = equilibrium phase A is present after nucleation from the metastable phase C.

3.1.4. Challenges of Cyclic Processes

A number of the issues discussed in the previous examples apply to both cyclic and line-type processes (where the starting point and end point of the path are different, and the path does not intersect itself), since any phase transition with inconclusive outcomes will require the introduction of a probabilistic setup for the optimization of the finite-time process. Similarly, any phase transition that is encountered along the path will introduce possibly large non-equilibrium states of the system. Thus, it becomes problematic to use the straightforward intuitive (linear response type) models of the relaxation to approximate equilibrium, which underlie our ability to perform, e.g., analytical calculations when optimizing the finite-time process. These aspects are present for both cyclic and line-type paths in thermodynamic space. Cyclic processes employing real materials instead of abstract thermodynamically perfect ones exhibit their own challenges when compared with a given line-type process in thermodynamic space, even if the line-type process also involves various kinds of phase transitions. As pointed out earlier, the major issue for cyclic paths is the ability to return to the original starting point with the properties of the material that serves as the working fluid being intact and identical to the state at the beginning of the cycle.

Several detrimental outcomes have already been discussed, such as the material being in a different (meta)stable phase when returning to the starting point of the cycle in thermodynamic space. Examples we have considered are a metastable phase B or C instead of the original equilibrium phase A, the equilibrium phase A instead of the original metastable phase C, the metastable phase B instead of the metastable phase C, or some kind of glassy state. We have also noted that if we were to start with the material in a glassy state, then aging processes during the cycle can result in the material being in a slightly different aged (or rejuvenated) glassy state when returning to the starting point at the end of the cycle, since the glassy material is not in a locally ergodic equilibrium state but in an ever-evolving marginally ergodic quasi-equilibrium state.

In fact, this issue, raised by the somewhat ill-defined glassy state where the material is permanently in a slowly evolving quasi-equilibrium, can also appear when employing metastable and thermodynamically stable crystalline modifications of the material as a working fluid. The reason is that in these systems we are often dealing with the accumulation of defects in each cycle, which are not eliminated or brought back to their equilibrium concentrations and spatial distributions after the cycle has ended: the material is back in its original modification (or has never left it), but many long-lived defects have been created while undergoing the cyclic thermodynamic process. If these defects affect the physical or chemical properties of the working fluid, then subsequent cycles will yield slightly different outcomes even for the same path in thermodynamic space with the same allocation of time along this path. Of particular importance are long-lasting non-equilibrium defects on the atomic or mesoscopic level, such as defect clusters, dislocations, domain boundaries, or grain boundaries. In the last case, we have usually already crossed the line to essentially permanent changes on most time scales and for most systems of interest, since now we are dealing with a polycrystalline material instead of the single-crystal material in thermodynamic equilibrium that we started with.

Here, we remark that sometimes these non-equilibrium states are preferable as working fluids because of their specific macroscopic properties. For example, if a glass is replaced by a poly- or single-crystalline material as the working fluid, it might no longer have the desired property spectrum, e.g., it might be more brittle, less transparent, etc. In addition to glasses, an extreme case of such materials with a desirable high "defect" density on the atomic level might be so-called "high-entropy" materials [140,141], where we aim for a state with a high degree of so-called "controlled disorder" [7], such as a super-sized version of a solid solution. As mentioned earlier, solid solutions are equilibrium or long-lived near-equilibrium states; often, a material called a "solid solution" is not yet fully relaxed into thermodynamic equilibrium, and "emits"/"absorbs" heat while settling into the maximum entropy state. However, these high-entropy solid (solutions) do not necessarily correspond (and actually are unlikely to correspond) to thermodynamic equilibrium phases/states at the temperatures and pressures at which the material is being used in devices. As a consequence, (T, p) cycles performed using such a material as a working fluid can very well lead to a change in structure (on the atomic level and on the mesoscopic level), in particular regarding the compositional atom distribution throughout the material, and properties of the material can even be different after the cycle. Another example for a type of quasi-equilibrium material where thermodynamic cycles (in finite time) are important regarding its properties are battery materials, since each cycle adds a certain number of more or less permanent "defects". Here, each discharge–recharge cycle (at constant temperature and pressure but at different external applied voltages) leaves the material slightly changed (in a thermodynamically lower metastable state), until no further cycles are possible and the battery needs to be replaced and the material recycled.

The accumulation of defects raises issues that are also present when we consider the finite size of the system or when we are dealing with an inhomogeneous material. In order for an inhomogeneous material to be the equilibrium phase, the material usually needs to be a composite material or agglomeration, otherwise the system will slowly evolve towards a homogeneous new (!) material. If we are dealing with such a composite, then a contact boundary is present, with its own physics and chemistry; in particular, we encounter possible mixing on the atomic level via diffusion of atoms between the phases in the boundary zone. At finite times, we cannot achieve a smooth distribution of the "foreign" atoms as minority "defect"/"solid solution"-type atoms inside the other phase. Thus, in each of the separate phases there is a gradient of the minority atoms in the "other" phase, violating the condition of homogeneity in that subset of the composite material. The equilibrium amount of minority atoms depends on the (T, p) values of the phase diagram; thus, at least locally in the interface zone, we deal with changes in this concentration as function of (T, p). Further "irreversibilities" can appear in the quasi-permanence of cracks and grain boundaries, which are typical mesoscopic features of a material.

This brings up the issue of the finite size of the material employed as a working fluid: if we are down to mesoscopic sizes, e.g., employing single crystalline grains of a material, then this might allow the system to fully equilibrate without appealing to the self-averaging principle. In particular, we could have single crystal-to-single crystal transformations, i.e., we do not end up with some of the essentially irreversible changes mentioned earlier, such as crystal-to-powder-like structure changes on the mesoscopic level, which destroy the macroscopic reversibility and prohibit the return of the material to its state at the beginning, thereby violating the assumption underlying every thermodynamic cycle. Another basic advantage of mesoscopic-size working fluids is that we do not need to worry too much about the speed with which external changes in the thermodynamic conditions propagate and spread through the material serving as the working fluid. On the other hand, the size of the surface is quite large compared to the volume of the material; thus, surface terms should be taken into account when computing thermodynamic functions for the system. Furthermore, every phase becomes metastable for mesoscopic size materials—including the thermodynamically stable one, in principle—and the number of metastable phases in the system can increase considerably. Finally, many thermodynamic engines must employ working fluids of macroscopic size in order to be useful in real-life situations.

3.1.5. Probabilistic Optimizations

In the previous subsections, we have already encountered one general way to deal with those cycles where the system ends up with a final material that is different from the original one. First, it is necessary to add a certain amount of work after the actual work cycle has been performed in order to rejuvenate the material and transform it back to its original state. This work, and in principle the time required to perform this work, needs to be taken into account when formulating the finite-time optimization problem, e.g., by subtracting the time needed for restoring the original phase t_{rejuv} from the total time t_{total} available for the cycle, such that we can start the next cycle directly after the material is restored.

As mentioned in the previous subsections, when dealing with competing metastable phases, one obtains a probability (as function of temperature and pressure) that indicates the likelihood of the metastable phase i leaving its locally ergodic region $\mathcal{R}_i$ on the energy landscape within the observation time spent at this point in thermodynamic space. In the definition of the escape time, this probability is provided in the form of the characteristic parameter b_{esc}, which indicates that an atomistic trajectory of length τ_{esc} will leave the region $\mathcal{R}_i$ with a probability exceeding b_{esc}. In the definition of the escape time, b_{esc} is usually assumed to be very small, such that even a small likelihood of leaving the system counts for defining the time scale $\tau_{esc}(i; T, p; b_{esc})$. Conversely, for a given time scale of interest t_{obs} which we want to spend at the point (T, p), we can set this time equal to the escape time, allowing us to compute the corresponding probability b_{esc}. As a consequence, once in every $1/b_{esc}$ cycles on average, the system will end up leaving phase i and end up in the correct (if we wanted to leave phase i) or incorrect (if we wanted to stay in phase i) phase.

Now, we can construct a "decision"-like tree of outcomes for the cycle as far as the phases that occur during the trajectory in thermodynamic space are concerned, where each decision (i.e., whether a transition occurs or not along the path) is noted. For each transition that takes place, we can include the times needed for the transition in the list of legs of the path and also add the amount of excess entropy production/loss of work associated with this transition to the cost function of the finite-time optimization problem. For each path on the decision tree, we can compute the probability that the system will follow this decision tree path. Furthermore, we can treat each such path of the decision tree as a separate finite-time optimization problem that needs to be analyzed, taking into account the constraints due to the transitions that are defined to occur.

In the next step, we optimize the time allocation/distribution for every given path on the decision tree with respect to the thermodynamic cost function of the cycle. Here, we must decide whether, for a decision path where the "wrong" state of the material is found after the regular thermodynamic cycle has been finished, we are willing to "rejuvenate" the material to its original phase as part of the cycle or not. In the former case, we need to subtract this rejuvenation time t_{rejuv} from t_{total} and add the associated thermodynamic cost to the cost associated with this decision tree path. Alternatively, we could instead "replace" the material that ended up in the undesired phase with the material in the correct original phase obtained from a large reservoir/storage, such as that provided by, e.g., a recycling company. In the latter case, it is not quite clear which cost and time should be associated with such a replacement; a compromise might be that we do not subtract the replacement time from t_{total}—after all, the recycling company will rejuvenate enormous amounts of spent material in one fell swoop—yet still add the thermodynamic work needed in this external recycling process to this path.

Furthermore, we note that when optimizing the cycle following such a decision tree path we might end up influencing the occurrence probability of this path by varying the time we spend within the region in thermodynamic space where this transition occurs. This kind of feedback should be automatically included in the complete optimization problem until a self-consistent solution has been reached. In this discussion, we assume that the probabilities of the transitions occurring in the first place do not change by very much when we optimize the cycle for a given sequence of transitions assumed to occur along this decision tree path.

In a second step, we would now consider an ensemble of cycles representing all the decision tree paths and assign their appropriate probabilities. Because these probabilities will change depending on the choice of path in thermodynamic (T, p)-space, we can now perform an optimization on the path the cycle should traverse in thermodynamic space. Of course, for each choice of cycle path we need to perform the time allocation optimizations for all the decision tree paths together with a re-determination of the probabilities with which such decision tree paths will occur.

3.2. Optimal Schedule of Processes Aimed at Synthesis/Production of Materials

In the previous Section 3.1, we have discussed many aspects of thermodynamic cycles in finite time that involve a working fluid with a complex energy landscape where multiple locally ergodic and/or glassy regions are present on the energy landscape for the area of thermodynamic (T, p)-space in which the process takes place. The typical targets of the optimization are, e.g., minimization of the excess entropy/heat production or loss of work/generation of extra work due to the finite time available for completion of the cycle. Another important class of optimization problems associated with thermodynamic processes following some trajectory in thermodynamic space consists of the production of chemical compounds and materials in finite time. Here, the goal is the maximization of the desired product, where the target compound does not necessarily constitute the thermodynamic equilibrium phase of the system. Alternatively, for a given amount of product, the goal might be to minimize the amount of work or heat production if the product needs to be generated within a certain finite time.

Of course, the efficient production of chemicals has a long tradition in chemistry and chemical engineering, with many processes having been analyzed and optimized in the past [142–145]. We also find examples where such analyses have been performed by employing the viewpoint of finite-time thermodynamics, such as studies of distillation processes [146,147], various chemical reactions [148,149], or maximization of the desired product obtained upon cooling from a melt by a nucleation and growth process [128]. However, such analyses have usually not taken the complex energy landscape of the underlying chemical system into account.

Designing syntheses that target specific molecules that would never form by themselves from a given initial set of atoms or small molecules have been highly successful [150–152], with molecular chemists developing a plethora of so-called elementary name reactions [153] that describe specific reaction steps in the multi-step reaction paths starting from very simple and widely available educt molecules. We note that many of these target molecules correspond to high-lying minima on the energy landscape of the chemical system that is defined by the set of atoms making up the molecule. Due to the high degree of kinetic control of the chemical reactions involved in building up these specific molecules, moving from one minimum to the next on the energy landscape can be done in a controlled fashion.

In contrast, aiming for the synthesis of specific solid phases and designing efficient synthesis routes for this purpose is a much greater challenge for the experimentalist, even if the goal is only to obtain the thermodynamically stable phase [154,155]. One problem is that for many compositions in a chemical system even the equilibrium phase is not known with certainty. Trying to solve the task of identifying all stable solid phases of interest in a chemical system spawned the field of crystal structure prediction over thirty years ago (for reviews, see, e.g., [154,156,157]), which is now in the process of adding machine learning to its toolbox [158,159]. However, the major reason for the difficulty in systematically synthesizing such metastable solid compounds in the experiment is probably the lack of atom-level kinetic control of the chemical processes. Instead, thermodynamics-based tools such as variations of temperature, pressure, and attempts to (locally) vary the concentration of starting atoms or precursor materials (crystalline, amorphous, or layers of films) are mainly employed. Trying to alleviate this issue has led to various film- or atom layer-based methods [134,160,161], where systematic quantitative analyses of the outcome of the growth of various crystalline phases (e.g., from an amorphous precursor) have been performed as function of the applied (thermodynamic parameter-based) synthesis schedule [135,136]. However, there is clearly still a long way to go in this regard.

From the optimization point of view, the major challenge is to guide the system into the right metastable target phase within finite time, where we assume that we have full information about the set of relevant local minima and locally ergodic regions on the energy landscape together with information about the generalized barrier structure including both energetic and entropic barriers. The issues that arise are in many ways quite analogous to those discussed in Section 3.1: the many time scales associated with the movement on the energy landscape when exiting from or equilibrating in (intermediary) metastable phases or amorphous precursors, the amount of excess entropy/loss of work when trying to accelerate nucleation and growth processes, the persistence of quasi-equilibrium states such as glasses, and the probabilistic nature of the outcome of the synthesis even for identical schedules in (T, p)-space when metastable phases are targeted or need to be passed through on the route to the final solid compound of interest.

To address these issues, it is possible to employ empirical theoretical models for the various stages of the envisioned synthesis route. For example, the maximization of the amount of a crystalline solid phase generated via cooling from the melt within a finite time t_{total} has been investigated [128], where the temperature served as the thermodynamic control parameter. Using standard elementary models to describe the nucleation and growth stages of the process, it was found that the solution to the optimization problem consisted of a bang–bang-type schedule, with the temperature abruptly switching between a minimum value where the nucleation rate is high for the super-cooled melt and a maximum value where the growth rate of these nuclei is maximized (note that this maximum temperature must stay below the melting temperature of the material to avoid a re-melting). The same study investigated the competition between two different metastable phases, where the optimization goal was to preferentially generate one of these two phases during their growth from the super-cooled melt. Here, a bang–bang-type solution for the control parameter (in this case the temperature) was again obtained.

In the above study of crystallization from a melt, it was implicitly assumed that empirical models for both the nucleation and growth processes were available, together with values of the parameters in these models for the chemical system of interest. Furthermore, it was assumed that all relevant metastable phases in the system were known, such that the predictions would be able to guide the experimentalist in their syntheses. However, for many systems no such experiments have yet been performed, or none of the metastable phases that compete with the known equilibrium phase have yet been synthesized; thus, the empirical laws guiding the processes involved are not available to perform optimization with.

While distressing on the one hand, on the other it is clear that even some qualitative guidance in thermodynamic space would already be of great help for the experimentalist in finding ways to synthesize any of the metastable solid phases that have been predicted or intuited to exist in the system, thereby spurring theoretical work in the field even using very rough approximations. To address this challenge of synthesis route prediction, it is necessary to perform the finite-time optimization directly on the energy landscape of the system of interest [7]. A prerequisite is detailed knowledge of the structure of this landscape, including all local minima of relevance that individually (in the case of crystalline modifications) or in structurally related groups (in the case of solid solution phases) are at the center of locally ergodic regions corresponding to these (meta)stable phases, the (effective) local densities of states for these minima and associated locally ergodic regions, and the energetic and entropic barriers separating these regions. These generalized barriers are captured in form of probability flows between the locally ergodic regions as function of energy during the global landscape explorations [21,94,109]. Such flows, together with the local densities of states, can be determined using the so-called threshold algorithm [38,124,162], which explores the regions of the energy landscape which are accessible from all minima of relevance below a sequence of energy lids.

This information can be used to construct a Markov model description of the dynamics of the system on the level of the locally ergodic regions [7,163], i.e., we model the probabilistic dynamics on the energy landscape in a coarsened picture that considers the movement of the walker between locally ergodic regions instead of from microstate to microstate, as would be done in, e.g., a molecular dynamics simulation. The thermodynamic parameters such as temperature and pressure influence the transition probability entries in these Markov matrices via the Boltzmann factors and shifts in the enthalpy levels; we recall that a change in pressure from p_1 to p_2 corresponds to a shift $(p_2 - p_1)V$ in the potential enthalpy of the microstates of the system. As a result of the finite-time optimization, we find an optimal temperature–pressure schedule in thermodynamic space for each target phase; these schedules can then serve as a rough guideline for experimentalist to design their synthesis routes.

Examples of such thermodynamic optimizations on the energy landscape level include studies by Hoffmann et al. [163–165], who employed the results of energy landscape investigations of periodic approximants of the magnesium difluoride system [94,166,167] to construct such Markov matrices for the probability flows as function of temperature and pressure. Starting the Markovian time evolution from the system at very high temperatures, the optimal temperature-pressure schedules are computed, enabling to obtain not only the experimentally-known rutile structure but also predicted metastable alternative phases such as the anatase and the CdI_2 modification to be obtained with a certain probability.

Such in-principle studies constitute only the beginning of the applicability of such synthesis optimization to realistic systems. The main problem when constructing simulations on the level of the metagraph of the locally ergodic regions for solid state chemical systems is the number of atoms involved. Thus far, it is only possible for small periodic approximants to obtain the detailed landscape information required for the construction of the Markov model—single molecules or clusters are clearly much easier to deal with in this regard! The problem is that unless the transformation between the metastable phases occurs via, e.g., a second-order phase transition (as discussed above), real solid materials typically undergo nucleation and growth processes or grow from glassy or amorphous precursors. Thus, obtaining the appropriate time scales for the probability flows that allow us to model the Markovian evolution on the metagraph of the locally ergodic regions in a quantitatively realistic fashion requires information from landscape explorations for state spaces consisting of hundreds or thousands of atoms/variable periodic simulation cell, ideally on the ab initio level of energy. Nevertheless, such explorations are expected to become feasible with the availability of machine learning (ML) potentials for multi-atom systems, as the ML energies of these systems for multi-atom configurations in solids are reasonable approximations of the ab initio energies for the same configuration but can be computed orders of magnitude faster [168].

In this context, we comment on the issue of computing free energy differences and free energy barriers on the atomic level via molecular dynamics or Monte Carlo simulations of single walkers (or ensembles thereof) for systems that exhibit several locally ergodic regions that might compete with each other along the path in thermodynamic space. Examples of classic approaches to computing free energy differences, e.g., between two systems that can be transferred into each other by some change in characteristic parameters (such as the strength in the atom–atom interactions) or between the same system but at two different points in thermodynamic space, include the thermodynamic integration method and the thermodynamic perturbation method [169,170]. The basic assumption behind such approaches is the observation that the work needed to perform such a transformation/movement on the energy landscape constitutes a lower or upper bound on the free energy difference [171]. The closer the system can stay to (local) equilibrium during the procedure, the tighter these bounds will be. Usually, the transformation is performed in both the forward and backward direction, at least as long as we are moving between two global equilibrium states. The same considerations apply when the transfer is supposed to take place between two (meta)stable modifications A and B at the same location or at different locations in thermodynamic space.

In a practical realization of such a computation, one would employ an ensemble of walkers on the energy landscape. The movement of these walkers as the thermodynamic parameters are changed to drive the system from phase A to phase B then corresponds to a moving ensemble average along the path in thermodynamic space. If the transformation takes place in finite time t_{total}, either extra work needs to be expended or excess heat generated, as the system will always be slightly out of equilibrium. As a consequence, we are again dealing with a finite-time optimization problem, i.e., attempting to find the optimal path in thermodynamic space to move from phase A to phase B while keeping the ensemble representing the system close to equilibrium everywhere along the path. In addition, we need to identify the allocation of the available time along the path.

The general question of optimally moving a system in thermodynamic space on the level of the energy landscape, and thereby computing the free energy differences, has been addressed in the literature [30]; however, those derivations assume that no bifurcations will be encountered. If other locally ergodic regions or marginally ergodic regions—corresponding to metastable phases or glassy solids, respectively—can be accessed along with the target phase B with non-vanishing probability, then the ensemble will no longer stay in or close to thermodynamic equilibrium, since some of the walkers will visit or even end up in other locally ergodic or marginally ergodic regions of the landscape. Ensuring that all walkers reach phase B requires additional work performed on the system, adding to the uncertainty in the free energy calculation already present due to the finite-time effects along the "correct" route through the energy landscape.

In addition to accepting this handicap and paying the price of extra work or entropy/heat production to force the walkers to stay on the direct route between phases A and B, one can in principle again employ a decision-tree approach, as mentioned in Section 3.1.5, where a probabilistic formulation of the many possible outcomes of the cyclic thermodynamic paths was presented. The advantage of such an approach is that one does not add essentially uncontrollable forcing terms into the algorithm; furthermore, as a positive side-benefit, estimates for free energy differences among many metastable phases of the system can be obtained. The disadvantage is the enormous number of walkers needed to probe the energy landscape in a locally equilibrated way along many possible pathways through the landscape between the locally ergodic regions. The possible appearance of glassy states is another serious handicap of an unbiased approach, since such non- or at best quasi-equilibrium states might not be exited on realistic simulation time scales, possibly requiring the system to be heated close to the melting point. Such a deviation from the original thermodynamic path will quite likely result in many additional uncertainties in the calculation of the free energy. Nevertheless, addressing such issues is an important task in the context of free energy computations for systems with complex energy landscapes.

3.3. *Systems with Complex Energy Landscapes Outside Physics and Chemistry*

Systems with complex energy landscapes are also found outside the fields of chemistry and physics, ranging from mathematics and biology over engineering and economics to the humanities; for an overview, we refer to [8]. In the latter cases, the high-dimensional single or vector-valued function over a large space of microstates is often no longer called an energy function; instead, we speak of a generalized cost function. To provide some specific examples, this function is called the fitness function (which is to be maximized) in biological systems when discussing evolution; the welfare function in the context of thermo-economics for multi-agent systems; the happiness function for social systems; the cost function for planning problems in business level economics; and the objective function in abstract or practical combinatorial and global optimization problems in mathematics.

In many of these systems, one is mostly interested in identifying the local and global minima and maxima of the generalized cost function, i.e., most of the effort is devoted to developing or applying suitable global optimization techniques and algorithms. Because these algorithms must explore the landscape in an efficient fashion, many of them are inspired by the way in which physical and chemical systems proceed in a natural way to explore their energy landscapes in order to reach thermodynamically stable phases. Examples of such algorithms are the simulated annealing method [172] and genetic and evolutionary algorithms [173], which have spawned a multitude of variants. Here, the picture drawn from classical mechanics of a system rolling downhill under the force of gravity to reach a state of lower potential energy leads to the deterministic gradient descent approach, while stochastic methods involving random walks on the cost function landscape reflect the statistical nature of the approach to low-energy minima associated with equilibrium phases in statistical thermodynamics. Such algorithms have been analyzed by employing the analogy of a glass transition or a glassy intermediary region on the energy landscape, which must be passed through before the desired low-energy cost function minima can be identified [174,175].

The challenges faced in the design and optimization of global optimization algorithms to explore such multi-minima cost function landscapes with a limited amount of computational resources are very similar to those encountered in thermodynamic space when moving from a phase that is stable at high temperatures to the thermodynamically stable equilibrium phase at low temperatures discussed above. In this context, we note that the energy landscapes of such combinatorial optimization problems frequently do not exhibit a well-separated ground state basin that would be analogous to the well-defined crystalline thermodynamic equilibrium phase at low temperatures. Instead, many of the landscapes of such optimization problems are more similar to those of spin glasses, which by definition or construction do not have a well-defined locally ergodic region surrounding the global energy minimum. Instead, many minima with nearly the same energy as the global minimum energy exist on the complex energy landscape, and these are located far away from the global minimum. This constitutes a qualitative difference from the energy landscape of a crystalline material modeled with a realistically large periodic approximant such that isolated defect configurations can be included; in the latter case, all low-energy minima that have nearly the same energy as the global minimum correspond to equilibrium defect configurations of the thermodynamically stable zero temperature equilibrium phase, and as such belong to the same locally ergodic region.

Quite generally, employing a whole ensemble of walkers on the cost function landscape as opposed to only a single walker allows us to compare the evolution of a thermodynamic system toward a (meta)stable phase with the gradual establishment of (meta)stable states of biological, ecological, social, or economic systems, which can reach an equilibrium-like state regarding the exchange of (biological or economic) goods and resources with a hypothetical external environment. In particular, when moving from one metastable biological, ecological, or economic state to another, we encounter problems similar to those we have discussed for the movement between two metastable solid phases: we need to invest a large amount of "work" or resources to accomplish the transformation into the desired biological, ecological, or economic target state. Doing this while minimizing the extra work or loss of resources within a finite time is clearly analogous to a finite-time optimization problem. We also note that chaotic and nearly unpredictable changes between two stable biological, ecological, or economic states can occur in such systems. This can result in the system being in a non-equilibrium or quasi-equilibrium situation which can persist for very long times, in analogy to the glassy states of chemical materials.

More concretely, in biological systems we can consider scenarios such as the attempt to breed certain traits into farm animals or to change the resistance of plants against "pests" within as few generations as possible or while employing a minimal number of intermediary breeding animals to be analogues of the scenario involving transformation of a given solid phase into a different metastable material in an as efficient a manner as possible. Similarly, we can consider the recovery of an ecosystem [176,177] after, e.g., a destructive volcanic eruption. Typically, the ecosystem observed in the region V around the volcano must proceed through a series of metastable ecosystems featuring pioneering and other intermediary plant and animal generations before a stable ecosystem is reestablished. Achieving this with a minimal amount of effort within a finite time, perhaps measured in few decades, requires careful fine-tuning of the environmental conditions experienced by the region V as function of time, which can strongly influence the types of plants and animals that will grow and settle in region V after the disturbance. This recovery process is quite analogous to a series of phase transitions when moving from, e.g., the melt to the low-temperature equilibrium phase via several intermediary (high-temperature) phases after the system has experienced an abrupt change in its thermodynamic environment, such as a quench in temperature and/or exposure to a cycle of high and low pressures.

Such environmental boundary conditions can include the general climate of the region, the plants or animals introduced by humans in region V after the volcanic eruption, and/or the (fixed) distribution of plants and animals in the geographic region G surrounding the region of interest near the volcano V. While the local climate or weather are difficult to influence by human intervention, the plant and animal populations in the surrounding region G can be controlled by humans. For example, if region G exhibits an agricultural monoculture or if predatory animals (wolves, bears, etc.) are systematically eliminated in G, then this will have a different influence on the final ecological state in region V compared to if region G were a wild forest. Note that the ecosystem reached in the long-time limit may be different from the original one before the volcano erupted; there can be many feasible metastable ecosystems that are stable on long time scales in region V, and the one the system settles into might depend on the environmental boundary conditions. We remark that there is actually no reason to assume that these boundary conditions completely determine the final ecosystem in region V; in principle, many long-time metastable ecosystems could exist for the same set of environmental boundary conditions.

While it is clear that there are many fascinating examples of (thermodynamic-like) processes for systems that have complex energy landscapes in fields of science outside the realm of physics and chemistry, we do not want to go into greater detail here as far as these biological, ecological, social, and economic systems are concerned. Considering, e.g., the mathematical formulation of thermo-economics and expounding the correspondence of its variables with those of thermodynamics would require a lengthy presentation which is beyond the purview of this perspective. Nevertheless, it should be clear that the concepts of finite-time thermodynamics are applicable to many of these non-physical systems, and can provide guidance about the optimal route toward the establishment of the desired states in these systems. Conversely, insights gained from dealing with systems in biology, ecology, or economics may inspire new work in the finite-time thermodynamics of chemical and physical systems.

4. Conclusions

After a short presentation of some fundamental concepts of finite-time thermodynamics and complex energy landscapes, we have discussed a number of prototypical processes and cycles that involve phase transitions of the working fluid material or the target material of a synthesis process. A major focus has been on cycles that include a simple back-and-forth transition between two equilibrium phases, at least in the limit of infinite time available for performing the cycle. To study such systems in the context of finite-time thermodynamics, it is necessary to employ appropriate models for the nucleation and growth processes involved in the phase transitions, in addition to models describing the relaxation to equilibrium along those parts of the thermodynamic path where the system stays within an equilibrium phase. Similar considerations arise when we try to optimize synthesis routes to achieve specific (meta)stable phases in a given chemical system, where the synthesis includes standard first-order phase transitions or where various nucleation and growth processes occur, starting from thermodynamically stable materials and also from those that are not in thermodynamic equilibrium, such as amorphous precursors or glasses.

In principle, this task is straightforward as long as the available time for the whole process t_{total} is sufficiently large to allow the system to remain close to equilibrium along the whole path in thermodynamic space and no bifurcations occur. In that case, the main challenge is to adapt and derive appropriate models for the above mentioned relaxations and transitions, and then to implement them as part of the optimization problem.

However, when analyzing the time scales involved, in particular those of the nucleation and growth processes during the phase transitions, the equilibration and escape times associated with the many competing metastable phases, and the time scale on which glassy materials very slowly convert into the crystalline equilibrium phase, it becomes clear that such cycles can run into major problems in practical applications. The reason for this is that the finite time t_{total} available for finishing the thermodynamic work cycle or line-type path is often not sufficient to guarantee that the material employed as the working fluid is transformed into precisely the specified phase or state via the planned phase transitions along the path. In addition to encountering difficulties in reaching the intended phases during the process itself, the system might become stranded in alternative (meta)stable phases or a glassy state at the end of the process, i.e., the working fluid might not be returned to its original thermodynamic and structural state upon return to the origin of the cycle in thermodynamic space or the desired product material might not be obtained in sufficient amounts or purity. As a consequence, for a non-negligible fraction of the cycles, large amounts of undesired (metastable) phases or glassy states can appear along the thermodynamic path or non-equilibrium defects may accumulate in an essentially irreversible fashion.

Performing (and optimizing) these processes will presumably require quite radical measures; a large amount of additional work or time t_{rejuv} may have to be invested after the end of the cycle in order to rejuvenate the material back into its original state, especially if one wants to repeat the cycle many times as part of a long-time process, in which case it is not usually possible to discard the working fluid at the end of the cycle. Essentially, one would have to add a recycling stage for the material in its final state or add a generic loss term to the final accounting of the process. Nevertheless, finite-time optimizations will be useful even here in trying to minimize the number of cycles where irreversible changes in the working fluid take place. Alternatively, it would be necessary to substantially alter the work cycle or synthesis path in thermodynamic space, avoiding all possible appearances of undesired phases and states of the material.

In trying to address this problem in full generality, we must account for all possible paths that the system might follow as far as the phases and states it can display are concerned. This requires the construction of probabilistic trees for a systematic analysis of the possible bifurcations during the thermodynamic process under study: in this approach, we note the possible outcomes together with their probability of occurrence at each point or leg along the path in thermodynamic phase space where a transition to a (meta)stable phase or quasi-equilibrium state such as a glass can occur. Combining this information with the models for the nucleation and growth processes and for the relaxation processes as functions of temperature and pressure, we can formulate and investigate the probabilistic optimal control problem for performing finite-time processes involving bifurcations regarding both the choice of path in thermodynamic space and the time allocation along this path.

In addition to the use of materials as working fluids that can experience phase transitions during a thermodynamic cycle in finite time or designing efficient synthesis routes towards new materials that exhibit metastable phases, there exist a multitude of other systems that also exhibit complex energy landscapes with many long-time stable states, analogous to the metastable phases in chemical systems; such systems appear in economics, where many different business strategies and economic systems can be stable on reasonably long time scales, as well as in biology and ecology, where many different kinds of animals and plants can survive in similar ecological niches and where whole ecosystems can be realized for the same external geographical and climate conditions. Studying and optimizing the dynamics of these systems that exhibit stable states which can be transformed into alternative ones—but possibly with high costs if the transition is forced to take place on short time scales—can potentially benefit from the use of methods developed for the study of finite-time thermodynamics and the optimization of thermodynamic cycles in finite time.

Funding: This research received no external funding.

Institutional Review Board Statement: Not applicable.

Data Availability Statement: Data is contained within the article.

Acknowledgments: Many thanks go to all the participants in the many workshops on thermodynamics in the Telluride Summer Research Center (now Telluride Science and Innovation Center) from

1988 to the present for their valuable discussions and inspirations. Particular thanks belong to Peter Salamon, Jim Nulton, Bjarne Andresen, Karl Heinz Hoffmann, Stan Sieniutycz, Ryszard Mrugala, Steve Berry, and Chris Essex for introducing me to the field of finite-time thermodynamics over 35 years ago, and for always being willing to provide feedback on unusual ideas thrown at them.

Conflicts of Interest: The author declares no conflicts of interest.

References

1. Tisza, L. *Generalized Thermodynamics*; MIT Press: Cambridge, MA, USA, 1966.
2. DeHoff, R.T. *Thermodynamics in Materials Science*; McGraw-Hill: New York, NY, USA, 1993.
3. Swendsen, R.H. *An Introduction to Statistical Mechanics and Thermodynamics*; Oxford University Press: Oxford, UK, 2012.
4. Goldstein, H. *Klassische Mechanik*; Akad. Verlagges.: Wiesbaden, Germany, 1978.
5. Simonyi, K. *Kulturgeschichte der Physik*, 3rd ed.; Wissenschaftlicher Verlag Harri Deutsch: Frankfurt, Germany, 2004.
6. Lautrup, B. *Physics of Continuous Matter*; IOP: London, UK, 2005.
7. Schön, J.C. Energy landscapes in inorganic chemistry. In *Comprehensive Inorganic Chemistry III*; Reedijk, J., Poeppelmeier, K., Eds.; Elsevier: Amsterdam, The Netherlands, 2023; Volume 3, pp. 262–392. [CrossRef]
8. Schön, J.C. Energy landscapes—Past, present, and future: A perspective. *J. Chem. Phys.* **2024**, *161*, 050901. [CrossRef]
9. Pathria, R.K. *Statistical Mechanics*, 2nd ed.; Butterworth-Heinemann: Oxford, UK, 1996.
10. Sieniutycz, S.; Salamon, P. *Finite Time Thermodynamics and Thermoeconomics*; Taylor and Francis: New York, NY, USA, 1990.
11. Bejan, A. Entropy generation minimization: The new thermodynamics of finite-size devices and finite-time processes. *J. Appl. Phys.* **1996**, *79*, 1191–1218. [CrossRef]
12. Hoffmann, K.H.; Burzler, J.M.; Schubert, S. Endoreversible Thermodynamics. *J. Non-Equil. Thermodyn.* **1997**, *22*, 311–355.
13. Andresen, B. Current trends in Finite-Time Thermodynamics. *Angew. Chem. Int. Ed. Engl.* **2011**, *50*, 2690–2704. [CrossRef] [PubMed]
14. Curzon, F.L.; Ahlborn, B. Efficiency of a Carnot engine at maximum power output. *Amer. J. Phys.* **1975**, *43*, 22–24. [CrossRef]
15. Andresen, B.; Berry, R.S.; Nitzan, A.; Salamon, P. On lumped models for thermodynamic properties of simulated annealing problems. *Phys. Rev. A* **1977**, *15*, 2086. [CrossRef]
16. Reitlinger, H.B. *Sur l'Utilisation de la Chaleur dans le Machine de Feu*; Vaillant-Carmane: Liege, Belgium, 1929; p. 25.
17. Novikov, I.I. The efficiency of atomic power stations. *J. Nucl. Energy* **1958**, *7*, 125–128.
18. Bejan, A. *Entropy Generation through Heat and Fluid Flow*; Wiley-Interscience: New York, NY, USA, 1982.
19. Sciubba, E. Beyond thermoeconomics? The concept of extended exergy accounting and its application to the analysis and design of thermal systems. *Exergy Int. J.* **2001**, *1*, 68–84. [CrossRef]
20. Schön, J.C.; Jansen, M. Determination, Prediction, and Understanding of Structures Using the Energy Landscape Approach—Part I. *Z. Krist.* **2001**, *216*, 307–325.
21. Schön, J.C.; Wevers, M.A.C.; Jansen, M. Entropically stabilized region on the energy landscape of an ionic solid. *J. Phys. Cond. Matter* **2003**, *15*, 5479–5486. [CrossRef]
22. Farquhar, I.E. *Ergodic Theory in Statistical Mechanics*; Interscience: New York, NY, USA, 1964.
23. Palmer, R. Broken ergodicity. *Adv. Phys.* **1982**, *31*, 669–735. [CrossRef]
24. Thirumalai, D.; Mountain, R.D.; Kirkpatrick, T.R. Ergodic Behaviour in Supercooled Liquids and Glasses. *Phys. Rev. A* **1989**, *39*, 3563–3574. [CrossRef] [PubMed]
25. Schön, J.C. Structure Prediction and Modelling of Solids: An Energy Landscape Point of View. In *Proceedings of RIGI-Workshop 1998*; Schreuer, J., Ed.; ETH Zürich: Zürich, Switzerland, 1998; pp. 75–93.
26. Jackson, J.D. *Classical Electrodynamics*; Wiley: New York, NY, USA, 1975.
27. Lebon, G.; Jou, D.; Casas-Vazquez, J. *Finite-Time Thermodynamics: Economy, Ecology, and Heat Engines*; Springer: Berlin, Germany, 2008.
28. Sieniutycz, S.; Shiner, J.S. Thermodynamics of Irreversible Processes and its Relation to Chemical Engineering: Second Law Analyses and Finite Time Thermodynamics. *J. Non-Equil. Thermodyn.* **1994**, *19*, 303–348.
29. Nulton, J.D.; Salamon, P.; Andresen, B.; Anmin, Q. Quasistatic processes as step equilibrations. *J. Chem. Phys* **1985**, *83*, 334–338. [CrossRef]
30. Schön, J.C. A thermodynamic distance criterion of optimality for the calculation of free energy changes from computer simulations. *J. Chem. Phys.* **1996**, *105*, 10072–10083. [CrossRef]
31. Rozonoer, L.I.; Tsirlin, A.M. Optimal control of thermodynamic processes. *Automat. Remote Control* **1983**, *44*, 55–62; 209–220; 314–326.

32. Andresen, B.; Salamon, P.; Berry, R.S. Thermodynamics in finite time: Extremals for imperfect heat engines. *J. Phys. Chem.* **1977**, *66*, 1571–1577. [CrossRef]
33. Muschik, W.; Hoffmann, K.H. Endoreversible Thermodynamics: A Tool for Simulating and Comparing Processes of Discrete Systems. *J. Non-Equil. Thermodyn.* **2006**, *31*, 293–317. [CrossRef]
34. Wagner, K.; Hoffmann, K.H. Endoreversible modeling of a PEM fuel cell. *J. Non-Equil. Thermodyn.* **2015**, *40*, 183–294. [CrossRef]
35. Jansen, M.; Pentin, I.V.; Schön, J.C. A universal representation of the states of chemical matter including metastable configurations in phase diagrams. *Angew. Chem. Int. Ed.* **2012**, *51*, 132–135. [CrossRef]
36. Schön, J.C.; Doll, K.; Jansen, M. Predicting solid compounds via global exploration of the energy landscape of solids on the ab initio level without recourse to experimental information. *Phys. Stat. Sol. (b)* **2010**, *247*, 23–39. [CrossRef]
37. Jansen, M.; Doll, K.; Schön, J.C. Addressing chemical diversity by employing the energy landscape concept. *Acta. Cryst. A* **2010**, *66*, 518–534. [CrossRef] [PubMed]
38. Neelamraju, S.; Oligschleger, C.; Schön, J.C. The threshold algorithm: Description of the methodology and new developments. *J. Chem. Phys.* **2017**, *147*, 152713. [CrossRef]
39. Struik, L.C.E. *Physical Aging in Amorphous Polymers and Other Materials*; Elsevier: Amsterdam, The Netherlands, 1978.
40. Lundgren, L.; Svedlindh, P.; Nordblad, P.; Beckman, O. Dynamics of the relaxation time spectrum in a CuMn spin glass. *Phys. Rev. Lett.* **1983**, *51*, 911–914. [CrossRef]
41. Sibani, P.; Hoffmann, K.H. Aging and relaxation dynamics in free-energy landscapes with multiple minima. *Phys. A* **1997**, *234*, 751–763. [CrossRef]
42. Kob, W.; Barrat, J.; Sciortino, F.; Tartaglia, P. Aging in a Simple Glass Former. *J. Phys. Cond. Matter* **2000**, *12*, 6385–6394. [CrossRef]
43. Hannemann, A.; Schön, J.C.; Jansen, M.; Sibani, P. Non-equilibrium Dynamics in Amorphous $Si_3B_3N_7$. *J. Phys. Chem. B* **2005**, *109*, 11770–11776. [CrossRef]
44. Milchev, A.; Binder, K.; Heermann, D.W. Fluctuations and Lack of Self-Averaging in the Kinetics of Domain Growth. *Z. Phys. B Cond. Matt.* **1986**, *63*, 521–535. [CrossRef]
45. Huitema, H.; Van der Eerden, J.P. Can Monte-Carlo simulations describe dynamics? A test on Lennard-Jones systems. *J. Chem. Phys.* **1999**, *110*, 3267–3274. [CrossRef]
46. Schön, J.C. Energy landscape concepts for chemical systems under extreme conditions. *J. Innov. Mater. Extrem. Cond.* **2021**, *2*, 5–57.
47. Balian, R. *From Microphysics to Macrophysics: Volume 1*; Springer: Berlin/Heidelberg, Germany, 1991.
48. Schön, J.C. Enthalpy landscapes of the earth alkali oxides. *Z. Anorg. Allg. Chem.* **2004**, *630*, 2354–2366. [CrossRef]
49. Stillinger, F.H.; Weber, T.A. Hidden Structure in Liquids. *Phys. Rev. A* **1982**, *25*, 978–989. [CrossRef]
50. Bengtzelius, U.; Götze, W.; Sjölander, A. Dynamics of supercooled liquids and the glass transition. *J. Phys. C* **1984**, *17*, 5915–5934. [CrossRef]
51. Buchenau, U. Energy landscape—A key concept in the dynamics of liquids and glasses. *J. Phys. Cond. Matter* **2003**, *15*, S955–S966. [CrossRef]
52. Doliwa, B.; Heuer, A. What Does the Potential Energy Landscape Tell Us about the Dynamics of Supercooled Liquids and Glasses? *Phys. Rev. Lett.* **2003**, *91*, 235501. [CrossRef]
53. Goldstein, M. Viscous Liquids and the Glass Transition: A Potential Energy Barrier Picture. *J. Chem. Phys.* **1969**, *51*, 3728–3739. [CrossRef]
54. Heuer, A. Properties of a Glass-Forming System as Derived from its Potential Energy Landscape. *Phys. Rev. Lett.* **1997**, *78*, 4051–4054. [CrossRef]
55. Schön, J.C.; Sibani, P. Energy and entropy of metastable states in glassy systems. *Europhys. Lett.* **2000**, *49*, 196–202. [CrossRef]
56. Middleton, T.F.; Wales, D. Energy Landscapes of Some Model Glass Formers. *Phys. Rev. B* **2001**, *64*, 024205. [CrossRef]
57. Sastry, S. Glass-forming liquids and the glass transition: The energy landscape approach to dynamics and thermodynamics. *J. Indian Inst. Sci.* **2006**, *86*, 731–749.
58. Raza, Z.; Alling, B.; Abrikosov, I.A. Computer simulations of glasses: The potential energy landscape. *J. Phys. Cond. Matter* **2015**, *27*, 293201. [CrossRef] [PubMed]
59. Muthukumar, M. Entropic barrier model for polymer diffusion in concentrated polymer solutions and random media. *J. Non-Cryst. Solids* **1991**, *131–133*, 654–666. [CrossRef]
60. Angell, C.A. Formation of Glasses from Liquids and Biopolymers. *Science* **1995**, *267*, 1924–1934. [CrossRef]
61. Angell, C.A. Landscapes with Megabasins: Polyamorphism in Liquids and Biopolymers and the Role of Nucleation in Folding and Folding Diseases. *Phys. D Nonlinear Phenom.* **1997**, *107*, 122–142. [CrossRef]

62. Schön, J.C. Energy Landscape of two-dimensional Lattice Polymers. *J. Phys. Chem. A* **2002**, *106*, 10886–10892. [CrossRef]
63. Binder, K.; Baschnagel, J.; Paul, W. Glass Transition of Polymer Melts: Test of Theoretical Concepts by Computer Simulation. *Prog. Poly. Sci.* **2003**, *28*, 115–172. [CrossRef]
64. Muthukumar, M. Modeling polymer crystallization. *Adv. Polym. Sci.* **2005**, *191*, 241–274.
65. Alexa, P.; Oligschleger, C.; Gröger, P.; Morchutt, C.; Vyas, B.; Lotsch, B.V.; Schön, J.C.; Gutzler, R.; Kern, K. Short-range structural correlations in amorphous 2D polymers. *ChemPhysChem* **2019**, *20*, 2340–2347. [CrossRef] [PubMed]
66. Tsuji, Y.; Prasad, D.V.K.; Elatresh, S.F.; Hoffmann, R.; Ashcroft, N.W. Structural Diversity and Electron Confinement in Li_4N: Potential for 0-D, 2-D, and 3-D Electrides. *J. Amer. Chem. Soc.* **2016**, *138*, 14108–14120. [CrossRef]
67. Niblett, S.P.; Biedermann, M.; Wales, D.J.; de Souza, V.K. Pathways for diffusion in the potential energy landscape of the network glass former SiO_2. *J. Chem. Phys.* **2017**, *147*, 152726. [CrossRef] [PubMed]
68. Sherrington, D.; Kirkpatrick, S. Solvable Model of a Spin Glass. *Phys. Rev. Lett.* **1975**, *35*, 1792–1796. [CrossRef]
69. Mezard, M.; Parisi, G.; Virasoro, M.A. *Spin Glass Theory and Beyond*; World Scientific: Singapore, 1987.
70. Klotz, T.; Kobe, S. Exact low energy landscape and relaxation phenomena in Ising spin glasses. *Acta Phys. Slov.* **1994**, *44*, 347–356.
71. Sibani, P.; Schriver, P. Local phase-space structure and low-temperature dynamics of short-range Ising spin glasses. *Phys. Rev. B* **1994**, *49*, 6667–6671. [CrossRef] [PubMed]
72. Levinthal, C. Are there pathways for protein folding? *J. Chim. Phys.* **1968**, *65*, 44–45. [CrossRef]
73. Chelvanayagam, G.; Reich, Z.; Bringas, R.; Argos, P. Prediction of Protein Folding Pathways. *J. Mol. Biol.* **1992**, *227*, 901–916. [CrossRef]
74. Fontana, W.; Stadler, P.F.; Bornberg-Bauer, E.G.; Griesmacher, T.; Hofacker, I.L.; Tacker, M.; Tarazona, P.; Weinberger, E.D.; Schuster, P. RNA folding and combinatory landscapes. *Phys. Rev. E* **1993**, *47*, 2083. [CrossRef] [PubMed]
75. Bryngelson, J.D.; Onuchic, J.N.; Socci, N.D.; Wolynes, P.G. Funnels, Pathways, and the Energy Landscape of Protein Folding: A Synthesis. *Proteins Struct. Funct. Gen.* **1995**, *21*, 167–195. [CrossRef] [PubMed]
76. Dill, K.A.; Bromberg, S.; Yue, K.; Fiebig, K.M.; Yee, D.P.; Thomas, P.D.; Chan, H.S. Principles of protein folding—A perspective from simple exact models. *Prot. Sci.* **1995**, *4*, 561–602. [CrossRef]
77. Onuchic, J.N.; Luthey-Schulten, Z.; Wolynes, P.G. Theory of protein folding: The energy landscape perspective. *Ann. Rev. Phys. Chem.* **1997**, *48*, 545–600. [CrossRef]
78. Dobson, C.M.; Sali, A.; Karplus, M. Protein folding: A perspective from theory and experiment. *Angew. Chem. Int. Ed. Engl.* **1998**, *37*, 868–893. [CrossRef]
79. Honig, B. Protein Folding: From the Levinthal Paradox to Structure Prediction. *J. Mol. Biol.* **1999**, *293*, 283–293. [CrossRef]
80. Hodge, I.M. Physical Aging in Polymer Glasses. *Science* **1995**, *267*, 1945–1947. [CrossRef]
81. Li, M.S.; Nordblad, P.; Kawamura, H. Aging Effects in Ceramic Superconductors. *Phys. Rev. Lett.* **2001**, *86*, 1339–1342. [CrossRef]
82. Gutzow, I.; Schmelzer, J. *The Vitreous State*; Springer: Berlin/Heidelberg, Germany, 1995.
83. Sibani, P.; Schön, J.C.; Salamon, P.; Andersson, J.O. Emergent hierarchies in complex systems. *Europhys. Lett.* **1993**, *22*, 479–485. [CrossRef]
84. Schön, J.C.; Sibani, P. Properties of the energy landscape of network models for covalent glasses. *J. Phys. A Math. Gen.* **1998**, *31*, 8165–8178. [CrossRef]
85. Schön, J.C. Preferential trapping on energy landscapes in regions containing deep-lying minima - the reason for the success of simulated annealing? *J. Phys. A Math. Gen.* **1997**, *30*, 2367–2389. [CrossRef]
86. Hoffmann, K.H.; Schön, J.C. Kinetic Features of Preferential Trapping on Energy Landscapes. *Found. Phys. Lett.* **2005**, *18*, 171–182. [CrossRef]
87. Bryngelson, J.D.; Wolynes, P.G. Spin glasses and the statistical mechanics of protein folding. *Proc. Nat. Acad. Sci. USA* **1987**, *84*, 7524–7528. [CrossRef] [PubMed]
88. Pannetier, J.; Bassas-Alsina, J.; Rodriguez-Carvajal, J.; Caignaert, V. Prediction of crystal structures from crystal chemistry rules by simulated annealing. *Nature* **1990**, *346*, 343–345. [CrossRef]
89. Freeman, C.M.; Newsam, J.M.; Levine, S.M.; Catlow, C.R.A. Inorganic Crystal Structure Prediction Using Simplified Potentials and Experimental Unit Cells—Application to the Polymorphs of Titanium-Dioxide. *J. Mater. Chem.* **1993**, *3*, 531–535. [CrossRef]
90. Schön, J.C.; Jansen, M. Determination of Candidate Structures for Lennard-Jones-Crystals through Cell Optimisation. *Ber. Bunsenges.* **1994**, *98*, 1541–1544. [CrossRef]
91. Bush, T.S.; Catlow, C.R.A.; Battle, P.D. Evolutionary programming technique for predicting inorganic crystal structures. *J. Mater. Chem.* **1995**, *5*, 1269. [CrossRef]

92. Schön, J.C.; Jansen, M. Determination of Candidate Structures for Simple Ionic Compounds through Cell Optimisation. *Comp. Mater. Sci.* **1995**, *4*, 43–58. [CrossRef]
93. Takada, A.; Catlow, C.R.A.; Price, G.D. Computer modelling of B_2O_3: Part I. New interatomic potentials, crystalline phases and predicted polymorphs. *J. Phys. Cond. Mat.* **1995**, *7*, 8659–8692. [CrossRef]
94. Wevers, M.A.C.; Schön, J.C.; Jansen, M. Global Aspects of the Energy Landscape of Metastable Crystal Structures in Ionic Compounds. *J. Phys. Cond. Matter* **1999**, *11*, 6487–6499. [CrossRef]
95. Woodley, S.M.; Battle, P.D.; Gale, J.D.; Catlow, C.R.A. The prediction of inorganic crystal structures using a genetic algorithm and energy minimisation. *Phys. Chem. Chem. Phys.* **1999**, *1*, 2535–2542. [CrossRef]
96. Wevers, M.A.C.; Schön, J.C.; Jansen, M. Characteristic regions on energy landscapes of complex systems. *J. Phys. A Math. Gen.* **2001**, *34*, 4041–4052. [CrossRef]
97. Oganov, A.R.; Glass, C.W. Crystal structure prediction using ab initio evolutionary techniques: Principles and applications. *J. Chem. Phys.* **2006**, *124*, 244704. [CrossRef] [PubMed]
98. Schön, J.C.; Jansen, M. Prediction, determination and validation of phase diagrams via the global study of energy landscapes. *Int. J. Mat. Res.* **2009**, *100*, 135–152. [CrossRef]
99. Zagorac, D.; Schön, J.C.; Doll, K.; Jansen, M. Structure prediction for PbS and ZnO at different pressures and visualization of the energy landscape. *Acta Phys. Pol. A* **2011**, *120*, 215–220. [CrossRef]
100. Heard, C.J.; Schön, J.C.; Johnston, R.L. Energy Landscape Exploration of Sub-Nanometre Copper-Silver Clusters. *Chem. Phys. Chem.* **2015**, *16*, 1461–1469. [CrossRef]
101. Skundric, T.; Zagorac, D.; Schön, J.C.; Pejic, M.; Matovic, B. Crystal Structure Prediction of the Novel Cr_2SiN_4 Compound via Global Optimization, Data Mining, and the PCAE Method. *Crystals* **2021**, *11*, 891. [CrossRef]
102. Berry, R.S. Potential Surfaces and Dynamics: What Clusters Tell us. *Chem. Rev.* **1993**, *93*, 2379–2394. [CrossRef]
103. Wales, D.J.; Doye, J.P.K. Global Optimization by Basin Hopping and the Lowest Energy Structures of Lennard-Jones Clusters Containing up to 110 Atoms. *J. Phys. Chem. A* **1997**, *101*, 5111–5116. [CrossRef]
104. Wales, D.J.; Doye, J.P.K.; Miller, M.A.; Mortenson, P.N.; Walsh, T.R. Energy Landscapes: From Clusters to Biomolecules. In *Advances in Chemical Physics, Volume 115*; Prigogine, I., Rice, S.A., Eds.; Wiley: New York, NY, USA, 2000; pp. 1–111.
105. Johnston, R.L. *Atomic and Molecular Clusters*; CRC Press: London, UK, 2002.
106. Ferrando, R.; Jellinek, J.; Johnston, R.L. Nanoalloys: From theory to applications of alloy clusters and nanoparticles. *Chem. Rev.* **2008**, *108*, 845–910. [CrossRef] [PubMed]
107. Woodley, S.M.; Hamad, S.; Catlow, C.R.A. Exploration of multiple energy landscapes for zirconia nanoclusters. *Phys. Chem. Chem. Phys.* **2010**, *12*, 8454–8465. [CrossRef]
108. Pacheco-Contreras, R.; Dessens-Felix, M.; Borbon-Gonzalez, D.J.; Paz-Borbon, L.O.; Johnston, R.L.; Schön, J.C.; Posada-Amarillas, A. Tetrahelix conformations and transformation pathways in Pt_1Pd_{12} clusters. *J. Phys. Chem. A* **2012**, *116*, 5235–5239. [CrossRef] [PubMed]
109. Neelamraju, S.; Schön, J.C.; Doll, K.; Jansen, M. Ab initio and empirical energy landscapes of $(MgF_2)_n$ clusters (n = 3, 4). *Phys. Chem. Chem. Phys.* **2012**, *14*, 1223–1234. [CrossRef]
110. Pacheco-Contreras, R.; Borbon-Gonzalez, D.J.; Dessens-Felix, M.; Paz-Borbon, L.O.; Johnston, R.L.; Schön, J.C.; Jansen, M.; Posada-Amarillas, A. Determination of the energy landscape of $Pd_{12}Pt_1$ using a combined genetic algorithm and threshold energy method. *RSC Adv.* **2013**, *3*, 11571–11579. [CrossRef]
111. Neelamraju, S.; Oakley, M.T.; Johnston, R.L. Chiral effects on helicity studied via the energy landscape of short (d, l)-alanine peptides. *J. Chem. Phys.* **2015**, *143*, 165103. [CrossRef]
112. Rapacioli, M.; Tarrat, N.; Schön, J.C. Exploring energy landscapes at the DFTB quantum level using the threshold algorithm: The case of the anionic metal cluster Au_{20}^-. *Theor. Chem. Acc.* **2021**, *140*, 85. [CrossRef]
113. Miller, M.A.; Wales, D.J. Energy landscape of a model protein. *J. Chem. Phys.* **1999**, *111*, 6610–6616. [CrossRef]
114. Komatsuzaki, T.; Hoshino, K.; Matsunaga, Y.; Rylance, G.J.; Johnston, R.L.; Wales, D.J. How many dimensions are required to approximate the potential energy landscape of a model protein. *J. Chem. Phys.* **2005**, *122*, 084714. [CrossRef]
115. Vendruscolo, M.; Dobson, C.M. Towards complete descriptions of the free-energy landscapes of proteins. *Phil. Trans. Roy. Soc. A* **2005**, *363*, 433–452. [CrossRef] [PubMed]
116. Neelamraju, S.; Johnston, R.L.; Schön, J.C. A Threshold-Minimization Scheme for Exploring the Energy Landscape of Biomolecules: Application to a Cyclic Peptide and a Disaccharide. *J. Chem. Theo. Comp.* **2016**, *12*, 2471–2479. [CrossRef] [PubMed]
117. Margerit, W.; Charpentier, A.; Maugis-Rabusseau, C.; Schön, J.C.; Tarrat, N.; Cortes, J. IGLOO: An Iterative Global Exploration and Local Optimization Algorithm to Find Diverse Low-Energy Conformations of Flexible Molecules. *Algorithms* **2023**, *16*, 476. [CrossRef]

118. Salomon, G.; Tarrat, N.; Schön, J.C.; Rapacioli, M. Low-Energy Transformation Pathways between Naphtalene Isomers. *Molecules* **2023**, *28*, 5778. [CrossRef]
119. Naaz, F.; Chauhan, M.S.; Yadav, K.; Singh, S.; Kumar, A.; Prasad, D.L.V.K. Structural Phase Transitions in Perovskite $BaCeO_3$ with Data Mining and First Principles theoretical Calculations. *J. Phys. Chem. C* **2024**, *128*, 4766–4778. [CrossRef]
120. Babin, V.; Roland, C.; Darden, T.A.; Sagui, C. The free energy landscape of small peptides as obtained from metadynamics with umbrella sampling corrections. *J. Chem. Phys.* **2006**, *125*, 204909. [CrossRef] [PubMed]
121. Wales, D.J.; Bogdan, T.V. Potential energy and free energy landscapes. *J. Phys. Chem. B* **2006**, *110*, 20765–20776. [CrossRef]
122. Schön, J.C.; Čančarević, Ž.P.; Hannemann, A.; Jansen, M. Free enthalpy landscape of SrO. *J. Chem. Phys.* **2008**, *128*, 194712. [CrossRef]
123. Salamon, P.; Wales, D.; Segall, A.; Lai, Y.A.; Schön, J.C.; Hoffmann, K.H.; Andresen, B. Rate constants, timescales, and free energy barriers. *J. Non-Equil. Thermodyn.* **2016**, *41*, 13–18. [CrossRef]
124. Schön, J.C.; Putz, H.; Jansen, M. Investigating the energy landscape of continuous systems—The threshold algorithm. *J. Phys. Cond. Matter* **1996**, *8*, 143–156. [CrossRef]
125. Berry, R.S.; Kazakov, V.A.; Sieniutycz, S.; Szwast, Z.; Tsirlin, A.M. *Thermodynamic Optimization of Finite Time Processes*; Wiley: Chichester, UK, 1999.
126. Kaushik, S.C.; Tyagi, S.K.; Kumar, P. *Finite Time Thermodynamics of Power and Refrigeration Cycles*; Springer: Berlin/Heidelberg, Germany, 2018.
127. Santoro, M.; Schön, J.C.; Jansen, M. Finite-time thermodynamics and the gas-liquid phase transition. *Phys. Rev. E* **2007**, *76*, 061120. [CrossRef] [PubMed]
128. Schön, J.C. Finite-Time Thermodynamics and the Optimal Control of Chemical Syntheses. *Z. Anorg. Allg. Chem.* **2009**, *635*, 1794–1806. [CrossRef]
129. Zettlemoyer, A.C. *Nucleation*; Dekker: New York, NY, USA, 1969.
130. Catlow, C.R.A.; DeLeeuw, N.H.; Anwar, J.; Davey, R.J.; Roberts, K.J.; Unwin, P.R. (Eds.) *Faraday Discussions 136: Crystal Growth and Nucleation*; The Royal Society of Chemistry: London, UK, 2007.
131. Anwar, J.; Zahn, D. Uncovering molecular processes in crystal nucleation and growth by using molecular simulation. *Angew. Chem. Int. Ed.* **2011**, *50*, 1996–2013. [CrossRef]
132. Baldus, H.P.; Jansen, M. Novel high-performance ceramics—Amorphous inorganic networks from molecular precursors. *Angew. Chem. Int. Ed.* **1997**, *36*, 328–343. [CrossRef]
133. Hannemann, A.; Schön, J.C.; Jansen, M. Modeling the sol-gel synthesis route of amorphous $Si_3B_3N_7$. *J. Mater. Chem.* **2005**, *15*, 1167–1178. [CrossRef]
134. Fischer, D.; Jansen, M. Synthesis and Structure of Na_3N. *Angew. Chem. Int. Ed.* **2002**, *41*, 1755–1756. [CrossRef]
135. Liebold-Ribeiro, Y.; Fischer, D.; Jansen, M. Experimental substantiation of the "Energy Landscape Concept" for solids: Synthesis of a new modification of LiBr. *Angew. Chem. Int. Ed.* **2008**, *47*, 4428–4431. [CrossRef]
136. Schön, J.C.; Fischer, D. Thin Films and Monolayers—Prediction, Modeling and Experiments. *J. Innov. Mater. Extr. Cond.* **2023**, *4*, 52–76.
137. Fischer, D.; Andriyevsky, B.; Schön, J.C. Systematics of the allotrope formation in elemental gallium. *Mater. Res. Expr.* **2019**, *6*, 116401. [CrossRef]
138. Papon, P.; Leblond, J.; Meijer, P.H.E. *The Physics of Phase Transitions*; Springer: New York, NY, USA, 2002.
139. Klesnil, M.; Lukas, P. *Fatigue of Metallic Materials*; Elsevier: Amsterdam, 1992.
140. George, E.P.; Raabe, D.; Ritchie, R.O. High-entropy alloys. *Nat. Rev. Mater.* **2019**, *4*, 515–534. [CrossRef]
141. Matovic, B.; Zagorac, D.; Cvijovic-Alagic, I.; Zagorac, J.; Butulija, S.; Ercic, J.; Hanzel, O.; Sedlak, R.; Lisnichuk, M.; Tatarko, P. Fabrication and characterization of high-entropy pyrochlore ceramic. *Bol. Soc. Esp. Ceram. Y Vidr.* **2023**, *62*, 66–76. [CrossRef]
142. Denbigh, K.G. Optimum temperature sequence in reactors. *Chem. Eng. Sci.* **1958**, *8*, 125–132. [CrossRef]
143. Horn, F.; Troltenier, U. Über den optimalen Temperaturverlauf im Reaktionsrohr. *Chem. Ing. Techn.* **1960**, *32*, 382–393. [CrossRef]
144. Aris, R. The determination of optimum operating conditions by the methods of dynamic programming. *Z. Elektrochem.* **1961**, *65*, 229–244. [CrossRef]
145. Mansson, B.; Andresen, B. Optimal temperature profile for an ammonia reactor. *Ind. Eng. Chem. Proc. Des. Develop.* **1986**, *25*, 59–65. [CrossRef]
146. Schön, J.C.; Andresen, B. Multiple Modes for the Operation of a Distillation Column. *Industr. Engin. Chem. Res.* **1996**, *35*, 2327–2333. [CrossRef]
147. Schaller, M.; Hoffmann, K.H.; Siragusa, G.; Salamon, P.; Andresen, B. Numerically optimized performance of diabatic distillation columns. *Comp. Chem. Engin.* **2001**, *25*, 1537–1548. [CrossRef]

148. Schön, J.C.; Andresen, B. Finite-Time Optimization of Chemical Reactions: $nA \rightleftharpoons nB$. *J. Phys. Chem.* **1996**, *100*, 8843–8853. [CrossRef]
149. Bak, T.A.; Salamon, P.; Andresen, B. Optimal Behavior of Consecutive Chemical Reactions $A \rightleftharpoons B \rightleftharpoons C$. *J. Phys. Chem. A* **2002**, *106*, 10961–10964. [CrossRef]
150. Corey, E.J. General Methods for the Construction of Complex Molecules. *Pure Appl. Chem.* **1967**, *14*, 19–37. [CrossRef]
151. Corey, E.J. The Logic of Chemical Synthesis—Multi-Step Synthesis of Complex Carbogenic Molecules. *Angew. Chem. Int. Ed. Eng.* **1991**, *30*, 455–465. [CrossRef]
152. Ugi, I.; Bauer, J.; Bley, K.; Dengler, A.; Dietz, A.; Fontain, E.; Gruber, B.; Herges, R.; Knauer, M.; Reitsam, K.; et al. Computer-Assisted Solution of Chemical Problems - The Historical Development and the Present State of the Art of a New Discipline of Chemistry. *Angew. Chem. Int. Ed. Eng.* **1993**, *32*, 201–227. [CrossRef]
153. Li, J.J. *Name Reactions: A Collection of Detailed Mechanisms and Synthetic Applications*, 4th ed.; Springer: Berlin/Heidelberg, Germany, 2009.
154. Schön, J.C.; Jansen, M. A First Step towards Planning of Syntheses in Solid State Chemistry: Determination of Promising Structure Candidates using Global Optimization. *Angew. Chem. Int. Ed. Eng.* **1996**, *35*, 1286–1304. [CrossRef]
155. Schön, J.C. On the way to a theory of solid state synthesis: Issues and open questions. *Adv. Chem. Phys.* **2015**, *157*, 125–134.
156. Woodley, S.M.; Catlow, C.R.A. Crystal structure prediction from first principles. *Nat. Mater.* **2008**, *7*, 937. [CrossRef]
157. Zurek, E. Discovering New Materials via A Priori Crystal Structure Prediction. In *Reviews in Computational Chemistry, Volume 29*; Parrill, A.L., Lipkowitz, K.B., Eds.; Wiley: New York, NY, USA, 2016; Chapter 5, pp. 274–326.
158. Aykol, M.; Merchant, A.; Batzner, S.; Wei, J.N.; Cubuk, E.D. Predicting emergence of crystals from amorphous precursors with deep learning potentials. *Nat. Comput. Sci.* **2025**, *5*, 105–111. [CrossRef]
159. Schön, J.C. Enhancing synthesis prediction via machine learning. *Nat. Comput. Sci.* **2025**, *5*, 95–96. [CrossRef]
160. Noh, M.; Johnson, D.C. Designed Synthesis of $[TiSe_2]_m[NbSe_2]_n$ Superlattices from Modulated Reactants. *Angew. Chem. Int. Ed.* **1996**, *35*, 2666–2669. [CrossRef]
161. Esters, M.; Alemayehu, M.B.; Jones, Z.; Nguyen, N.T.; Anderson, M.D.; Grosse, C.; Fischer, S.F.; Johnson, D.C. Synthesis of Inorganic Structural Isomers By Diffusion-Constrained Self-Assembly of Designed Precursors: A Novel Type of Isomerism. *Angew. Chem. Int. Ed.* **2015**, *54*, 1130–1134. [CrossRef]
162. Schön, J.C. Studying the Energy Hypersurface of Multi-Minima Systems—The Threshold and the Lid Algorithm. *Ber. Bunsenges.* **1996**, *100*, 1388–1391. [CrossRef]
163. Hoffmann, K.H.; Fischer, A.; Schön, J.C. Controlled Dynamics and Preferential Trapping on Energy Landscapes. In *Energy Landscapes of Nanoscale Systems*; Wales, D.J., Ed.; Elsevier: Amsterdam, The Netherlands, 2022; Volume 21, pp. 211–245.
164. Hoffmann, K.H.; Schön, J.C. Controlled dynamics on energy landscapes. *Eur. Phys. J.* **2013**, *86*, 220. [CrossRef]
165. Hoffmann, K.H.; Schön, J.C. Combining pressure and temperature control in dynamics on energy landscapes. *Eur. Phys. J. B* **2017**, *90*, 84. [CrossRef]
166. Wevers, M.A.C.; Schön, J.C.; Jansen, M. Determination of structure candidates of simple crystalline AB_2-systems. *J. Solid State Chem.* **1998**, *136*, 223–246. [CrossRef]
167. Wevers, M.A.C. Energetische und Entropische Aspekte der Energielandschaften von MgF_2, CaF_2 und $Li_xNa_{6-x}N_2$ (X = 0, 1, . . . , 6) Sowie ein Vergleich mit ab initio Rechnungen. Ph.D Thesis, University of Bonn, Bonn, Germany, 1999.
168. Chen, C.; Ong, S.P. A universal graph deep learning interatomic potential for the periodic table. *Nat. Comput. Sci.* **2022**, *2*, 718–728. [CrossRef]
169. Kirkwood, J.G. Statistical Mechanics of Fluid Mixtures. *J. Chem. Phys.* **1935**, *3*, 300–313. [CrossRef]
170. Zwanzig, R. High-Temperature Equation of State by a Perturbation Method. I. Nonpolar Gases. *J. Chem. Phys.* **1954**, *22*, 1420–1426. [CrossRef]
171. Watanabe, M.; Reinhardt, W.P. Direct Dynamical Calculation of Entropy and Free Energy by Adiabatic Switching. *Phys. Rev. Lett.* **1990**, *65*, 3301–3304. [CrossRef]
172. Kirkpatrick, S.; Gelatt, C.D., Jr.; Vecchi, M.P. Optimization by Simulated Annealing. *Science* **1983**, *220*, 671–680. [CrossRef] [PubMed]
173. Holland, J.H. *Adaptation in Natural and Artificial Systems*; University of Michigan Press: Ann Arbor, MI, USA, 1975.
174. Andresen, B.; Hoffmann, K.H.; Mosegaard, K.; Nulton, J.; Pedersen, J.M.; Salamon, P. On lumped models for thermodynamic properties of simulated annealing problems. *J. Phys. I* **1988**, *49*, 1485–1492. [CrossRef]
175. Salamon, P.; Sibani, P.; Frost, R. *Facts, Conjectures, and Improvements for Simulated Annealing*; SIAM Monographs: Philadelphia, PA, USA, 2002.

176. del Moral, R.; Grishin, S. Volcanic disturbances and ecosystem recovery. In *Ecosystems of Disturbed Ground*; Walker, L.R., Ed.; Elsevier: Amsterdam, The Netherlands, 1999; Chapter 5, pp. 137–160.
177. Carrillo, U.; Diaz-Villanueva, V. Impacts of volcanic eruptions and early recovery in freshwater environments and organisms. *Biol. Rev.* **2021**, *96*, 2546–2560. [CrossRef] [PubMed]

entropy

MDPI

Article

(Finite-Time) Thermodynamics, Hyperbolicity, Lorentz Invariance: Study of an Example

Bernard Guy

Mines Saint-Etienne, Institut Mines Télécom, 42100 Saint-Étienne, France; bernard.guy15@wanadoo.fr

Abstract: Our study lies at the intersection of three fields: finite-time thermodynamics, relativity theory, and the theory of hyperbolic conservation laws. Each of these fields has its own requirements and richness, and in order to link them together as effectively as possible, we have simplified each one, reducing it to its fundamental principles. The example chosen concerns the propagation of chemical changes in a very large reactor, as found in geology. We ask ourselves two sets of questions: (1) How do the finiteness of propagation speeds modeled by hyperbolic problems (diffusion is neglected) and the finiteness of the time allocated to transformations interact? (2) How do the finiteness of time and that of resources interact? The similarity in the behavior of the pairs of variables (x, t and resources, resource flows) in Lorentz relativistic transformations allows us to put them on the same level and propose complementary-type relationships between the two classes of finiteness. If times are finite, so are resources, which can be neither zero nor infinite. In hyperbolic problems, a condition is necessary to select solutions with a physical sense among the multiplicity of weak solutions: this is given by the entropy production, which is Lorentz invariant (and not entropy alone).

Keywords: finite-time thermodynamics; finite resources; Lorentz invariance; entropy production; optimization; complementarity relations; hyperbolic problem; conservation of matter; four vectors; non-equilibrium

1. Introduction

"Finite-time thermodynamics" studies the evolution of non-equilibrium systems subject to finiteness constraints: finite resources, finite time allocated to transformations, etc. A criterion is used to define the values of physical quantities that make a given objective optimal (see, among many references, [1–7]).

It should be emphasized that the established term "finite-time thermodynamics" requires the incorporation of a minimum dissipation criterion. We put (finite-time) in parentheses if we emphasize the first part (finite-time) at the expense of the second (minimum dissipation), but, for us, the second part is indeed present.

Studies on the subject are generally based on a single space-time reference frame. They do not consider the case of another reference frame moving relative to the first. A relativistic approach raises a number of questions: How do finiteness constraints change? How do quantities change and accommodate the contraction/dilation effects established for space and time? How do optimization criteria change? What compatibility is there between different points of view?

We will not address these questions in general terms but rather in a specific manner, shifting them in two new directions that constitute the two objectives of our research:

(1) *How does thermodynamics in finite time work for large systems in terms of spatial dimensions?* We are thinking of geological systems measuring tens or hundreds of meters, or even more. In mathematical models, the transport of constituents within such systems is not driven by diffusion and its infinite propagation speeds but by advection with finite propagation speeds (diffusion is negligible at the scale considered). It is, therefore, appropriate to consider weak solutions in the mathematical sense for the propagation equations, simulating the displacement of discontinuities (hyperbolic problems). The finiteness of the velocities will lead to the finiteness of the times through simple relationships. The word geology attached to the example we will discuss simply indicates that we are dealing with large scales of space and time, where propagations, regardless of their actual speeds, have time to become noticeable.

(2) *What is the link between the finite space and time of (finite-time) thermodynamics and the finite resources it discusses?* Indeed, thermodynamics in finite time encompasses studies in which finiteness concerns both time and resources using the same methods. Does this not point us towards a deeper organic link expressing a form of equivalence between time and resources? Addressing this question brings us back to the fundamental philosophical question of the nature and existence of time. We have addressed this question in our works using relational logic: time and space have no definition in themselves; they are names for relationships between phenomena, or in this case, between resources. This is one way of interpreting the theory of relativity, which is expressed mathematically in Lorentz's relations. The pairs of quantities (x, t and resources, resource flows) behave in the same way with respect to Lorentz transformations. The mathematical development proposed in this study will, thus, lead us to propose organic relationships between time and finite resources.

These two directions are, a priori, independent of each other. Their possible intersections constitute complementary research directions that will be outlined in the conclusions of this article.

Addressing these research directions requires articulating three fields: thermodynamics, in particular finite-time thermodynamics; the theory of relativity; and the theory of hyperbolic conservation laws. Each of these fields is very rich, and we will not go into their details here. Therefore, we will present a very simplified view of each of them, focusing on their philosophical foundations (that have been renewed to a certain extent) without exploiting their full richness.

Thermodynamics, in its elementary structure, studies evolving systems (second law) subject to conservation constraints (first law). Similarly, thermodynamics in finite time is understood here in its simplest form, based on the very words that define it: finite time, i.e., the study of systems for which a finite evolution time is considered, whether imposed or derived from the solution of the problem. In this context, the question of temperature, although essential in finite-time thermodynamics, will not be addressed directly, although it is hidden in entropy, and entropy production will be the dissipation criterion to be minimized. The duality of conservation and evolution, which is at the heart of thermodynamics, refers to the duality of space and time discussed in the theory of relativity. We look at the latter theory with a philosophical perspective (see below) that is also fruitful in thermodynamics.

For us, the theory of relativity is fundamentally a theory that proposes to construct and discuss the concepts of space and time by choosing a standard of motion that allows different systems in relative motion to interrelate. This interpretation, which is an extension of classical interpretations, is set out in our numerous works. It is based on a relational logic in which objects of thought (space, time) are the names of comparisons between material entities, which we will refer to here as the (finite) resources studied, as we have said. It is not so much the high value of the speed of light that matters as the relationships between the different speeds involved (in our relational understanding, important in relativity are the v/c ratios, whatever the value of the standard c, as if it could be considered on its own, which cannot be done in general terms: for us, the expression "speed of light" has no meaning by itself).

To gain this broader view of relativity, we draw on philosophy. The strength of physics lies in giving full value to the tools it uses (measurements, concepts, mathematical symbols, equations). It assumes correspondence with the compartments of reality that are supposed to exist separately (we can speak of realism or, in a slightly different sense, positivism or pragmatism). This allows for remarkable efficiency. Philosophy, on the other hand, confronts an infinitely complex reality where everything is connected to everything else. The world infinitely exceeds all representations we can make of it. Philosophy never stops reflecting on the conditions that make objective knowledge possible and on the paths that lead us to adopt one or another concept of physics. There are moments in the history of physics when it is essential to take a philosophical step back from what has been constructed, in order to possibly shift our choices in understanding reality. This, therefore, concerns relativity but, also, in another way, thermodynamics, as we have shown in our works. This allows us to accommodate various difficulties (conceptual, experimental, mathematical) that have arisen since the last construction. But efficiency in our actions will require a new pause in fundamental questioning.

Finally, the theory of hyperbolic problems is provided by the example chosen. This field is not well known to thermodynamicists (it is present in its own way in extended irreversible thermodynamics [8,9]; see also [10]). The term "hyperbolic problem" refers to the classification of partial differential equations (parabolic, elliptic, hyperbolic). A hyperbolic problem exhibits solutions with finite propagation speeds. It gives rise to discontinuous solutions and must, therefore, be placed in an appropriate mathematical framework, that of distributions. We then face a multiplicity of solutions (known as weak solutions), and a criterion must be added to select those that make physical sense. This is entropy production, and this structure (equation + selection criterion), thus, *shows a similarity to finite-time thermodynamics*. The fact that the example chosen comes from geology is not important. What is important is the mathematical structure of the hyperbolic problem, its concern with discussing different contrasting scales of space and time, and its ability to deal with real discontinuities in the sense of distribution theory.

In return for these simplified/modified views on thermodynamics, relativity, and the theory of hyperbolic conservation laws, we will provide original results, instructive answers to the two specific questions specified above, and research proposals for the scientific community. The logical structure of our approach is summarized in Figure 1.

Four boxes are distinguished, as variants of thermodynamics in finite time in the broad sense. We make a distinction, along the vertical axis, between the treatment of continuous physical quantities (bottom) and discrete quantities (top); along the horizontal axis, between the finiteness of time in the strict sense (left) and the finiteness of resources (right). The transition from the lower level to the upper level corresponds to larger scales of space and time where propagation speeds are finite and diffusion is neglected (we mark this with $D \to 0$; discontinuous solutions; hyperbolicity of the problem). The transition from the left level to the right level amounts to exchanging the reference parameters, i.e., time and space (left) and flow of resources (right), according to a relational interpretation of the theory of relativity (they behave in the same way in Lorentz transformations). Arrows 1 and 2 refer to topics discussed in this study, as mentioned above, while arrows 3 and 4 represent research directions briefly discussed in the conclusions.

Our plan will be as follows. In the first part (theoretical background), we will set out the useful formulations for placing ourselves in a relativistic framework (in particular, using four vectors, writing the transformation of quantities and requesting Lorentz invariance). In the second part (results), we present the example studied and its solution: the finiteness relations express the conservation of matter, and the criterion for optimizing or choosing physical quantities is the production of entropy. We will look at how the various quantities and equations used (conservation, optimization criterion) are transformed. We will highlight the complementary-type relationships between time finiteness and resource finiteness. We conclude with a discussion section.

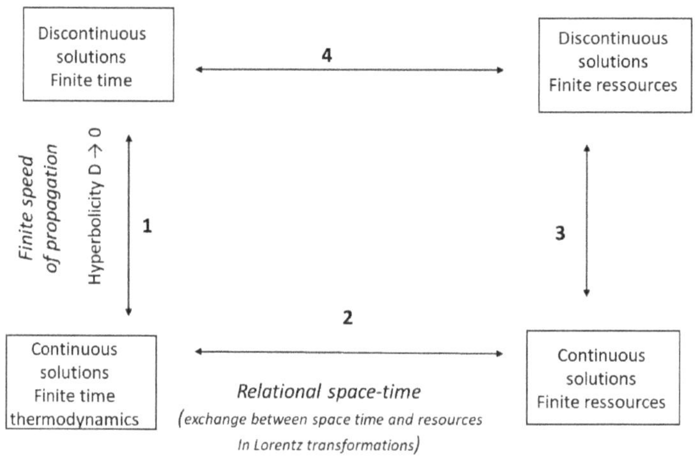

Figure 1. Logical structure of our research.

2. Theoretical Background

2.1. *Revision of Mathematical Tools*

If we place ourselves in a framework compatible with the relativist approach, we must revise our conceptual equipment. It is useful to deal with physical quantities not "on their own" but by associating them in pairs that allow us to define four vectors: it is these that will be the subject of relativistic transformations. The pairs (position, time; energy, momentum; electric charge, current) are among the best known. In relativity, we extend this formalism to three-component vector pairs such as electric field, magnetic field in appropriate tensorial mathematical entities.

In the present thermodynamic approach, we will have to deal with entropy in this way. What pair should we consider for it? For reasons that will become clear later, we propose the pair (entropy, entropy flow), denoted (S, F), where entropy S is a scalar depending on the variables f of the system, and F is a three-dimensional vector. This pair is already used as such in the solution of the hyperbolic-type equations we will be dealing with in our example (part 3) Entropy is a non-conserved quantity but this does not affect our choice. In the Lorentz transformation (see below), we will retain the entropy production associating the two vectors S and F (in their temporal and spatial derivatives, respectively): this mathematical form is conserved regardless of its sign. As we will see it is remarkable that Kruzhkov's formulation of hyperbolic problems puts equalities and inequalities on the same footing. It is possible to choose another time reference based on the pair (f, g) that satisfies the entropy condition (Section 2.4), and this will not change the behavior with respect to Lorentz invariance. It is tantamount to placing ourselves in a potentially open system, which is in line with the conclusions of quantum thermodynamics: there is no rigorously closed physical system that has no relationship with the rest of the universe [11].

There is still controversy surrounding the choice of an entropic optimization criterion in relativistic thermodynamics, as we will discuss further below.

For the concentration variables used in our example, we must similarly consider a pair (concentration, concentration flux) for a small, open rock system; we will call it (f, g).

2.2. Links Between the Concepts of Space and Time

Our work on relativity has led us to renew our interpretation of the concepts of space and time (see, for example, [12–14]). Our approach is not one of substantial epistemology, in which the concepts of space and time are each grasped for their own sake; it is of relational epistemology, in which they are grasped in opposition/composition to each other, based on a comparison between phenomena or movements (these being envisaged in a primary way before words, through experience, experimental protocols, embodied cognition, designation). To space correspond the relative movements that are barely perceptible or "stopped", compared to faster movements on which time is defined. Quantitative measurements are made by giving a particular movement the role of a standard: this is the meaning of the second postulate of the theory of relativity. In this context, it is interesting to take up the mathematization of time using three intermediate parameters: the coordinates of the position of the mobile that defines time (the photon in the atomic clock has replaced the relative position of the sun and the earth), i.e., t_x, t_y, t_z. From these three coordinates, we can define the scalar t in various ways, for example, by $t = \sqrt{t_x^2 + t_y^2 + t_z^2}$. This formula is a way of constructing the usual time as a scalar. The invariance of the relativistic distance is another topic; it may also be written with the three intermediate pre-time coordinates. This approach offers greater symmetry between spatial and temporal variables and opens up new possibilities compared with the standard approach. For entropy, we need to define an intermediate vector with three components, S_x, S_y, S_z, measuring the inhomogeneity of a system along the three directions of space and enabling us to define the entropy scalar. This is useful for generalizing the results that follow, but, no more than for time, we will not need this generalization, as we will be reasoning on a simple one-dimensional example. We will be dealing with pairs of scalars: (x, t) or (S, F). Generalization to several space and time variables is not a priori a problem, apart from the cumbersomeness of the formalism. For now, this provides a framework for placing the pairs (x, t) and (f, g) on the same footing.

2.3. General Conservation Constraints

The systems studied in (finite-time) thermodynamics are subject to finiteness constraints of various kinds. Here, we shall consider constraints on the conservation of matter: together with the equivalent conservation of energy, these are the most fundamental constraints that can be expressed for common physical systems. In the general case, we must write them in the following local form; f is a physical quantity defined by its volume concentration, and $\vec{g}$ is a flux vector in three-dimensional space:

$$\frac{\partial f}{\partial t} + div\,\vec{g} = 0 \tag{1}$$

Knowing that we will restrict ourselves here to laws of the form

$$\frac{\partial f}{\partial t} + \frac{\partial g}{\partial x} = 0 \tag{2}$$

where f and g are two scalars. We will assume that, due to the granularity of the representation adopted, there is no production term in the mass balances. The preceding relationship can be understood another way: in a relational spirit, the space and time variables x and

t, and more generally the physical quantities f and g, are not known in themselves but only in their reciprocal relationships: we express the latter by the equality of their coupled variations (partial derivatives) of the type

$$\frac{\partial f}{\partial t} = -\frac{\partial g}{\partial x} \tag{3}$$

(we keep the—sign to respect the first conservation formulation). Knowing that the dual expression

$$\frac{\partial f}{\partial x} = -\frac{\partial g}{\partial t} \tag{4}$$

would also be an admissible law. In the general framework proposed in our work, complete formulations would be of the type

$$\frac{\partial f_x}{\partial t_x} + \frac{\partial f_y}{\partial t_y} + \frac{\partial f_z}{\partial t_z} + \frac{\partial g_x}{\partial x} + \frac{\partial g_y}{\partial y} + \frac{\partial g_z}{\partial z} = 0$$

which is derived by summation of elementary equalities of the type $\frac{\partial f_i}{\partial t_j} = -\frac{\partial g_j}{\partial x_i}$. It can be shown [15,16] that the equations of mechanics and Maxwell's equations can be put into this form through various changes of variables. See also [17]. The nullity of a transport derivative along an appropriate motion would be yet another way of understanding Equation (2).

For the pair (entropy, entropy flow), we do not have a strict equality in a conservation equation, but the inequality expressing the second law in an open system; the entropy balance is written:

$$I = \frac{\partial S}{\partial t} + \frac{\partial F}{\partial x} \geq 0 \tag{5}$$

And we also write (even if this rarer notation is confusing) the following:

$$I = P(S) = \frac{d_i S}{dt}$$

where $d_i S$ refers to the change in entropy due to the phenomena internal (i for internal) to the system (as opposed to the processes in relation with the fluxes).

2.4. Lorentz Invariance and Conservation Laws

Lorentz transformation will play hereafter. It allows us to pass from the values (x, t) (we use one dimension of space and one of time), evaluated in a first frame of reference, to those (x', t') in a second frame of reference moving relative to the first. Subject to various constraints (linearity, isotropy of space, homogeneity of space and time, compliance with the two relativity postulates), we obtain the following results:

$$x' = ax + bt$$

$$t' = bx + at \tag{6}$$

with only two coefficients a and b and not four (expressing a formal symmetry between t and x; we have taken $c = 1$), which depend on the speed v of displacement between the two reference frames, according to

$$a = \frac{1}{\sqrt{1 - \frac{v^2}{c^2}}} \qquad b = \frac{-v}{\sqrt{1 - \frac{v^2}{c^2}}} \tag{7}$$

The expressions of the Lorentz transformations in relations (6) and (7) depend on the choice of clocks made. In our new perspective on the theory of relativity, mentioned in the introduction, we highlight the importance of the hidden movements of photons in clocks (which cannot be punctual). Depending on the choices made, various expressions of the Lorentz transformations are possible, as we show in [18], commenting on various results in the literature. Here, we used the standard formulations.

As we shall see, there is a special link between Lorentz invariance and the general form of linear conservation laws, explained above for pairs (f, g) of two scalars; in Equation (2), we have written:

$$\frac{\partial f}{\partial t}+\frac{\partial g}{\partial x}=0$$

Let us now write down the Lorentz transformation for the quantities f and g linked by law (2). We want this law to be Lorentz invariant, i.e., to be conserved in a change of Galilean reference frame R → R′ (Einstein's first postulate), with reference frame R′ moving at speed v relative to reference frame R. It is shown in [16] that the law

$$\frac{\partial f\prime}{\partial t\prime}+\frac{\partial g\prime}{\partial x\prime}=0 \tag{8}$$

is also verified, provided we write:

$$f' = af + bg$$

$$g' = bf + ag \tag{9}$$

where the coefficients a and b describe the Lorentz transformations for x and t. Equations (2), (6) and (7) have been used (the Lorentz transformation assumes the second postulate of relativity). We, therefore, obtain transformation laws for physical quantities that are identical to the transformations for space and time coordinates (these results can be found in the developments of standard relativity, even if it is a little hidden; we give this result a general value. Expressed in the standard covariant notation adopted in relativity, any conservation law can be expressed as: $\partial_i J_i = 0$, where J is the suitable 4-vector. When expressing it in this form, it is immediately clear that a Lorentz transformation would not change the form of the law).

The reciprocal holds. If the transformation laws of type (6), (7) and (9) are verified, then the laws (2) linking f to g and (8) linking f' to g' are such that

$$\frac{\partial f}{\partial t}+\frac{\partial g}{\partial x}=\frac{\partial f\prime}{\partial t\prime}+\frac{\partial g\prime}{\partial x\prime} \tag{10}$$

This last result expresses both the form of the law sought and its invariance under the action of the change of reference frame; if we note that relations (10) are valid for all forms of functions f and g, they are particularly verified for functions f and g equal to zero, and the common value of the two expressions in relation (10) is zero. This is written as:

$$\frac{\partial f}{\partial t}+\frac{\partial g}{\partial x}=\frac{\partial f\prime}{\partial t\prime}+\frac{\partial g\prime}{\partial x\prime}=0 \tag{11}$$

Thus, *a necessary and sufficient condition for a law to be Lorentz invariant* (the Lorentz transformation being given for x and t) *is that it has the specified form of type (2)* or a mathematically equivalent form (at orders derived or integrated with respect to the original formulation). The characteristic of this form is to equalize derivatives of quantities with respect to time with derivatives of coupled quantities with respect to space. This is the case for what we call the basic laws of physics. As mentioned above (note (3)), this is the case for Maxwell's equations and the equations of mechanics usually tested for Lorentz invariance. If the form of the equations confers Lorentz invariance, as is well known, we emphasize here the reciprocal proposal and its generality. This result expresses *a strong link between relational thinking (we spoke of relational epistemology) and Lorentz invariance.* It also shows that the pair (x, t) is on the same footing as pairs of physical quantities in duality and behaves in the same way or, conversely, that pairs of type (f, g) behave like pairs (x, t). This is a way of saying that space and time can be defined by the measurements of f and g. This will allow us to understand the expressions "finite time" and "finite resources" as having a real mathematical equivalence (and not just a practical one), which is one of the objectives of our study (see Section 3.5).

2.5. Entropy Balance

The same applies to entropy (by formally replacing f and g by S and F in the linear conservation relation)

$$\frac{\partial S}{\partial t}+\frac{\partial F}{\partial x}=\frac{\partial S\prime}{\partial t\prime}+\frac{\partial F\prime}{\partial x\prime} \tag{12}$$

In particular, if

$$\frac{\partial S}{\partial t}+\frac{\partial F}{\partial x}\geq 0 \tag{13}$$

We will have

$$\frac{\partial S\prime}{\partial t\prime}+\frac{\partial F\prime}{\partial x\prime}\geq 0 \tag{14}$$

With again

$$S' = aS + bF$$

$$F' = bS + aF \tag{15}$$

where a and b are the coefficients of the Lorentz transformations given by (7). In the relativistic thermodynamics literature, it is generally accepted that the entropy scalar function S is a relativistic invariant, i.e., $S' = S$, although it is pointed out that consensus on the issue does not seem to have been reached (see in the numerous references [19–44] along with general considerations on relativistic thermodynamics).

For the general reasons we have given we think it useful to embed S in four vectors (S, F). In short, we mustn't separate physical quantities from the laws in which they are involved, in this case laws involving the simultaneous time and space derivatives of dualistic quantities. It's a way of joining up with the requirements of quantum mechanics concerning the absence of rigorously isolated systems. We are then led to

$$P(S)=\frac{\partial S}{\partial t}+\frac{\partial F}{\partial x}=P(S')=\frac{\partial S\prime}{\partial t\prime}+\frac{\partial F\prime}{\partial x\prime}\geq 0 \tag{16}$$

and relativistic invariance concerns entropy production, not entropy alone. The optimization criterion we are interested in here (see below) is that given by (13), and, in the study of reference frame changes, we will rely on relation (16).

3. Results

3.1. Study of an Example

Let us turn now to the concrete example already suggested. We assume that the balance of a physical quantity f is governed by the following conservation equation, analogous to (2) seen above:

$$\frac{\partial f}{\partial t}+\frac{\partial g}{\partial x}=0$$

where g represents the flux of the quantity f (recall that the problem is posed in one dimension of space x and one of time t). This equation simulates the behavior of a rock defined by the concentration f (related to the unit length) of a single chemical constituent C (e.g., iron) in the solid phase ([45,46]; ion-exchange columns provide an analogous problem [47]). The rock is traversed in its porosity p (assumed to be low) by an *aqueous fluid* in disequilibrium, characterized by the concentration g in the fluid. The flow involves the velocity w of the percolating fluid and the porosity p in the complete expression pwg. We will assume that we have normalized the quantities so that pw (Darcy velocity) = 1, and a velocity is, thus, hidden in g. Then, the velocity w of the fluid is assumed to be uniform, constant, and non-dimensional. This somewhat surprising choice is made for practical reasons, in order to respect the simple form of the continuity equation of type (2).

Another surprising choice on our part is the inclusion of our example in a relativistic thermodynamics framework. This physical example is not likely to involve anything moving at relativistic speeds. Without fast particles, it is not immediately clear how "relativistic thermodynamics", the theoretical structure used here, could be tested. Would ordinary Newtonian/Galilean mechanics, perhaps in its diffusion form, not suffice? In response to this natural objection, we would say that we are using the word relativistic in a broad sense, certainly in continuity with and consistent with the standard meaning: we speak of relativity to refer to the general structure of our understanding of space and time, extending beyond the realm of physics and into the human and social sciences. It points the way to a close relationship between time and space. Relativity allows us to discuss the behavior of space and time, which manifest themselves in interesting ways when we consider relative movements between reference frames (refer to our introductory remarks).

We will also assume that a local equilibrium is achieved between the solid and the fluid, according to a law, called an isotherm, not necessarily linear, $g = g(f)$. By normalizing the quantities appropriately, we assume that $g(0) = 0$ and $g(1) = 1$ (Figure 2).

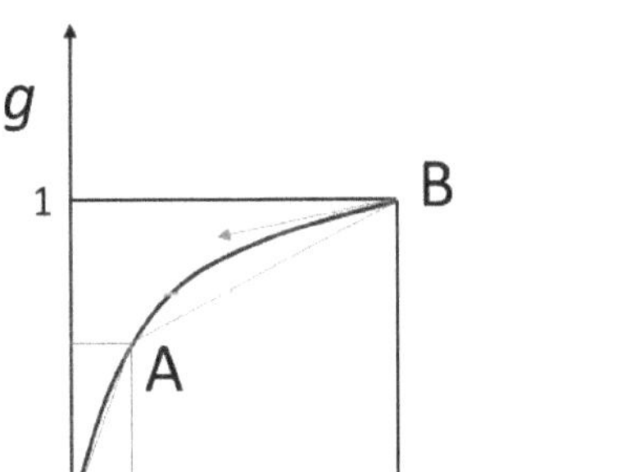

Figure 2. Isotherm linking rock and fluid concentrations at equilibrium. The concentration f of the chemical constituent in the solid is shown on the abscissa, that in the fluid, g, on the ordinate. Concentrations are normalized so that the amplitude of variation is between 0 and 1 in the solid and in the fluid respectively. The isotherm is concave downwards and connects points O (0, 0) and B (1, 1). An intermediate point A is shown. Hyperbolic problem theory teaches that the velocities of progression of different concentrations in space are proportional to the slopes of the isotherm at the corresponding points. So, the arrow drawn from point B, tangent to the isotherm at this point, gives the speed of progression of concentration (1, 1). Similarly, shocks or discontinuities between various points have velocities proportional to the slopes connecting the points concerned [48]. So, a shock between point (0, 0) and point A will have a velocity along OA, and the shock between point B and point (1, 1) will have a velocity along AB.

We impose a disequilibrium on the system with the following contrasted initial and boundary conditions (Figure 3): the fluid arrives at $x = 0$ with $g = 1$ in a rock characterized by $g = 0$ for t = 0 from abscissa 0 to infinity to the right. We are considering a reactor of size L. We are interested in the finite time T for complete transformation of the reactor to the value 1 of the concentrations. This will correspond to a finite content Φ of the component inside the reactor and an integrated flux necessary for transformation Γ.

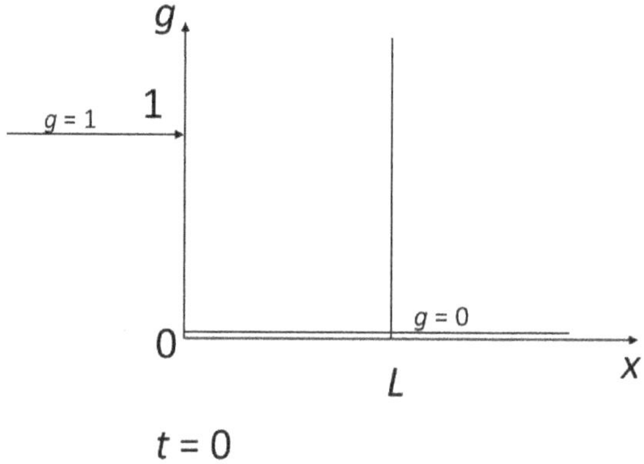

Figure 3. Representation of the reactor studied. The reactor of finite length L is represented along the x-axis. The fluid enters from its left side at $x = 0$, and leaves to the right from $x = L$. The initial concentration in the fluid contained in the reactor (and along the x-axis) is $g = 0$, the incoming concentration at $x = 0$ is $g = 1$. The corresponding equilibrium concentrations for the solid are $f = 0$ and $f = 1$, respectively.

Various scenarios are possible, described by $g(x, t)$ distributions. For reasons of generality and because it is interesting and fruitful to do so (see introduction), we pose the problem in terms of distributions in the mathematical sense (weak solutions). Multiple, discontinuous solutions (also called shocks) may indeed be encountered; continuous solutions are called rarefaction waves or *détentes*. We need to impose an optimization criterion that selects the weak solutions that make physical sense. Let us insist: it is an important property of hyperbolic problems when looking for weak solutions in the mathematical sense, i.e., representing true discontinuities in the sense of distribution theory, to present multiple solutions (non-unique). Solutions are complete spatio-temporal evolutions $g(x, t)$ linking boundary conditions as specified (Riemann problem). To do this, we postulate the existence of a (entropy, entropy flux) pair, i.e., (S, F), where S depends on the quantity g (or, what is equivalent, to f, thanks to local equilibrium), verifying (see the theory of hyperbolic problems [48,49]):

$$P(S) = \frac{\partial S}{\partial t} + \frac{\partial F}{\partial x} \geq 0$$

There are weak formulations of the previous entropy condition. It also corresponds to an extremum [50]. This is in line with [3]. The local equilibrium condition can be written as [45]:

$$\frac{\partial S}{\partial f} = \frac{\partial F}{\partial g}$$

Hyperbolic problem theory teaches that the velocities of progression of different concentrations in space are proportional to the slopes of the isotherm at the corresponding points (scalar case). Similarly, shocks or discontinuities between various points have velocities proportional to the slopes connecting the points concerned (cf. [48,51]). Refer to the caption in Figure 2.

3.2. Solving the Problem

Figures 4–6 show various weak solutions to the problem posed in Figure 3. In Figure 4, two compositional shocks or discontinuities propagate through the system. The shock leading from the starting concentration ($f = g = 0$) to that of point A moves more rapidly than the shock seeing the concentration change from that of point A to that of point B. The concentration at point A can have any value between 0 and 1. In Figure 5, after an initial shock similar to that in the previous figure, we see a rarefaction wave between A and B. Figure 6 shows the propagation of a single shock going directly from level 0 to level 1. The various profiles satisfy the shock conditions just described.

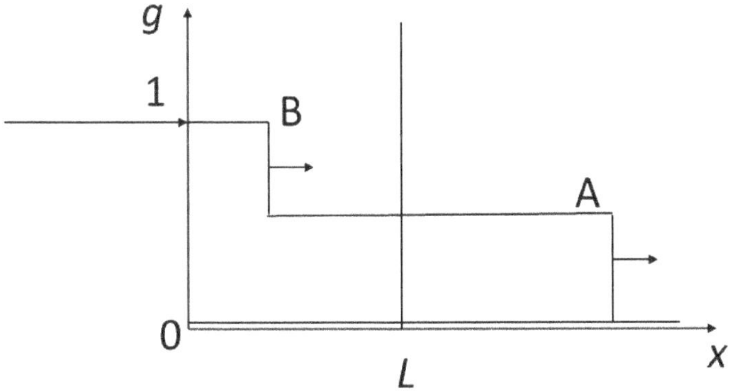

Figure 4. Representation of the reactor crossed by two composition discontinuities. The state of the reactor at a given moment has been represented, traversed by a composition discontinuity leading from $g = 0$ to point A, followed by another discontinuity connecting point A to point B; points A and B are shown in Figure 2. Applying the rules governing velocities recalled in the caption of Figure 2, we can see that the velocity of the first front is almost six times faster than that of the second. The arrow perpendicular to the vertical discontinuities indicates the direction of the corresponding fronts (to the right).

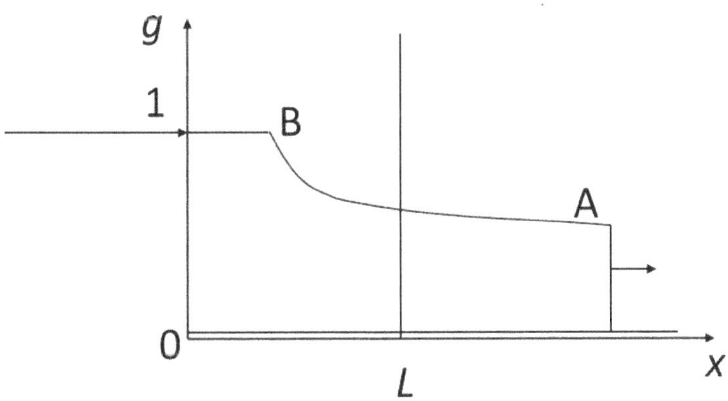

Figure 5. Representation of the reactor pervaded by a shock and a rarefaction wave. By comparison with the previous figure, the discontinuity between 0 and A is kept, but there is a rarefaction wave between A and B. The advance velocity of B is given by the tangent to the isotherm at the corresponding point (see Figure 2). The arrow perpendicular to the vertical discontinuity indicates the direction of the corresponding front (to the right).

Figure 7 shows a rarefaction wave between point O (0, 0) and point B (1, 1). Only this rarefaction wave meets the entropy condition. In fact, this profile can be understood as meeting a stability constraint: if a fluctuation causes an intermediate composition between 0 and 1 to appear, this composition is maintained and propagated in the sequence of compositions, because its speed is between the speeds of the upstream and downstream compositions (the profiles in the previous figures are not stable, as they cannot avoid the propagation of intermediate compositions between the compositions of the initial condition and the composition of the boundary condition, should these appear through fluctuation). More generally, practitioners of hyperbolic problems trace paths in the isotherms connecting the extreme points of the boundary conditions. In the general case, and for isotherms with more complicated shapes than that in this study, the optimal path in the sense of the entropy condition is given by the convex envelope to the isotherm between the extreme points.

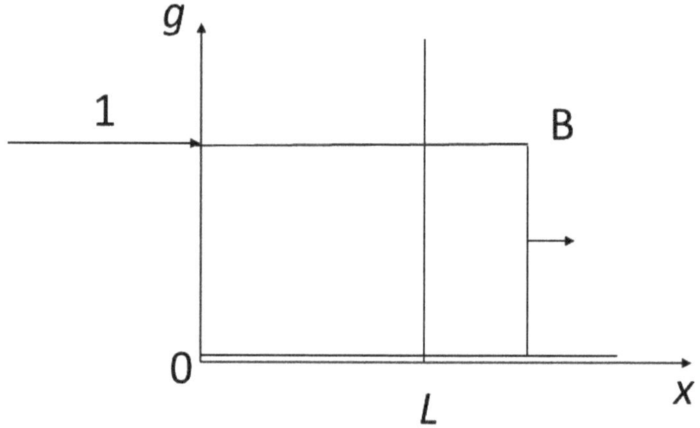

Figure 6. Representation of a reactor traversed by a single shock. Here, we have the case of a single discontinuity between point O (0, 0) and point B (1, 1) with a velocity equal to 1, i.e., the slope of the segment connecting the two points on the isotherm (Figure 2). The arrow perpendicular to the vertical discontinuity indicates the direction of the corresponding front (to the right).

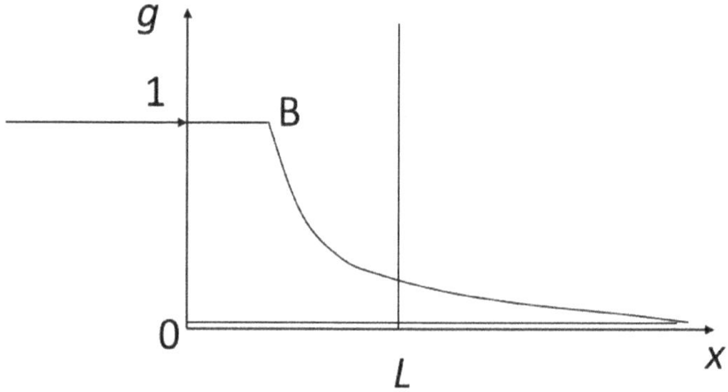

Figure 7. Representation of the reactor through which a composition rarefaction wave takes place. In contrast to Figures 4–6, the reactor here is traversed by a complete rarefaction wave linking point (0, 0) and point (1, 1) of the isotherm (Figure 2). The first velocities (for the smallest concentrations), visible on the slopes of the isotherm, are faster than those for the largest concentrations, which arrive behind (lower slopes) and are therefore ahead, generating the rarefaction wave. Only this configuration respects the entropy condition and makes physical sense.

3.3. "Finite" Quantities

On the basis of the above results, we can define various quantities characterizing the finiteness of the problem: time T for the complete transformation of the reactor of size L. We can see from the above that this is the time it takes for the composition B to pass through the reactor from abscissa 0 to abscissa L. Depending on the case, the speed of B is not the same, particularly if B is produced after a shock (non-physical solution, Figures 4 and 6) or at the end of a rarefaction wave (physical solution, Figure 7, or partially physical, Figure 5).

The times T are as follows, for Figures 4–7, respectively (we call T_4 the transformation time corresponding to the propagation shown in Figure 4, and similarly for other times):

$T_4 = L/v_{AB}$ where v_{AB} is the velocity of the shock passing from A to B

$T_5 = L/v_B$ where v_B is the velocity of point B (tangent to the isotherm)

$T_6 = L/v_{OB} = L$ where $v_{OB} = 1$

$T_7 = L/v_B = T_5$

We can also calculate the total quantity Φ of chemical component C contained in the reactor of dimension L and the finite flux Γ that was required for the transformation. Here, we see that $\Phi = \Gamma = L$. This particular result originates from the initial condition where $g(t = 0) = 0$.

3.4. Moving Frame: Transformation of Equations

Let us now consider a moving frame of reference R′, relative to the frame of reference R used to solve the problem described in the previous section. Let v be the relative velocity between the two reference frames.

In our example, there seems to be a clear preferred frame of reference in the form of the rock medium in which the water carrying the atoms of the chemical element we are interested in (iron, for example) moves. This is clearly a specific physical frame of reference, and with this frame of reference present, it is simply not clear how Lorentz invariance could be appropriate. There are two answers to this question. The first is that the reference frames used in standard relativity all have distinctive features that make each one privileged in its own way (in the example of Langevin's twins, the reference frame at rest relative to the Earth, the earth itself, is quite special). Second, we can say that we are operating at a mathematical level and that the two reference frames R and R′ have no other characteristic at this level than that of having a relative displacement. From a general point of view, in the geological example, we can easily imagine other reference frames in relative motion (such as the fluid in migration).

We will bring into play the relativistic approach described in the first section and ask ourselves a series of questions:

- What will be the new conservation equation linking the physical quantities in the moving frame of reference?
- What will be the new optimization criterion?
- What relationships can we write between the quantities x, t, f, g, S, F (frame R) and the quantities x', t', f', g', S' and F' (frame R′)?
- How will the finite quantities L and T, as well as Φ and Γ, which relate to the (finite-time) thermodynamics problem, be transformed?
- Can we write relationships between these different finite quantities?

These questions can be answered simply for this particular example, thanks to the results of the first part. The conservation Equation (2) is Lorentz invariant and the new equation is identical to (8) above

$$\frac{\partial f'}{\partial t'} + \frac{\partial g'}{\partial x'} = 0$$

And, as far as the optimization principle is concerned, it is also Lorentz invariant and we will have the condition given by (16)

$$P(S\prime) = \frac{\partial S\prime}{\partial t\prime} + \frac{\partial F\prime}{\partial x\prime} \geq 0$$

We are, therefore, in the simple situation of having the same Lorentz transform formulas for both space and time variables and for concentrations, as in (6) and (9).

$$x' = ax + bt$$

$$t' = bx + at$$

and

$$f' = af + bg$$

$$g' = bf + ag$$

where a and b are the coefficients given by relation (7). We will not dwell on what concerns the primary physical quantities involved in the Lorentz transformations just written. In our research on (finite-time) thermodynamics, we will take a closer look at what concerns the finite quantities L, T, Φ and Γ.

3.5. *Moving Reference Frame: Transformation of Finite Quantities*

The transformation of space amplitudes L is given by the following classical formula:

$$L = L'/a \tag{17}$$

with

$$a = \frac{1}{\sqrt{1 - \frac{v^2}{c^2}}}$$

where a is greater than 1. This relationship expresses that a length L' evaluated in the moving frame of reference is seen contracted from the fixed frame of reference. For durations T, we also have, according to the classic formula

$$T = aT' \tag{18}$$

where, this time, the durations evaluated T' in the moving frame of reference are seen dilated from the fixed frame of reference. From the above relationships, we derive

$$LT = L'T \tag{19}$$

or

$$\frac{L}{L\prime}\frac{T}{T\prime} = 1 \tag{20}$$

which expresses that the relative variations of lengths and durations are correlated. This relationship has been commented on in the literature. In the case of three dimensions of space, rather than "conservation of a hypervolume" [52], we prefer to see it as that of a norm $\Sigma t_i x_i$, see [53].

Let us now look at how the quantities Φ and Γ are transformed. To do this, let us return to the first postulate of relativity: the laws of physics are the same in the two reference frames in Galilean motion relative to each other. Just as quantities Φ are evaluated in proportions of lengths L, quantities Φ' will be evaluated in proportions of lengths L', in accordance with the following ratios

$$\Phi/L = \Phi'/L' \tag{21}$$

For Γ flows, we can similarly say that they are evaluated by T durations and that the following relationships must be respected

$$\Gamma/T = \Gamma'/T' \tag{22}$$

Relationships (21) and (22) could just as well be understood by saying that Φ and Γ provide the units of space and time. By multiplying the two previous relationships, we obtain

$$\Phi\,\Gamma/LT = \Phi'\,\Gamma'/L'T' \tag{23}$$

By virtue of $LT = L'T'$ (19), this gives

$$\Phi\,\Gamma = \Phi'\,\Gamma' \tag{24}$$

This relationship expresses that the relative variations of the finite quantities Φ and Γ are also correlated. This expression could be obtained independently if we decide to measure space and time using the quantities f and g. In both cases (by direct independent determination or indirectly via Equations (19), (21), and (22)), the expression $\Phi\,\Gamma = \Phi'\,\Gamma'$ remains the same, even if the specific values of Φ, Γ, Φ', Γ' that verify it differ.

Other similar relationships can be obtained. Carrying (17) into (21), we obtain

$$\Phi = \Phi'/a \tag{25}$$

Using (18), this gives

$$\Phi T = \Phi' T' \tag{26}$$

which we can also write

$$\frac{\Phi}{\Phi\prime}\frac{T}{T\prime} = 1 \tag{27}$$

Analogous reasoning yields the relationship

$$\Gamma L = \Gamma' L' \tag{28}$$

which we can also write

$$\frac{\Gamma}{\Gamma\prime}\frac{L}{L\prime} = 1 \tag{29}$$

This expresses a complementarity between finite resources (reactor contents, summation of the flux supplied) and finite space and time. The size of the systems mobilized and the possible durations of the reactions concerning them have something to do with the quantities of resources involved, directly or contributed. We believe that the results expressed in relations (26) to (29) are important, even from a purely qualitative point of view. They place (finite) time and space on the same level as (finite) material flows and resources. They were obtained thanks to the theory of relativity, but this theory could now be forgotten. However, it has fully played its role, which we see as fundamentally that of linking the construction of space and time variables to the comparison of concrete material processes at work in physical reality.

4. Discussion and Conclusions

At the end of this work, let us sketch out a few conclusions: the future will tell us what general value they may have, having been obtained by studying an example. We can use them as a basis for further fundamental research, to be carried out over the long term. In summary, the relationship between relativity and thermodynamics in this work is based on the fact that the thermodynamic quantities f and g are treated as relativistic, even though the velocities are low. It is also based on the combination of relationships involving the pairs (f, g) and (x, t), which highlight the organic relationships between finite resources and finite time. The relationship between finite-time thermodynamics and hyperbolic problems is based on the fact that there are multiple weak solutions and that a thermodynamic criterion is used to select them.

The constraints brought by the relativistic functioning as influencing the thermodynamic functioning, particularly in the use of four vectors, have oriented us, as far as the optimization criterion is concerned, towards the entropy production. Entropy alone is not a relativistic invariant.

We have seen the complementary-type relationships between finite resources and finite time and space. Finitudes are linked between space and time, on the one hand, and resources and their flows, on the other. *If time is finite, so are resources, which can be neither zero nor infinite.* This is a way of reiterating that space and time cannot be thought of on their own and are a way of understanding the phenomena associated with material resources. This is a way to explore dimension 2 in Figure 1. Along these lines, we can imagine posing the problem differently, by exchanging the roles of the variables (x, t), on the one hand, and (f, g), on the other, in the mathematical equations, in particular the conservation Equation (2) from which we start. The equivalence between the two approaches expresses the equivalence between time/space and flows/quantities of matter.

With regard to the role of hyperbolicity (dimension 1 in Figure 1) for large systems, we have seen that neglecting diffusion allows us to obtain finite propagation speeds and ensure a finite time that depends directly on them. The trade-off for this choice is the multiplicity of weak solutions for $g(x, t)$ evolutions.

We mentioned two research directions, labeled 3 and 4, in Figure 1. To proceed in direction 3, we must resume the hyperbolic approach by exchanging the variables (x, t) with the variables (f, g). These mark space and time, and the old scale of space and time is distorted. To proceed in direction 4, we must introduce weak solutions into the Lorentz transformation. A line of research into these questions is provided by the works of Andresen and Essex ([54] and other references therein). These authors are interested in what we are able to see of systems, depending on the comparative speeds of the processes taking place within them and the time we spend observing them. Thus, we can say that when scanning a spatial system, the abrupt variations found within it may or may not be considered as discontinuities depending on the speed of the scan (i.e., the temporal pixel). If the scan is slow or defined with a sufficient temporal resolution, we will see the intermediate states of the abrupt variations and will not consider them as real mathematical discontinuities. At faster observation speeds, we will not see the intermediate states (and with larger temporal pixels): we will then conclude that discontinuities are present. If we now compare the views we have of a system from two reference frames in relative motion, we conclude from relativistic effects relating to time dilation that the question of discontinuities will not arise in the same way in the two frames (the question may arise in one frame but not in another). All this must be clarified in appropriate research.

We have also seen that relativistic changes in conservation constraints and optimization criteria are analogous. Can we see these two types of mathematical relations as the expression of one and the same point of view? We can think of Kruzhkov's [55] formulation of hyperbolic problems, where conservation and entropy conditions derive from a single mathematical writing. We may also think of the formulations of the two principles of thermodynamics as forms of a single principle entitled "stable-equilibrium-state-principle" according to [56] (see also [57]) from which, out of equilibrium, formulations are derived.

Finally, the results we obtained give us food for thought on the relationships between thermodynamics and relativity. We can indeed write complementarity-type relations linking entropy variations and its flux to spatial and temporal quantities (identical to (26) and (27)). Is this a way of understanding the necessity for homogeneity, of both spatio-temporal origin (based on regularly spaced graduations, in the postulate c = cte) and thermodynamic origin in the second law? In [58], we laid propositions in this direction.

Funding: This research received no external funding.

Data Availability Statement: There are no data.

Acknowledgments: I would like to thank Bjarne Andresen for his invitation and his trust. The critical dialogue with him was very fruitful and was a great opportunity for me. This does not mean his complete agreement to the paper, as the imperfections of the paper are my sole responsibility. I would like to thank Michel Feidt for his helpful comments, and Marc Doumas for his help with bibliographical research. The critical comments made by the reviewers and editors greatly helped to improve an early version of the manuscript.

Conflicts of Interest: The author declares no conflicts of interest.

References

1. Andresen, B.; Berry, R.S.; Nitzan, A.; Salamon, P. Thermodynamics in finite time. I. The step-Carnot cycle. *Phys. Rev. A* **1977**, *15*, 2086. [CrossRef]
2. Andresen, B.; Salamon, P.; Berry, R.S. Thermodynamics in finite time. *Phys. Today* **1984**, *37*, 62. [CrossRef]
3. Bejan, A. Entropy generation minimization: The new thermodynamics of finite-size devices and finite-time processes. *J. Appl. Phys.* **1996**, *79*, 1191–1218. [CrossRef]
4. Feidt, M. *Optimal Thermodynamics in Finite Physical Dimensions*; Lavoisier: Paris, France, 2013; p. 428.
5. Petrescu, S.; Enache, V.; Feidt, M.; Stanciu, C.; Costea, M.; Boriaru, N. Unification perspective of finite physical dimensions thermodynamics and finite speed thermodynamics. *Int. J. Energy Environ. Engin.* **2015**, *6*, 245–254. [CrossRef]
6. Petrescu, S.; Banches, E.; Popescu, G.H.; Boriaru, N.; Borcila, B.; Stanciu, C.; Costea, M.; Dobre, C. Concepts and fundamental equations in Thermodynamics with Finite Speed. *IOP Conf. Ser. Mater. Sci. Eng.* **2016**, *147*, 012144. [CrossRef]
7. Haddad, W.M. Temporal Asymmetry, Entropic Irreversibility, and Finite-Time Thermodynamics: From Parmenides–Einstein Time-Reversal Symmetry to the Heraclitan Entropic Arrow of Time. *Entropy* **2012**, *14*, 407–455. [CrossRef]
8. Jou, D.; Casas-Vázquez, J.; Lebon, G. *Extended Irreversible Thermodynamics*; Springer Nature: Berlin/Heidelberg, Germany, 2001.
9. Dreyer, W.; Weiss, W. The classical limit of relativistic extended thermodynamics. *Ann. L'institut Henri Poincaré Phys. Théorique* **1986**, *45*, 401–418.
10. Ruggeri, T. The Entropy Principle from Continuum Mechanics to Hyperbolic Systems of Balance Laws: The Modern Theory of Extended Thermodynamics. *Entropy* **2008**, *10*, 319–333. [CrossRef]
11. Gemmer, J.; Michel, M.; Mahler, G. Quantum thermodynamics. In *Emergence of Thermodynamic Behavior Within Composite Quantum Systems*; Springer: Berlin/Heidelberg, Germany, 2004; p. 288.
12. Guy, B. Thinking time and space together. *Philos. Sci.* **2011**, *15*, 91–113.
13. Guy, B. On the cross-fertilization of the foundations of thermodynamics, quantum mechanics and relativity theory. In Proceedings of the 16th Joint European Thermodynamics Conference, Prague, Czech Republic, 14–18 June 2021; HAL Publication Repository, CCSD, CNRS: Paris, France, 2021. Available online: https://hal.science/hal-03259573v1 (accessed on 25 June 2025).
14. Guy, B. Refreshing the theory of relativity. *Preprints* **2025**. [CrossRef]
15. Tsabary, G.; Censor, A. An alternative mathematical model for special relativity. *arXiv* **2004**, arXiv:math-ph/0402054v1.
16. Guy, B. Degree zero of physical laws. In *Heuristic Considerations*; HAL Publication Repository, CCSD, CNRS: Paris, France, 2012. Available online: https://hal.science/hal-00723183v1 (accessed on 25 June 2025). (In French)

17. Chen, X. Three-dimensional time theory: To unify the principles of basic quantum physics and relativity. *arXiv* **2005**, arXiv:quant-ph/0510010v1.
18. Guy, B. The Lorentz transformation, time and space. In *Generalization of the Gamma Factor As a Function of the Direction of the Hidden Movement in the Clocks*; HAL Publication Repository, CCSD, CNRS: Paris, France, 2019. Available online: https://hal.science/hal-02279222v1 (accessed on 25 June 2025).
19. Ares de Parga, G.; Lopez-Carrera, B.; Angulo-Brown, F. A proposal for relativistic transformations in thermodynamics. *J. Phys. A Math. Gen.* **2005**, *38*, 2821. [CrossRef]
20. Gavassino, L. Proving the Lorentz invariance of the Entropy and the co-variance of thermodynamics. *Found. Phys.* **2022**, *52*, 11. [CrossRef]
21. Kluitenberg, G.A.; de Groot, S.R.; Mazur, P. Relativistic thermodynamics of irreversible processes. *Physica* **1953**, *19*, 689–704. [CrossRef]
22. Garcia-Colín, L.S.; Sandoval-Villalbazo, A. Relativistic Non-Equilibrium Thermodynamics Revisited. *arXiv* **2005**, arXiv:gr-qc/0503047v2. [CrossRef]
23. Güémez, J.; Mier, J.A. Relativistic mechanics and thermodynamics: IV. Thermodynamic processes. *Eur. J. Phys.* **2023**, *45*, 015701. [CrossRef]
24. Güémez, J. Relativistic Thermodynamics: A Modern 4-Vector Approach. *Phys. Res. Intern.* **2011**, *3D*, 387351. [CrossRef]
25. Yuen, C.K. Lorentz Transformation of Thermodynamic Quantities. *Am. J. Phys.* **1970**, *38*, 246–252. [CrossRef]
26. Ciancio, V.; Rogolino, P.; Farsaci, F.; Udrişte, C. On entropy production in relativistic thermodynamics. In Proceedings of the WSEAS International Conference Proceedings Mathematics and Computers in Science and Engineering, Faro, Portugal, 3–5 November 2010.
27. Söderholm, L.H. Some Remarks on Relativistic Thermodynamics. *Phys. Scr.* **1970**, *2*, 139. [CrossRef]
28. Parvan, A.S. Lorentz transformations of the thermodynamic quantities. *An. Phys.* **2019**, *401*, 130–138. [CrossRef]
29. de Groot, S.R. The laws of relativistic thermodynamics. *Phys. A Stat. Mech. Its Appl.* **1977**, *88*, 172–182. [CrossRef]
30. Farías, C.; Moya, P.S.; Pinto, A. On the Relationship between Thermodynamics and Special Relativity. *arXiv* **2007**, arXiv:07123793.
31. Eimerl, D. On relativistic thermodynamics. *An. Phys.* **1975**, *91*, 481–498. [CrossRef]
32. Muschik, W.; von Borzeszkowski, H.H. Entropy identity and equilibrium conditions in relativistic thermodynamics. *Gen. Rel. Grav.* **2008**, *41*, 1285–1304. [CrossRef]
33. Baig, M.W. Relativistic Thermodynamic State Functions from a Time Dilation Perspective. *Russ. J. Phys. Chem. B* **2021**, *15*, 53–55. [CrossRef]
34. Habeger, C.C. The second law of thermodynamics and special relativity. *An. Phys.* **1972**, *72*, 1–28. [CrossRef]
35. Tane, J.L. An Extended Interpretation of the Concept of Entropy Opening a Link between Thermodynamics and Relativity. *Nat. Sci.* **2014**, *6*, 503–513. [CrossRef]
36. Livadiotis, G.; McComas, D.J. The theory of thermodynamic relativity. *Sci. Rep.* **2024**, *14*, 22641. [CrossRef]
37. Callen, H.; Horwitz, G. Relativistic thermodynamics. *Am. J. Phys.* **1971**, *39*, 938–947. [CrossRef]
38. Van Kampen, N.G. Relativistic Thermodynamics of Moving Systems. *Phys. Rev.* **1968**, *173*, 295–301. [CrossRef]
39. Tolman, R. *Relativity, Thermodynamics and Cosmology*; Clarendon Press: Oxford, UK, 1934.
40. Rengui, Y. The logical connection between special relativity thermodynamics. *Eur. J. Phys.* **1996**, *17*, 265. [CrossRef]
41. Landsberg, P.T. Special relativistic thermodynamics. *Proc. Phys. Soc.* **1966**, *89*, 1007. [CrossRef]
42. Pathria, R.K. Lorentz transformation of thermodynamic quantities. *Proc. Phys. Soc.* **1966**, *88*, 791. [CrossRef]
43. Tane, J.L. Thermodynamics and Relativity, A Condensed Explanation of their Close Connection. *arXiv* **2005**, arXiv:physics/0503106.
44. He, L.; Zhuang, P. Lorentz Invariance at Finite Temperature and Its Effect on Production Rate and Equation of State. *Chin. Phys. Let.* **2004**, *21*, 1228.
45. Guy, B. Mathematical revision of Korzhinskii's theory of infiltration metasomatic zoning. *Eur. J. Mineral* **1993**, *5*, 317–339. [CrossRef]
46. Guy, B. The behavior of solid solutions in geological transport processes: The quantization of rock compositions by fluid-rock interaction. In *Complex Inorganic Solids, Structural, Stability and Magnetic Properties of Alloys*; Turchi, P., Gonis, A., Rajan, K., Meike, A., Eds.; Springer: Berlin/Heidelberg, Germany, 2005; pp. 265–273.
47. Tondeur, D.; Klein, G. Multicomponent ion exchange in fixed beds. *Ing. Eng. Chem. Fundam.* **1967**, *6*, 351–361. [CrossRef]
48. Lax, P.D. Shock waves and entropy. In *Contribution to Nonlinear Functional Analysis*; Zarantonello, E.M., Ed.; Academic Press: New York, NY, USA, 1971; pp. 603–634.
49. Lax, P.D. Hyperbolic systems of conservation laws and the mathematical theory of the shock waves. In *SIAM Regional Conference Series in Applied Mathematics*; Society for Industrial and Applied Mathematics: Philadelphia, PA, USA, 1973; 48p.

50. Dafermos, C.M. The entropy rate admissibility criterion for solutions of hyperbolic conservation laws. *J. Differ. Equat.* **1973**, *14*, 202–212. [CrossRef]
51. Oleinik, O.A. Uniqueness and stability of the generalized solutions of the Cauchy problem for a quasi-linear equation. *Amer. Math. Soc. Trans.* **1963**, *287*, 979–1007.
52. Bequerel, J. *Exposé Élémentaire de la Théorie D'einstein*; Payot: Paris, France, 1922; p. 206.
53. Guy, B. The duality of space and time and the theory of relativity. *Hadr. J. Sup.* **2001**, *16*, 369–412.
54. Essex, C.; Andresen, B. Diffusion and kinetic theory on very long time and large space scales. *Entropy* **2024**, *26*, 1037. [CrossRef] [PubMed]
55. Kruzhkov, S.N. Generalized solutions for the Cauchy problem in the large for non-linear equation of first order. *Soviet. Math. Dokl.* **1969**, *10*, 785–788.
56. Gyftopoulos, E.P.; Beretta, G.P. *Thermodynamics, Foundations and Applications*; Dover: Mineola, NY, USA, 2005; p. 756.
57. Haywood, R.W. Teaching thermodynamics to first-year students by the single axiom approach. In *Teaching Thermodynamics*; Lewins, J.D., Ed.; Springer: Boston, MA, USA, 1985.
58. Guy, B. Some reflections on the relations between thermodynamics and relativity theory. *Comm. to 17th Joint European Thermodynamics Conference, Salerno*; 2023. HAL Publication Repository, CCSD, CNRS, Paris, France. Available online: https://hal.science/hal-05132112v1 (accessed on 25 June 2025).

MDPI AG
Grosspeteranlage 5
4052 Basel
Switzerland
Tel.: +41 61 683 77 34

Entropy Editorial Office
E-mail: entropy@mdpi.com
www.mdpi.com/journal/entropy

www.ingramcontent.com/pod-product-compliance
Lightning Source LLC
Chambersburg PA
CBHW040817250926
51850CB00007BA/1192
9783725872862